THE ROYAL HORTICULTURAL SOCIETY

# PESTS &
# DISEASES

# THE ROYAL HORTICULTURAL SOCIETY

# PESTS & DISEASES

PIPPA GREENWOOD
ANDREW HALSTEAD

DORLING KINDERSLEY
LONDON • NEW YORK • STUTTGART • MOSCOW

# A DORLING KINDERSLEY BOOK

**PROJECT EDITORS** *Laura Langley, Martha Swift*
**DESIGNER** *Rachael Parfitt*

**MANAGING EDITOR** *Louise Abbott*
**MANAGING ART EDITOR** *Lee Griffiths*

**DTP DESIGNERS** *Chris Clark, Matthew Greenfield*
**PRODUCTION CONTROLLER** *Martin Croshaw*

**ILLUSTRATIONS** *Sandra Pond and Will Giles*

First published in Great Britain in 1997
by Dorling Kindersley Limited,
9 Henrietta Street, London WC2E 8PS
Visit us on the World Wide Web at http: / / www.dk.com

Copyright © 1997 Dorling Kindersley Limited, London

A CIP catalogue record for this book is available
from the British Library

ISBN 0 7513 0366 6

Colour reproduction by Bright Arts, Hong Kong
Printed and bound by Star Standard, Singapore

# CONTENTS

Introduction 7

How To Use This Book 8

## GALLERY OF SYMPTOMS
Leaves 12 • Stems & Buds 33 • Flowers 42
Fruits, Berries, Nuts & Vegetables 47 • Soil, Roots, Tubers & Bulbs 53
Lawns 58 • Beneficial Garden Creatures 60

## GARDEN HEALTH AND PROBLEMS
Ecology of the Garden 66 • Types of Pests and Diseases 72
Controlling Pests and Diseases 78

## A–Z of Pests, Diseases and Disorders
97-192

## INDIVIDUAL PLANT PROBLEMS
Garden Trees 194 • Shrubs & Climbers 196 • Herbaceous Perennials 198
Lawns 199 • Bulbous Plants 200 • Annuals & Biennials 201 • Greenhouse & House Plants 202
Fruits & Nuts 204 • Vegetables & Herbs 205

## REFERENCE SECTION
Garden Insecticides 208
Garden Fungicides 210
Brand Names 212
Glossary 214 • Index 218
Conversion Tables 223
Addresses & Acknowledgments 224

# INTRODUCTION

The identification and diagnosis of a problem is halfway to solving it. This book helps the gardener to consider all the factors in the garden that affect growing plants, so that prompt action can be taken if needed.

It is impossible to garden without coming across pests, diseases and disorders on your plants. They are unwelcome in any garden, large or small. At the very least they spoil a plant's appearance, but if left to spread and develop unchecked, the more serious ones may dramatically reduce the plant's vigour, impairing growth, flowering ability or yield. Some may create wounds, which are of little significance in their own right, but which allow the entry to secondary organisms which then wreak havoc.

The key to a flourishing garden lies not only in quick identification and treatment of pests and diseases, but also in understanding how to keep plants healthy, thereby preventing problems occurring in the first place. Many common problems originate in cultural disorders (the conditions in which a plant is growing), or are only capable of serious damage on a plant which is already stressed by having to grow in an unsuitable site, or by poor cultivation. Once a cultural problem has been identified and understood, it is often quite simple to overcome it and so avoid a whole host of possible additional problems.

It is vital to identify correctly the pests, diseases and disorders themselves. If you are uncertain about what is attacking a plant it is impossible to tell whether or not it is something you need worry about, and what, if anything, you should do about it. *RHS Pests and Diseases* makes a wealth of information on all aspects of the problem side of gardening available to gardeners and professional horticulturists alike. By looking at the garden as a whole and then identifying the specific problem, it allows you to stay one step ahead of pests, diseases and disorders in the garden.

# HOW TO USE THIS BOOK

THE EASY TO USE format provides three ways to help you identify a plant pest, disease or disorder: by symptom, by plant and by problem. Symptoms of pests and diseases are shown in colour photographs in the *Gallery of Symptoms*, which is cross-referenced to the *A–Z of Pests, Diseases and Disorders*, a dictionary of most possible garden problems. *Individual Plant Problems* is a plant-by-plant listing, enabling you to look up the problems your particular plant could be suffering from and find their reference in the A–Z and Gallery. A section on *Garden Health and Problems* looks at the garden and plant cultivation as a whole, discussing the background to problems that might occur, while at the back of the book, the comprehensive *Reference Section* contains a glossary and index, and tables that relate to the use and correct application of all the most common commercially available garden chemicals.

## IDENTIFYING AND CONTROLLING PROBLEMS:

## BY SYMPTOM

*See* Gallery of Symptoms, *pp.10–63*

The Gallery section contains hundreds of photographs, enabling you to identify the cause of the symptoms on your plant and then referring you to the A–Z section. It is divided into six parts: Leaves; Stems and Buds; Flowers; Fruits, Berries, Nuts and Vegetables; Soil, Tubers, Roots and Bulbs; and Lawns. At the end are four pages enabling the identification of Beneficial Garden Creatures.

Colour photograph of affected plant part with symptoms clearly visible

Section heading, colour-coded for easy reference

Whether problem is a pest, a disease or a disorder

Name of problem, corresponding with detailed A–Z entry

Page reference in A–Z

### APPLE CAPSID
PEST • *See p.100*
This pest sucks sap from the fruitlets and causes yellowish-brown, raised, corky swellings. The damage is only skin deep and does not affect the fruits' keeping qualities.

**PLANTS AFFECTED** Apple trees
**SEASON** Late spring to early autumn

Plants affected by that particular problem

Brief description of visible and other symptoms

Season when problem is most likely to occur

---

## FRUITS, BERRIES, NUTS AND VEGETABLES

IF THE EDIBLE PARTS of a plant are attacked by a pest or pathogen, the whole reason for growing the plant may well be lost. Fruits and berries are particularly prone to attack, as they usually have a high sugar and moisture content. Nuts, peas and beans are also valuable food sources for many creatures.

In addition, some ornamental trees and shrubs are very commonly grown mainly for their attractive show of berries, and when these are eaten, removed or spoiled by disease, the plant's appearance is ruined.

In many cases prolonged protection of the plant may be needed, perhaps by using barriers or traps to keep pests away from the flowers or the developing and ripening fruits. Chemicals can be used with great success, but again these may sometimes need to be applied before the first signs of damage are noticed.

### FRUIT PROBLEMS

**APPLE SAWFLY** PEST • *See p.101*
Most infested fruits drop off in early summer. Some do develop as ripe fruit but have distinctive ribbon scars where the young larvae fed under the skin earlier in the summer.
**PLANTS AFFECTED** Apples
**SEASON** Mid-spring to early autumn

**APPLE SCALD** DISORDER • *See p.101*
Reddish-brown patches of discoloration develop on the skin of the fruit. The flesh beneath is usually undamaged, but it may be discoloured. Exposed surfaces are most badly affected.
**PLANTS AFFECTED** Apples
**SEASON** Mainly summer

**BIRD DAMAGE** PEST • *See p.107*
Blackbirds, starlings and other birds will peck holes in ripening fruits. Wasps may enlarge this damage, and brown rot frequently develops once the skin has been broken.
**PLANTS AFFECTED** Apples, pears, plums
**SEASON** Mid-summer to early autumn

**APPLE CAPSID** PEST • *See p.100*
This pest sucks sap from the fruitlets and causes yellowish-brown, raised, corky swellings. The damage is only skin deep and does not affect the fruits' keeping qualities.
**PLANTS AFFECTED** Apples
**SEASON** Late spring to early autumn

**CODLING MOTH** PEST • *See p.118*
This is the cause of maggoty apples in late summer. The caterpillar feeds at the core and tunnels out through the side or the eye end of the ripening fruit when it is fully developed.
**PLANTS AFFECTED** Apples, pears
**SEASON** Mid-summer to early autumn

**WASPS** PEST • *See p.188*
These familiar insects enlarge damage started by birds, but can also initiate feeding on softer-skinned fruits (see p.49). Their presence can make fruit picking hazardous.
**PLANTS AFFECTED** Tree fruits, grapes
**SEASON** Mid-summer to early autumn

**BROWN ROT** DISEASE • *See p.111*
Soft, brown rot develops on fruit (here on apple) and soon becomes studded with raised, creamy-white pustules (see also p.49). These are often arranged in concentric rings.
**PLANTS AFFECTED** Tree fruits
**SEASON** Late summer and autumn

47

# BY PLANT

*See* Individual Plant Problems, *pp.193–206*

A comprehensive listing of garden plants by type, with listings of the general and specific ailments to which each is liable. The problem can then be identified by referring to the *Gallery of Symptoms* or to the A–Z section.

Pests, diseases or disorders that the plant might suffer from, listed alphabetically

APPLE CANKER **36**, 100
APPLE CAPSID **47**, 100
APPLE LEAF MINER **19**, 101
APPLE SAWFLY **47, 48**, 101
APPLE SCAB **16**, 101
APPLE SOOTY BLOTCH 102
APPLE WATER CORE 102

Page reference for colour photograph in Gallery of Symptoms

Page reference for entry in A–Z of Pests, Diseases and Disorders

---

FRUIT AND NUT PROBLEMS

INDIVIDUAL PLANT PROBLEMS

# FRUITS AND NUTS

MANY GARDENERS are happy to use chemicals on ornamentals, but when it comes to cropping plants, most prefer to avoid them wherever possible. Good cultivation, preventive measures taken at the correct time, and the use of resistant cultivars will all help. Some problems affect many fruits (*see right*). Others are more host-specific; plants so affected are listed below. More resistant species or cultivars may be suggested in the entry for the problem in question in the *A–Z of Pests, Diseases and Disorders.*

**COMMON PROBLEMS**

The most common problems affecting plants and trees bearing edible fruits and nuts are:

APHIDS 99 BIRDS 106
BIRD DAMAGE TO FRUITS 107 SQUIRRELS 178
CORAL SPOT 119 WASPS 188
GREY MOULD 133
HONEY FUNGUS 136
MAGNESIUM DEFICIENCY 146
POWDERY MILDEWS 165

## A

**Apple**
APPLE BITTER PIT **48**, 100
APPLE BLOSSOM WEEVIL **46**, 100
APPLE CANKER **36**, 100
APPLE CAPSID **47**, 100
APPLE LEAF MINER **19**, 101
APPLE SAWFLY **47, 48**, 101
APPLE SCAB **16**, 101
APPLE SOOTY BLOTCH 102
APPLE WATER CORE 102
BLOSSOM WILT **41**, 108
BROWN ROT **41, 49**, 111
BROWN SCALE **47**, 118
CORAL SPOT **41**, 119
CROWN GALL **58**, 119
FIREBLIGHT **35**, 127
FRUIT TREE RED SPIDER MITE **20, 46**, 129
FRUIT TREE TORTRIX MOTH **45**
HONEY FUNGUS **54**, 136
IRON DEFICIENCY AND JIM-INDUCED CHLOROSIS **24**, 139
LEOPARD MOTH **44**, 142
MAGNESIUM DEFICIENCY **22**, 146
MONKEY SCALE **36**, 149
NITROGEN DEFICIENCY 151
PHONY PEAR **40**, 154
PHYTOPHTHORA 159
POTASSIUM DEFICIENCY **24**, 162
POWDERY MILDEWS **49**, 165
PH-GENERATION **44**, 169
ROSY APPLE APHID **48**, 173
SILVER LEAF 176
SCOTTING **3**, 
VIRUSES **23**, 187
WINTER MOTH **44**, 191
WOOLLY APHID **35**, 192

**Apricot** *see* Peach

## B

**Currant, black**
AMERICAN GOOSEBERRY
MILDEW **25, 49**, 98
BIG-BUD MITE **41**, 107
BLACKCURRANT GALL MIDGE
**26**, 108
BROWN SCALE **29, 46**, 111
CAPSID BUGS **29, 46**, 115
CURRANT BLISTER APHID 122
RASPBERRY BEETLE 167
RED BERRY MITE **49**, 168
RUSTS 173
VIRUSES **23**, 187

**Cherry**
APPLE LEAF MINER **19**, 101
BACTERIAL CANKER 104
BLOSSOM WILT **41**, 108
BROWN ROT **47**, 111
CHERRY BLACKFLY **24**, 116
CROWN GALL **58**, 119
FROST DAMAGE (BUDS AND FLOWERS) **28**, 127
FRUIT DAMAGE (CHERRIES AND STONE) **28**, 129
HONEY FUNGUS **54**, 136
IRON DEFICIENCY AND JIM-INDUCED CHLOROSIS **24**, 139
LEOPARD MOTH **44**, 142
MAGNESIUM DEFICIENCY **22**, 146
NITROGEN DEFICIENCY 151
PEAR AND CHERRY SLUGWORM **45**, 155
POWDERY MILDEWS **49**, 165
PH-GENERATION **44**, 169
SILVER LEAF 176
SPLITTING 177
VIRUSES **23**, 187
WINTER MOTH **44**, 191
WOOLLY APHID 35, 192

## C

## F

**Fig**
CORAL SPOT **41**, 119
DIEBACK 125
FROST DAMAGE **28, 44**, 129
GREY MOULD **41**, 133
LEAF MOTH **(FUNGAL) 12, 13**, 145

**Currant, red/white**
BROWN SCALE **36**, 111
CAPSID BUGS **29, 46**, 115
CORAL SPOT **41**, 119
CURRANT AND GOOSEBERRY
LEAF SPOT 123
CURRANT BLISTER APHID **27**, 122
CURRANT CLEARWING MOTH
**53**, 121
GOOSEBERRY SAWFLY **51**, 133
GREY MOULD **50**, 133
LEAF SPOT **(FUNGAL) 12, 13**, 145
MAGNESIUM DEFICIENCY 151
PEAR MIDGE **22**, 132
PEAR LEAF **(FUNGAL) 51**, 
POWDERY MILDEWS **49**, 165
PH-GENERATION **18, 50**, 178
SHOTHOLE **18, 50**, 176
SILVER LEAF 176

## G

**Gooseberry**
AMERICAN GOOSEBERRY
MILDEW **25, 49**, 98
APHIDS 99
BROWN SCALE **36**, 111
CAPSID BUGS **46**, 115
CORAL SPOT **58**, 119
CURRANT AND GOOSEBERRY
LEAF SPOT 123
GLASSHOUSE RED SPIDER
MITE **22, 23**, 132
HONEY FUNGUS **54**, 136
IRON DEFICIENCY AND JIM-INDUCED CHLOROSIS **24**, 139
RUSTS 173
VIRUSES **23**, 187

**Grape** *see* Vine

## H

**Hazelnut**
APHIDS 99
BIG-BUD MITE **41**, 107
LEAF SPOT **(FUNGAL) 12, 13**, 145
POWDERY MILDEWS **49**, 165
SQUIRRELS **52**, 178
VIRUSES **23**, 187
WINTER MOTH **51**, 191

## M

**Mulberry**
CORAL SPOT **58**, 119
DIEBACK 125
MILDEWY CANKER **58**, 110

## P

**Peach, nectarine, apricot**
BACTERIAL CANKER **35**, 104
BLOSSOM WILT **41**, 108
BROWN SCALE **36**, 111
CROWN GALL **58**, 119
FROST DAMAGE (BUDS AND FLOWERS) **28**, 129
FRUIT DAMAGE (FERTILE AND STONE) **28**, 129
MITE **22, 23**, 132
HONEY FUNGUS **54**, 136
GLASSHOUSE RED SPIDER
LEAF SPOT (FUNGAL) 145
NITROGEN DEFICIENCY 151
PEACH APHIDS 155
PEACH LEAF CURL **26**, 156
PHYTOPHTHORA DEFICIENCY 159
PHYTOPHTHORA 159
POTASSIUM DEFICIENCY 162
POWDERY MILDEWS **49**, 165
SHOTHOLE **18, 50**, 176
SILVER LEAF 176

**Pear**
APPLE AND PEAR CANKER **36**,
101
APPLE AND POWDERY MILDEW **25**,
101
APPLE SCAB **16**, 101
BLOSSOM WILT **41**, 108
BROWN ROT **47**, 111
BULLFINCHES **46**, 112
CORAL MOTH **41**, 118
CORAL SPOT **41**, 119
FIREBLIGHT **35**, 127
FROST DAMAGE (BUDS AND FLOWERS) **28**, 129
HONEY FUNGUS **54**, 136
MAGNESIUM DEFICIENCY **22**, 146
NITROGEN DEFICIENCY 151
PEAR AND CHERRY 155
PEAR BEDSTRAW APHID 157
PEAR LEAF BLISTER MITE **16**, 157
PEAR MIDGE **48**, 132

---

A–Z OF PESTS, DISEASES AND DISORDERS

APPLE AND PEAR CANKER

*p.115*, CHERRY BLACKFLY *p.116*, CONIFER APHIDS *p.119*, CURRANT BLISTER APHID *p.121*, CYPRESS APHID *p.121*, GREEN SPRUCE APHID *p.134*, HONEYSUCKLE APHID *p.137*, JUNIPER APHID *p.140*, LETTUCE ROOT APHID *p.145*, LUPIN APHID *p.146*, MEALY CABBAGE APHID *p.146*, MEALY PLUM APHID *p.146*, MELON COTTON APHID *p.147*, PEACH APHID *p.156*, PEAR BLISTRAW APHID *p.157*, PLUM LEAF-CURLING APHID *p.161*, ROOT APHIDS *p.170*, ROSE APHIDS *p.170*, ROSY APPLE APHID *p.173*, WATER LILY APHID *p.186*, WILLOW BARK APHID *p.190*, WOOLLY APHID *p.192*

### APPLE AND PEAR CANKER

**SYMPTOMS** Areas of bark sink inwards (*see p.36*), usually starting near a wound or bud. The bark becomes discoloured and then shrinks and cracks, forming concentric rings of flaky bark. The branch may become swollen around the cankered area. As the canker enlarges, the shoot may be girdled and so foliage and growth above it starts to die back. Stems and trunks may be attacked and very occasionally semi-mature fruits are infected and show rotting. In summer raised, white fungal pustules develop on the canker, and in winter raised, red fruiting bodies develop. Apples and pears are frequently affected; poplars (*Populus*), beeches (*Fagus*), willows (*Salix*), hawthorn (*Crataegus*) and sorbus may also succumb.
**CAUSE** The fungus *Nectria galligena*, which is spread mainly during the spring as wind-borne spores enter through wounded areas such as leaf scars, pruning cuts, scab infection, frost crack, attack by woolly aphid or other wounds.

**CONTROL** Prune out whole spurs or branches where possible. On larger branch or trunk infections, carefully pare away all infected and damaged bark and wood, cutting back to completely clean tissue, and treat the wound with a proprietary wound paint, such as one containing cresylic acid. Dispose of prunings and parings carefully. Sprays used for the control of apple scab and apple powdery mildew (*see p.101*) should also help suppress canker. Improve general growing conditions, in particular soil drainage, since unsuitable conditions such as wet soil may aggravate the disease. Avoid growing apples known to be particularly susceptible, such as 'Cox's Orange Pippin', 'James Grieve', 'Worcester Pearmain' and 'Spartan'. The following cultivars seem to show a degree of resistance under most circumstances: 'Laxton's Superb', 'Newton Wonder', 'Bramley's Seedling' and 'Lane's Prince Albert'.

### APPLE BITTER PIT

**SYMPTOMS** The skin of the apple fruit may develop sunken brown spots, each less than 1mm in diameter (*see p.48*). The flesh is spoiled by numerous pale brown spots or freckles. Affected apples may have an unpleasant, slightly bitter taste. Large fruits are especially susceptible, as are fruits produced by heavy-cropping trees. The symptoms may appear while the fruits are on the tree, but often only develop during storage.
**CAUSE** Calcium deficiency in the fruit. This results in the collapse of small groups of cells within the flesh, so that

brown flecks develop. Calcium levels in the soil may be perfectly adequate, but in dry conditions, the tree cannot take up the amount it needs. Large fruits and heavy-cropping trees are more prone to this disorder because of their greater demand for calcium. Too high a concentration of magnesium or calcium in fruits may also be a cause of bitter pit.
**CONTROL** Keep apple trees well watered and mulched to maintain soil moisture levels. Maintain even growth by feeding with a balanced fertilizer; avoid excessive use of high-nitrogen feeds. Spray developing fruits with calcium nitrate solution from early summer until early autumn. The cultivars 'Bramley's Seedling', 'Discovery' and 'Crispin' may be damaged by this chemical, so care should be taken when treating them.

### APPLE BLOSSOM WEEVIL

**SYMPTOMS** Apple flower buds start to develop but the petals fail to open and turn brown (*see p.46*). If these 'capped' blossoms are pulled apart, a creamy-white grub, up to 7mm long, or the pupal stage, will be seen. Damaged blossoms cannot produce fruits.
**CAUSE** A weevil, *Anthonomus pomorum*, which lays eggs in the unopened buds during mid- and late spring. The adult weevil is brown with whitish markings on its upper surface and is 4–6mm long.

ACTUAL LENGTH 4–6MM

ADULT APPLE BLOSSOM WEEVIL

**CONTROL** Light infestations have little impact on the yield of fruits. Spraying with pirimiphos-methyl or fenitrothion when flowers are at the green cluster stage is only worthwhile if heavy attacks have occurred in previous years.

### APPLE CAPSID

**SYMPTOMS** Ripe apples have raised yellowish-brown corky bumps or scabby areas on the skin (*see p.47*). Leaves at shoot tips may be distorted, with many small holes.
**CAUSE** Apple capsid, *Plesiocoris rugicollis*, which is a bright green sap-feeding insect up to 7mm long. It injects a toxic saliva into the plant as it feeds on the foliage and immature fruits during the early summer. This kills some plant cells and, as the developing leaves grow and expand, these dead areas tear into small holes. Fruits respond to the capsid's saliva by developing raised bumps or scabby areas.
**CONTROL** The fruit damage is only skin-deep and light attacks can be tolerated. If damage has been extensive in previous years, spray with dimethoate, heptenophos with permethrin, fenitrothion or pirimiphos-methyl at petal fall.

### APPLE FROST DAMAGE

**SYMPTOMS** Russeting of the skin, usually on the most exposed face of the fruit. The more protected faces are usually unharmed. Damaged areas may develop a corky, roughened appearance and the fruit may become deformed and uneven in growth. Flowers and young foliage subjected to frost wither and die off rapidly.

Page reference in bold type indicates that symptoms are illustrated in the Gallery section

# BY PROBLEM

*See* A–Z of Pests, Diseases and Disorders, *pp.97–192*

Comprehensive entries for the pests, diseases and disorders that may be found on garden plants clearly describe the symptoms, cause and control of each problem. Preventive and control measures include organic, chemical and biological methods as appropriate. In addition, suggestions are given for techniques for avoiding and minimizing further damage, and for resistant cultivars, if they are available. It is cross-referenced to the gallery section and to other associated entries. Over fifty of the entries have black and white illustrations of their own.

# GALLERY OF SYMPTOMS

Even for experienced gardeners, the correct identification of the exact cause of a plant problem, be it a pest, disease or disorder, can be difficult. For ease of comparison with affected plants, the detailed photographs in this section are grouped by the parts of plants attacked, and show similar types of symptom together so that differences can be seen.

# LEAVES

DAMAGE TO THE FOLIAGE can spoil a plant's appearance and, if any part of the plant is edible, the potential crop may be reduced. Leaf markings may take the form of spots, either flat (*pp.12–14*) or raised (*pp.14–17*) to the touch, often with the pycnidia (fruiting bodies) of fungi visible; blotches (*pp.17–20*); mottling (*pp.20–22*) or discoloration (*pp.22–25*). Leaves may also be distorted (*pp.25–27*), dying (*pp.28–29*) or damaged (*pp.29–31*). Pests may be visible (*p.32*).

The majority of leaf markings are caused by diseases. Weaker fungal diseases are only likely to cause extensive damage to a plant already lacking in vigour. Plants are particularly vulnerable if they are growing in unsuitable conditions, have not been well cared for, or have suffered some other serious attack. Both cultural and chemical remedies may be necessary to both improve the general health of the plant and to control the specific problem.

Leaf marking can occur at any stage, but older foliage is usually worst affected, especially in the autumn. Variegated leaf areas are especially prone to damage.

## MARKINGS ON LEAVES

**ORCHID VIRUS** DISEASE • *See p.154*

Dark brown or black, or occasionally yellow, markings appear on the leaf. These are usually seen as streaks or spotting. Flowering is often affected and the plant's overall vigour will decline.

**PLANTS AFFECTED** Most orchids
**SEASON** All year

**STRAWBERRY LEAF SPOT** DISEASE • *See p.179*

Reddish-purple spots are found on the leaves, which may later enlarge (*see p.17*). Off-white fungal growth may develop over the surface. Flowers can be affected.

**PLANTS AFFECTED** Strawberries
**SEASON** Mainly mid-summer to autumn

**BEAN CHOCOLATE SPOT** DISEASE • *See p.105*

Chocolate-brown spots are seen on the leaves and there are similar coloured streaks on the stems, pods and flowers. The spots may join together.

**PLANTS AFFECTED** Broad beans
**SEASON** Mainly late spring to autumn

**DELPHINIUM BLACK BLOTCH** DISEASE • *See p.122*

Black blotches appear on the leaves, which enlarge and often spread to the leaf stalks and stems. The leaves may eventually die off.

**PLANTS AFFECTED** Delphiniums
**SEASON** Late spring to autumn

**LEAF SPOT (FUNGAL)** DISEASE • *See p.143*

Rounded, grey or brown spots develop on the leaves (here of a castor oil plant) and may join together. Numerous pycnidia are often apparent on the spots.

**PLANTS AFFECTED** Most plants
**SEASON** Varies with host plant

**CELERY LEAF SPOT** DISEASE • *See p.115*

Dark brown or black, pinprick-sized spots appear on the leaves, which may discolour and die off. The leaf stalks can also be affected and the crop reduced.

**PLANTS AFFECTED** Celery and celeriac
**SEASON** All season

**MAHONIA RUST** DISEASE • *See p.146*

Dark brown, spore-filled pustules are found on the lower leaf surface, with bright orange spots on the upper surface (*see p.24*). Leaves may fall early, but overall plant health is rarely affected.

**PLANTS AFFECTED** Mahonias
**SEASON** Late spring to early autumn

## MARKINGS ON LEAVES *CONTINUED*

### BRASSICA DARK LEAF SPOT
DISEASE • *See p. 110*

Grey-brown, round spots are seen on the leaf surface, often with concentric rings of pycnidia on them. Affected areas may drop away, leaving a ragged leaf.

**PLANTS AFFECTED** Brussels sprouts, cauliflowers
**SEASON** Mainly mid-summer to late autumn

### LEAF SPOT (BACTERIAL)
DISEASE • *See p. 143*

Circular or angular spots of dead tissue are found on the leaves (here of hibiscus), usually surrounded by a bright yellow "halo". No fungal bodies present.

**PLANTS AFFECTED** Many plants
**SEASON** Varies with host plant

### PANSY LEAF SPOT
DISEASE • *See p. 154*

Purple-brown or buff-coloured spots appear on leaves and may join together. The affected areas may fall away, leaving holes. Older plants are usually the most badly affected.

**PLANTS AFFECTED** Pansies
**SEASON** Mainly mid-summer to late autumn

### ROSE BLACK SPOT
DISEASE • *See p. 171*

Purple-black spots are seen on leaves. The spots may join together or remain separate. Leaf yellowing and premature fall are common.

**PLANTS AFFECTED** Roses
**SEASON** Mainly early summer to autumn

### LEAF SPOT (FUNGAL)
DISEASE • *See p. 143*

Brown spots are seen on the leaves (here of a primula), often with a grey centre and sometimes accompanied by leaf yellowing. Pycnidia may be visible.

**PLANTS AFFECTED** Most plants
**SEASON** Varies with host plant

*Main part of leaf tissue remains healthy green colour*

*Blotches appear shiny and may be up to 1.5cm in diameter*

*Surface of spot is rough and slightly raised*

### TAR SPOT OF ACER
DISEASE • *See p. 180*

Large, black blotches with bright yellow "halos" develop on upper leaf surfaces. Leaves may fall early and the tree can look disfigured, but overall vigour is not usually affected.

**PLANTS AFFECTED** Acers (usually, sycamore)
**SEASON** Late spring to autumn

## MARKINGS ON LEAVES *CONTINUED*

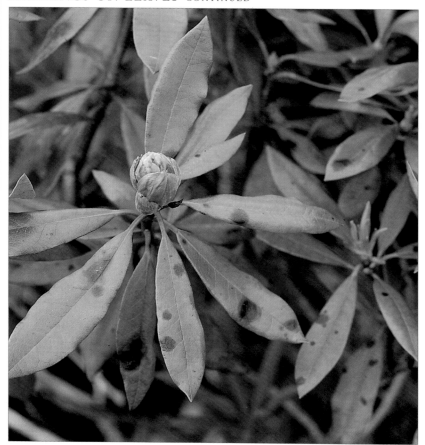

### RHODODENDRON LEAF SPOT
DISEASE • *See p.169*

Brownish-purple spots on leaves often have a pink or purple ring around the edge of each one. Concentric ringing is also seen, with pycnidia in the centre.

**PLANTS AFFECTED** Rhododendrons

**SEASON** All year (mainly late summer onwards)

### ARBUTUS LEAF SPOT
DISEASE • *See p.102*

Small spots appear, each with a dry, pale grey centre and a purple edge. These may enlarge but will rarely join together. Leaf fall may be slightly premature.

**PLANTS AFFECTED** Strawberry tree (*Arbutus*)

**SEASON** Mainly mid-summer to autumn

### PEA LEAF AND POD SPOT
DISEASE • *See p.155*

Yellow or brown, often sunken, spots are found on the leaves and pods. There is a general yellowing of adjacent tissues. Pycnidia are usually apparent.

**PLANTS AFFECTED** Peas

**SEASON** All season, but mainly summer

## RAISED MARKINGS ON LEAVES

### CHRYSANTHEMUM WHITE RUST
DISEASE • *See p.117*

Raised, warty, buff-coloured pustules appear beneath the leaf surface, corresponding with yellow, sunken areas above. Leaves may die back.

**PLANTS AFFECTED** Chrysanthemums

**SEASON** All season (mainly summer and autumn)

### WHITE BLISTER
DISEASE • *See p.189*

Shiny, white, raised spots are seen on the lower leaf surface of brassicas (here on a cabbage), often arranged in concentric rings. The upper leaf surface is yellowed and usually distorted.

**PLANTS AFFECTED** Edible brassicas

**SEASON** Mainly early summer to late autumn

### LIME NAIL GALL MITE
PEST • *See p.285*

Red or yellowish-green, tubular structures containing microscopic mites grow on the upper leaf surfaces. Heavily infested leaves may be distorted.

**PLANTS AFFECTED** Lime trees (*Tilia*)

**SEASON** Late spring to autumn

# RAISED MARKINGS ON LEAVES CONTINUED

### VIBURNUM WHITEFLY
PEST • *See p. 185*

Black, oval pupae encrusted with white wax are present on the lower leaf surface for much of the year. White-winged adults are found during the summer.

**PLANTS AFFECTED** Laurustinus; also *Arbutus*
**SEASON** All year

### CUSHION SCALE
PEST • *See p. 121*

Honeydew and sooty mould develop on the upper leaf surfaces. Oval, yellow or brown scales live on the underside (here of a camellia leaf) and in early summer white egg bands are deposited.

**PLANTS AFFECTED** Mainly camellias and hollies
**SEASON** All year

### PERIWINKLE RUST
DISEASE • *See p. 158*

Numerous small, dark brown, spore-filled pustules are found on the underside of the leaf. Leaves become pitted and distorted. The whole plant may become distorted and die back.

**PLANTS AFFECTED** Periwinkles (*Vinca*)
**SEASON** All season (mainly summer and autumn)

### PELARGONIUM RUST
DISEASE • *See p. 158*

Dark brown, fungal pustules are seen on the lower leaf surfaces, usually arranged in concentric rings. There is yellow discoloration on the upper surfaces.

**PLANTS AFFECTED** Zonal pelargoniums
**SEASON** All season

### IRIS RUST
DISEASE • *See p. 139*

Numerous raised, orange pustules are found on leaves. Leaves yellow and may die back prematurely. Older leaves are usually the worst affected. The plant's vigour is rarely affected.

**PLANTS AFFECTED** Irises
**SEASON** Mainly late summer to autumn

### ANTIRRHINUM RUST
DISEASE • *See p. 99*

Concentric rings of tiny, dark brown, spore-filled pustules are seen on lower leaf surfaces. A yellow "halo" appears around the edge of each spot.

**PLANTS AFFECTED** Snapdragons (*Antirrhinum*)
**SEASON** All season (mainly spring and summer)

### VINE ERINOSE MITE
PEST • *See p. 186*

Parts of the leaf bulge upwards; the underside of affected areas is covered in creamy-white hairs which turn brown. Microscopic mites live amongst the hairs.

**PLANTS AFFECTED** Grape vines
**SEASON** Late spring to autumn

### RHODODENDRON POWDERY MILDEW
DISEASE • *See p. 169*

Pale brown to buff, felty patches of fungal growth develop under leaves, with yellow blotches above. Leaves fall early and the plant may be weakened.

**PLANTS AFFECTED** Rhododendrons
**SEASON** All year on evergreen rhododendrons

### HYDRANGEA SCALE
PEST • *See p. 138*

Oval, white, waxy egg masses appear on the leaves and stems. Flat, oval, yellowish-brown scale insects are also present, and live on the underside of the leaves next to the leaf veins.

**PLANTS AFFECTED** Mainly hydrangeas
**SEASON** All year (eggs late spring to mid-summer)

## RAISED MARKINGS ON LEAVES CONTINUED

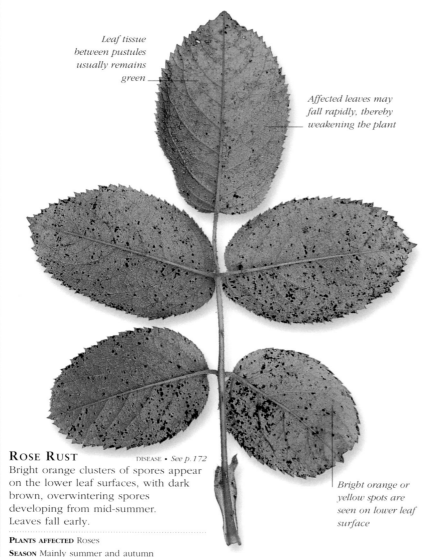

*Leaf tissue between pustules usually remains green*

*Affected leaves may fall rapidly, thereby weakening the plant*

### PLUM LEAF GALL MITE
PEST • See p. 161

Whitish-green structures containing microscopic mites develop on the upper leaf surface, especially around the leaf margins.

**PLANTS AFFECTED** Wild and cultivated plums
**SEASON** Late spring to autumn

### ROSE RUST
DISEASE • See p. 172

Bright orange clusters of spores appear on the lower leaf surfaces, with dark brown, overwintering spores developing from mid-summer. Leaves fall early.

**PLANTS AFFECTED** Roses
**SEASON** Mainly summer and autumn

*Bright orange or yellow spots are seen on lower leaf surface*

### APPLE SCAB
DISEASE • See p. 101

Greyish-green or khaki, scabby spots develop on the leaf surface. Slight blistering or puckering may also occur. Leaves yellow and fall prematurely and the tree is weakened.

**PLANTS AFFECTED** Edible and ornamental *Malus*
**SEASON** Mainly late spring and summer

### PEAR LEAF BLISTER MITE
PEST • See p. 157

The foliage develops many yellowish-green or pink, slightly raised patches, which later turn black. These patches form a band on each side of the midrib.

**PLANTS AFFECTED** Pear trees
**SEASON** Mid-spring to autumn

### HEMISPHERICAL SCALE
PEST • See p. 135

Convex, dark brown, hemispherical scale insects infest leaves (here of a fern) and stems. They are often accompanied by sticky honeydew and sooty mould.

**PLANTS AFFECTED** Many greenhouse plants
**SEASON** All year

### ELM GALL MITE
PEST • See p. 125

Hard, pale green swellings containing microscopic mites develop on the upper surface of elm leaves. Almost every leaf may be affected, but the tree's vigour is not usually reduced.

**PLANTS AFFECTED** Elms (*Ulmus*)
**SEASON** Mid-spring to autumn

# RAISED MARKINGS ON LEAVES CONTINUED

### LEEK RUST
DISEASE • *See p. 144*

Bright orange, oval pustules are found on the leaves, the outer leaves usually being the worst affected. The leaves may yellow and die off, usually becoming papery and brown.

**PLANTS AFFECTED** Leeks, chives, onions, shallots
**SEASON** Mainly mid-summer to late autumn

### HOLLYHOCK RUST
DISEASE • *See p. 136*

Orange, raised pustules appear on the lower leaf surface (here of a mallow), later turning buff-brown. The upper leaf surface has corresponding yellow or orange spots. Leaves may die back.

**PLANTS AFFECTED** Hollyhocks and mallows
**SEASON** All season

### ACER GALL MITES
PEST • *See p. 98*

Numerous small, round, reddish or yellowish-green structures (galls) form on the upper leaf surface, or felty hairs develop on the undersurface. Heavily infested leaves may become distorted.

**PLANTS AFFECTED** Maples and sycamores (*Acer*)
**SEASON** Late spring to autumn

### OAK GALL WASP (SPANGLE GALLS)
PEST • *See p. 152*

Flat, circular discs or galls containing tiny grubs form on the lower leaf surfaces. They are reddish at first, but yellowish-brown when mature.

**PLANTS AFFECTED** Deciduous oaks (*Quercus*)
**SEASON** Late summer to mid-autumn

### OAK GALL WASP (SILK BUTTON GALLS)
PEST • *See p. 152*

Golden-brown discs with central depressions develop on the underside of oak leaves. Each disc contains a single grub which will become a gall wasp.

**PLANTS AFFECTED** Deciduous oaks (*Quercus*)
**SEASON** Late summer to autumn

### WILLOW BEAN GALL SAWFLIES
PEST • *See p. 190*

Coffee bean-sized swellings, which are red or yellowish-green, form on willow leaves. Each gall contains a small, caterpillar-like larva.

**PLANTS AFFECTED** Willows (*Salix*)
**SEASON** Early summer to early autumn

# BLOTCHY MARKINGS ON LEAVES

### BEET LEAF MINER
PEST • *See p. 106*

Pale green blotches later dry up and become brown. White maggots live within the mined areas and later turn into brown pupae, from which adult flies will emerge.

**PLANTS AFFECTED** Beetroot and spinach beet
**SEASON** Late spring to late summer

### STRAWBERRY LEAF SPOT
DISEASE • *See 179*

Early spotting (*see p. 12*) develops into large blotches, each with a grey centre. These may appear slightly raised. Off-white fungal growth may develop.

**PLANTS AFFECTED** Strawberries
**SEASON** Mainly mid-summer to autumn

### HOLLY LEAF MINER
PEST • *See p. 136*

Irregular, yellowish-purple blotches occur on the upper leaf surfaces. There may also be linear tunnels in the leaf coming from the central blotch.

**PLANTS AFFECTED** Hollies (*Ilex*)
**SEASON** All year

## BLOTCHY MARKINGS ON LEAVES *CONTINUED*

### WILLOW LEAF BEETLES
PEST • *See p. 190*

Whitish or brown dried-up areas form where the adult beetles or their larvae have grazed away part of the leaves. The larvae often feed in groups of up to 20.

**PLANTS AFFECTED** Willows, poplars and aspens
**SEASON** Early summer to early autumn

### YUCCA LEAF SPOT
DISEASE • *See p. 192*

Circular, brown or grey-brown spots appear on leaf surfaces with concentric rings and central pycnidia. The spots may join together.

**PLANTS AFFECTED** Yuccas
**SEASON** All year (mainly mid-summer onwards)

### SHOTHOLE
DISEASE • *See p. 175*

Spots or blotches of discoloured tissues develop. The damaged areas then fall out, leaving holes (*see p. 30*). No insects are present, but fungal growth is sometimes visible.

**PLANTS AFFECTED** Ornamental and fruiting *Prunus*
**SEASON** Mainly mid- to late summer

### FUCHSIA RUST
DISEASE • *See p. 129*

Numerous tiny, pale orange pustules are found on the lower leaf surface, with the upper surface showing yellow-orange or purple discoloration. Leaves may wither and fall prematurely.

**PLANTS AFFECTED** Fuchsias and willowherbs
**SEASON** Mainly early summer to autumn

### DELPHINIUM LEAF MINER
PEST • *See p. 122*

Brown, dried-up areas extend from the leaf tips to the leaf stalk. Small, white grubs live within the mined areas. Adult flies cause pale spots on the leaves.

**PLANTS AFFECTED** Delphiniums and monkshood
**SEASON** Early to late summer

### LABURNUM LEAF MINERS
PEST • *See p. 141*

Roughly spherical, whitish-brown leaf mines are caused by small caterpillars. In heavy infestations the mines join together to form irregular blotches.

**PLANTS AFFECTED** Laburnum
**SEASON** Early summer to early autumn

### ROSE SLUGWORM
PEST • *See p. 172*

Pale green caterpillars with brown heads graze the leaves, causing white or brown dried-up areas. They usually feed on the lower leaf surface, but may occur on the upper surface in shaded areas.

**PLANTS AFFECTED** Roses
**SEASON** Early summer to early autumn

### RASPBERRY VIRUS
DISEASE • *See p. 168*

Yellow-green flecks, mosaics, mottles or ring spots are seen on the leaves. The leaves may become small and distorted and the plant will have a reduced crop. The plant's overall growth is very poor.

**PLANTS AFFECTED** Cane fruits
**SEASON** Spring to autumn

### RASPBERRY LEAF AND BUD MITE
PEST • *See p. 167*

Pale yellow blotches form on the upper leaf surface, with reduced hairiness of the lower leaf surfaces. Leaves at the shoot tips are often distorted.

**PLANTS AFFECTED** Raspberry
**SEASON** Late spring to early autumn

# BLOTCHY MARKINGS ON LEAVES CONTINUED

*Galls may form between the leaf veins or along them*

*Spots are usually creamy-white, but on purple-tinged leaves they are often red or pink*

*Felt galls disfigure but cause little real damage to plants*

## PEAR AND CHERRY SLUGWORM
PEST • *See p. 157*

Small caterpillars covered in black slime graze the leaf surface, causing it to dry up and turn brown. The larvae often feed on the upper leaf surface.

**PLANTS AFFECTED** Cherry, pear, plum, hawthorn
**SEASON** Early summer to mid-autumn

## FELT GALL MITES
PEST • *See p. 126*

Microscopic mites cause the abnormal growth of dense patches of hairs (here on beech), usually on the underside of the foliage. In late summer the hairs dry up and become brown.

**PLANTS AFFECTED** Some deciduous trees
**SEASON** Late spring to autumn

## APPLE LEAF MINER
PEST • *See p. 101*

Small caterpillars make white or brown lines where they have bored through the leaves. Pupation occurs in white silk cocoons spun on the outside of the leaf.

**PLANTS AFFECTED** Apples and cherries
**SEASON** Early to late summer

## CELERY LEAF MINER
PEST • *See p. 115*

Brown, dried-up patches develop on the leaves. White maggots live within the mined areas and later become brown pupae.

**PLANTS AFFECTED** Celery, parsnip, parsley
**SEASON** Early summer to early autumn

## LILAC LEAF MINER
PEST • *See p. 145*

Irregular, brown blotch mines develop where whitish-green caterpillars have fed inside the foliage. The caterpillars then roll up the leaf tips with silk threads and complete their feeding there.

**PLANTS AFFECTED** Lilac, privet and ash
**SEASON** Early summer to early autumn

## OEDEMA
DISORDER • *See p. 152*

Raised, pale green and, later, brown spots are found on the lower leaf surface (here of a pelargonium). The upper leaf surface may show corresponding yellow spotting.

**PLANTS AFFECTED** Many plants
**SEASON** All year

## BLOTCHY MARKINGS ON LEAVES CONTINUED

*Pycnidia can develop on lesions*

*Large areas of leaf may be killed as blotches join together*

*Affected areas become dry*

### HELLEBORE LEAF BLOTCH
DISEASE • *See p.135*

Slate-grey or grey-brown blotches are seen on the leaves. The lesions may eventually join together. Leaf and flower stems and flowers may also be affected.

**PLANTS AFFECTED** Hellebores
**SEASON** Mainly early summer to autumn

### ROSE POWDERY MILDEW
DISEASE • *See p.171*

White, powdery fungal growth is found on the leaves. They may become distorted, generally yellowed and fall early. Stems can also be affected.

**PLANTS AFFECTED** Roses
**SEASON** All season (mainly summer to autumn)

### HORSE CHESTNUT LEAF BLOTCH
DISEASE • *See p.137*

Tan-brown, irregular blotches appear, each with a bright yellow margin, on the leaves. The leaf margins and tips are most severely affected.

**PLANTS AFFECTED** Horse chestnuts (*Aesculus*)
**SEASON** Mid-summer to autumn

## MOTTLED MARKINGS ON LEAVES

### GLASSHOUSE LEAFHOPPER
PEST • *See p.132*

Coarse, pale spots are found on the upper leaf surface, caused by small, active insects feeding from the lower leaf surface.

**PLANTS AFFECTED** Many greenhouse plants
**SEASON** All year

### FRUIT TREE RED SPIDER MITE
PEST • *See p.129*

The foliage of apple and plum trees becomes increasingly dull and yellowish, usually with fine, pale mottling, and may fall prematurely.

**PLANTS AFFECTED** Apple and plum
**SEASON** Late spring to early autumn

### CARNATION TORTRIX MOTH
PEST • *See p.114*

Pale green caterpillars with brown heads bind leaves together with silk threads and graze away the inner surfaces. The leaves will then dry up and turn brown.

**PLANTS AFFECTED** Greenhouse and garden plants
**SEASON** All year

## MOTTLED MARKINGS ON LEAVES CONTINUED

### PELARGONIUM VIRUS
DISEASE • *See p. 158*

Yellow flecks, spots, mottles or sometimes ring spots are seen on the leaves. Leaves may be small and distorted, and overall growth poor.

**PLANTS AFFECTED** Pelargoniums
**SEASON** All season

### EUONYMUS SCALE
PEST • *See p. 126*

Pale yellow mottling and elongate white scale insects are found on the foliage. Stems become encrusted with the dark brown, pear-shaped female scales. Heavy infestations may cause dieback.

**PLANTS AFFECTED** Evergreen spindles (*Euonymus*)
**SEASON** All year

### ROSE LEAFHOPPER
PEST • *See p. 171*

A coarse, pale mottling develops on the upper leaf surface. Pale yellow insects live on the underside of the leaves. Roses growing in warm, sheltered places can become heavily infested by summer.

**PLANTS AFFECTED** Roses
**SEASON** Mid-spring to mid-autumn

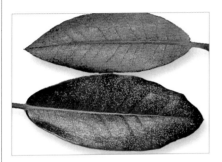

### RHODODENDRON LACEBUG
PEST • *See p. 168*

A coarse, pale yellow mottling develops on the upper leaf surface. A rusty brown deposit and insects are found on the lower leaf surface.

**PLANTS AFFECTED** Rhododendrons
**SEASON** Early spring to late summer

### ROSE VIRUS
DISEASE • *See p. 172*

Yellow flecks, spots, mottles, vein-banding or ring spots are found on the leaves. Flowers may show slight colour-breaking symptoms and growth may be affected, but these are both rare.

**PLANTS AFFECTED** Roses
**SEASON** All season

### CUCUMBER MOSAIC VIRUS
DISEASE • *See p. 120*

Yellow mottles, mosaicing and blotches are seen on leaves (here of a marrow), which may be distorted (*see p. 25*). The whole plant is stunted and dies early.

**PLANTS AFFECTED** Wide host range
**SEASON** All season

21

## MOTTLED MARKINGS ON LEAVES CONTINUED

### BANDED PALM THRIPS
PEST • *See p. 104*

Black, white-banded, narrow-bodied, elongate insects cause a silvery-brown discoloration of the upper leaf surfaces of the younger foliage.

**PLANTS AFFECTED** Some house plants
**SEASON** All year

### GLASSHOUSE RED SPIDER MITE
PEST • *See p. 132*

Early stages of spider mite attack show as a fine, pale mottling of the upper leaf surface. Tiny, yellow-green mites are found on the underside of the leaves.

**PLANTS AFFECTED** Greenhouse and garden plants
**SEASON** Mid-spring to late autumn

### PLUM RUST
DISEASE • *See p. 161*

Numerous dark brown, spore-filled pustules are found on the underside of the leaf. Corresponding bright yellow or orange spots are seen on the upper surface. Leaves fall prematurely.

**PLANTS AFFECTED** Plums and anemones
**SEASON** Mainly late summer and autumn

### CAMELLIA YELLOW MOTTLE
DISEASE • *See p. 114*

Yellow or white mottled markings and patches of discoloration form on leaves. General growth is not affected. Usually only isolated branches are affected.

**PLANTS AFFECTED** Camellias
**SEASON** All year

### TULIP FIRE
DISEASE • *See p. 183*

Buff-coloured or bleached flecks develop on the leaves. Early infection causes extensive leaf distortion and the development of fuzzy, grey fungal growth. Flowers may also be affected.

**PLANTS AFFECTED** Tulips
**SEASON** Mid- to late spring

### PRIVET THRIPS
PEST • *See p. 165*

Elongate, narrow-bodied insects cause green leaves (*above right*) to develop a silvery-brown discoloration (*above left*), especially during late summer. The thrips are found on the upper leaf surfaces.

**PLANTS AFFECTED** Privet and lilac
**SEASON** Early summer to early autumn

## DISCOLOURED LEAVES

### MAGNESIUM DEFICIENCY
DISORDER • *See p. 146*

Yellow, brown or otherwise discoloured areas form between the leaf veins and around the leaf edge (here on a tomato). Older leaves are the worst affected.

**PLANTS AFFECTED** Many plants
**SEASON** Mainly late spring to autumn

### LETTUCE DOWNY MILDEW
DISEASE • *See p. 144*

Angular, yellow discoloured areas on the leaves rapidly turn pale brown and papery or soft. Off-white fungal growth may develop beneath the leaf.

**PLANTS AFFECTED** Lettuces
**SEASON** All season (mainly summer and autumn)

### CACTUS CORKY SCAB
DISORDER • *See p. 113*

Pale, roughened areas appear on the surface and may become brown and sunken. In extreme cases holes develop as the affected tissue dies.

**PLANTS AFFECTED** Cacti, succulents
**SEASON** All year

## DISCOLOURED LEAVES CONTINUED

### VIRUS
DISEASE • *See p.187*

Virus symptoms include yellow vein-clearing (here on a sweet pea), mottling, mosaicing, spotting or ring spotting seen on the leaves. Leaves may be distorted (*see p.25*) and general growth is stunted.

**PLANTS AFFECTED** Wide host range
**SEASON** Mainly late spring to late summer

### BEGONIA POWDERY MILDEW
DISEASE • *See p.107*

White or pale grey, powdery fungal growth is found mainly on upper leaf surfaces. Affected leaves turn papery and soft. May affect stems and buds.

**PLANTS AFFECTED** Begonias
**SEASON** All year

### NARCISSUS VIRUS
DISEASE • *See p.150*

Yellow streaking or flecking is seen on leaves. Plants can become stunted and occasionally symptoms are apparent on flower stems or petals. Growth and performance may be unaffected.

**PLANTS AFFECTED** Daffodils, narcissi
**SEASON** Mainly early to late spring

### GLASSHOUSE RED SPIDER MITE
PEST • *See p.132*

If heavily infested, leaves lose much of their green colour and dry up. Fine silk webbing may appear between the leaves and the mites swarm over the plant.

**PLANTS AFFECTED** Greenhouse and garden plants
**SEASON** Mid-spring to late autumn

### WEEDKILLER DAMAGE – CONTACT
DISORDER • *See p.189*

Yellow spots and blotches appear (here on a rose), which may turn brown as the leaf tissue dies off. Plant growth can be stunted and the plant may die.

**PLANTS AFFECTED** Any plant
**SEASON** Mainly spring to summer

### LEEK MOTH
PEST • *See p.144*

Whitish-brown patches develop on the leaves where the leaf-mining caterpillars have been feeding. Stems and bulbs are also tunnelled. Small, oval, net-like cocoons may be found on the foliage.

**PLANTS AFFECTED** Leeks and onions
**SEASON** Early summer to mid-autumn

23

## DISCOLOURED LEAVES CONTINUED

### IRON AND MANGANESE DEFICIENCY

DISORDER • *See p.139*

Yellowing between the veins on leaves (here of a rhododendron) is seen. The youngest leaves are affected most badly. Overall growth may be affected.

**PLANTS AFFECTED** Many lime-hating plants
**SEASON** Varies with host plant

### OAK POWDERY MILDEW

DISEASE • *See p.152*

Powdery, white fungal growth appears on the leaf surface. Leaves, particularly young ones, may be distorted and die off prematurely.

**PLANTS AFFECTED** Oaks (*Quercus*)
**SEASON** Spring to autumn

### CHERRY BLACKFLY

PEST • *See p.116*

Leaves at the shoot tips become curled, with black aphids on the underside. Affected leaves often turn brown in mid-summer.

**PLANTS AFFECTED** Cherry trees
**SEASON** Late spring to late summer

*Circular holes eaten in foliage by adult weevils*

*Foliage becomes ragged*

### POTASSIUM DEFICIENCY

DISORDER • *See p.162*

Leaf tips and edges are scorched and may show brownish-purple spotting on the underside (here on a pelargonium). Flowering and fruiting will be poor.

**PLANTS AFFECTED** Many plants
**SEASON** Mainly mid-summer to autumn

### BEECH LEAF-MINING WEEVIL

PEST • *See p.106*

The leaf tips dry up and turn brown where the leaf-mining larvae have been feeding. The dried-up areas will then break off.

**PLANTS AFFECTED** Beeches (*Fagus*)
**SEASON** Mid-spring to late summer

### MAHONIA RUST

DISEASE • *See p.146*

Bright, discoloured spots appear on the upper leaf surface, with pustules on the underside (*see p.12*). Leaves may fall prematurely, but the overall health of the plant is rarely affected.

**PLANTS AFFECTED** Mahonias
**SEASON** Late spring to early autumn

## DISCOLOURED LEAVES CONTINUED

### GERANIUM DOWNY MILDEW
DISEASE • *See p. 131*

Pale green or light brown, angular patches of discoloration are seen on the leaves. Off-white, slightly fuzzy fungal patches develop beneath.

**PLANTS AFFECTED** *Geranium* species and cultivars
**SEASON** All season

### CUCUMBER MOSAIC VIRUS
DISEASE • *See p. 120*

Leaves may be distorted and reduced in size, and also discoloured, with blotches and mosaicing (*see p.21*). On some hosts, flowers show colour-breaking.

**PLANTS AFFECTED** Wide host range
**SEASON** All season

### CHRYSANTHEMUM LEAF MINER
PEST • *See p. 131*

White or brown sinuous lines develop where the leaf-mining grubs have been feeding. In heavy attacks most of the leaf area becomes discoloured.

**PLANTS AFFECTED** Chrysanthemums
**SEASON** Mid-spring to late autumn

### POWDERY MILDEW
DISEASE • *See p. 165*

White, powdery fungal growth develops on the upper leaf surface (here of a geranium). The leaf can become distorted and the leaf stalk may also be affected.

**PLANTS AFFECTED** Wide range of plants
**SEASON** Varies with host plant

### LEAF AND BUD EELWORMS
PEST • *See p. 142*

Brown areas enclosed by the larger leaf veins form islands or wedges in the leaves. Eventually the whole leaf will become discoloured.

**PLANTS AFFECTED** Many plants
**SEASON** Mainly late summer to late autumn

### CARROT MOTLEY DWARF VIRUS
DISEASE • *See p. 114*

Foliage turns yellow and pink and may become twisted and stunted. Leaf stalks are also twisted. Foliage dies off prematurely and the crop is reduced.

**PLANTS AFFECTED** Carrots and parsley
**SEASON** Mainly summer

## DISTORTED LEAVES

### AMERICAN GOOSEBERRY MILDEW
DISEASE • *See p. 98*

Powdery fungal growth is seen on puckered leaves, usually on the upper surfaces. Young leaves wither and die off. Stems and fruits may be infected.

**PLANTS AFFECTED** Gooseberries and blackcurrants
**SEASON** All season

### APPLE POWDERY MILDEW
DISEASE • *See p. 101*

Powdery, white fungal growth develops on leaves. On young leaves the upper and lower surfaces may be attacked, and the foliage withers and dies.

**PLANTS AFFECTED** Edible apples and crab apples
**SEASON** Spring to autumn

### VIRUS
DISEASE • *See p. 187*

Growth may be distorted and stunted (as on this tobacco plant) and the plant usually dies off early. Other symptoms include yellow mottling, spotting or mosaicing on the leaves.

**PLANTS AFFECTED** Most plants
**SEASON** Mainly summer and autumn

## DISTORTED LEAVES *CONTINUED*

*Withered remains of leaf still attached to gall*

*Texture is firm but compressible and becomes softer with age*

*Gall is rounded, but may also be forked with finger-like protrusions*

### CAMELLIA GALL
DISEASE • *See p. 113*

Large galls, up to 18cm long, develop in place of leaves. The surface becomes covered in spores and so may appear white in places. The plant's overall vigour is rarely affected.

**PLANTS AFFECTED** Camellias
**SEASON** Late summer to autumn

### PEACH LEAF CURL
DISEASE • *See p. 156*

Leaves become puckered; they are pale green at first, but soon turn red and purple. A powdery, white spore layer develops on the surface.

**PLANTS AFFECTED** Peaches, nectarines, almonds
**SEASON** Spring and early summer

### BAY SUCKER
PEST • *See p. 104*

Leaves at the shoot tips become yellow, with thickened, down-curled margins. Later, the damaged areas dry up and turn brown. Fluffy, white sucker nymphs may be seen on the lower leaf surface.

**PLANTS AFFECTED** Bay (*Laurus nobilis*)
**SEASON** Late spring to mid-autumn

### BLACKCURRANT GALL MIDGE
PEST • *See p. 108*

Leaves fail to expand, remaining tightly crumpled and later drying up. Small, white maggots may be found between the folds of infested leaves.

**PLANTS AFFECTED** Blackcurrant
**SEASON** Late spring to late summer

### BOX SUCKER
PEST • *See p. 109*

Shoot extension is severely restricted and the cupped leaves form cabbage-like structures. Flattened, pale green, aphid-like insects may be present and excrete a white, runny liquid.

**PLANTS AFFECTED** Box (*Buxus*)
**SEASON** Mid-spring to mid-summer

### IRREGULAR WATERING
DISORDER • *See p. 140*

Leaves (here of a rhododendron) become puckered and distorted, but retain their normal colour. Not all leaves will be affected.

**PLANTS AFFECTED** All plants
**SEASON** All season (mainly spring and summer)

## DISTORTED LEAVES *CONTINUED*

### VIRUS
DISEASE • *See p. 187*

The plant (here a *Passiflora*) may be distorted and stunted and will usually die off early. Leaves often also display yellow mottling, spotting, mosaicing or veining.

**PLANTS AFFECTED** Most plants
**SEASON** Mainly summer and autumn

### WEEDKILLER DAMAGE – BROADLEAVED
DISORDER • *See p. 188*

Leaves (here of a pelargonium) become cup-shaped, often with the veins close together. Stems and petioles may be curled or distorted (*see below right*).

**PLANTS AFFECTED** Any plant
**SEASON** Mainly spring and summer

### ROSE LEAF-ROLLING SAWFLY
PEST • *See p. 171*

The margins of affected leaflets are rolled inwards to form slender tubes in which small, pale green larvae feed. Rolled leaflets often hang downwards.

**PLANTS AFFECTED** Cultivated and wild roses
**SEASON** Late spring to mid-summer

### VIOLET GALL MIDGE
PEST • *See p. 186*

Infested leaves have greatly swollen and curled leaf margins. Pale orange maggots or white silk cocoons can be found inside. Galled leaves stay on the plant over the winter.

**PLANTS AFFECTED** Violets
**SEASON** All year

### PLUM LEAF-CURLING APHID
PEST • *See p. 161*

In the spring leaves become tightly curled and crinkled, due to greenfly feeding on the underside of the leaves. Leaves remain distorted all summer.

**PLANTS AFFECTED** Plums, damsons and gages
**SEASON** Early to late spring

### CURRANT BLISTER APHID
PEST • *See p. 121*

Leaves at the shoot tips develop raised, red or yellowish-green, puckered areas. Pale yellow aphids may be found on the underside of affected leaves.

**PLANTS AFFECTED** Red, white and blackcurrants
**SEASON** Mid-spring to early summer

### TARSONEMID MITES
PEST • *See p. 181*

Microscopic mites live and feed at shoot tips and in flower buds. Their feeding causes scarring on the stems and progressive distortion and stunting of the foliage. Growing tip may be killed.

**PLANTS AFFECTED** Many greenhouse plants
**SEASON** All year

### GLEDITSIA GALL MIDGE
PEST • *See p. 133*

Leaves at the shoot tip fail to expand and resemble small seed pods. Each galled leaflet contains several whitish-orange maggots.

**PLANTS AFFECTED** Honey locust (*Gleditsia*)
**SEASON** Early to late summer

### WEEDKILLER DAMAGE – BROADLEAVED
DISORDER • *See p. 188*

Stems and petioles (here of a tomato plant) may become curled and distorted. Leaves narrow or curl up (*see top, centre*) and stems may be roughened.

**PLANTS AFFECTED** Any plant
**SEASON** Mainly spring and summer

# LEAVES DYING OFF

### FROST DAMAGE
DISORDER · *See p.129*

The foliage (here of a hydrangea) becomes withered and hangs limply. It may become blackened or turn pale green or brown. The exposed parts of the plant are most severely affected.

**PLANTS AFFECTED** Many plants

**SEASON** Mainly winter and spring

### KEITHIA THUJINA NEEDLE BLIGHT
DISEASE · *See p.141*

Tiny, black, slightly raised spots are embedded in the foliage, leaving pits or holes as the spores are released. Foliage turns brown and dies back.

**PLANTS AFFECTED** *Thuja plicata*

**SEASON** Mainly late summer and early autumn

### POTATO BLIGHT
DISEASE · *See p.163*

Brown patches develop on the leaves, mainly around the tips and edges. Fluffy, white fungal growth can develop. Leaves eventually wither and die. Top growth may collapse.

**PLANTS AFFECTED** Potatoes and tomatoes

**SEASON** Mid- to late summer

### DROUGHT
DISORDER · *See p.124*

Leaves (here of a hosta) wilt and later dry out or shrivel, often turning brown and papery. In the early stages lowered temperatures and watering may revive the plant; later, dieback or death occurs.

**PLANTS AFFECTED** Many plants

**SEASON** Mainly summer

### PEA WILT
DISEASE · *See p.156*

Leaves turn yellow and die back. Internally the stem's vascular system is stained brown, but this is only visible if the outer stem is stripped away. The whole plant wilts and is killed rapidly.

**PLANTS AFFECTED** Peas and sweet peas

**SEASON** Mainly summer

### LETTUCE GREY MOULD
DISEASE · *See p.144*

Fluffy, grey fungal growth appears on the leaves. This is often combined with brown or orange slimy rotting of the stem and base of the plant.

**PLANTS AFFECTED** Lettuces

**SEASON** Mainly spring to autumn

## LEAVES DYING OFF *CONTINUED*

### NARCISSUS LEAF SCORCH
DISEASE • *See p. 150*

Leaves are red-brown and scorched as they emerge. Brown spots with pycnidia are found on the leaves. Bulbs do not rot or show discoloration.

**PLANTS AFFECTED** Narcissi, amaryllis, crinums
**SEASON** Spring

### VERTICILLIUM WILT
DISEASE • *See p. 185*

Foliage wilts and dies and soft stems wilt. Usually only part of the plant is affected initially. The leaf veins and vascular tissue are discoloured.

**PLANTS AFFECTED** Many plants
**SEASON** All year

### LEAF SPOT (BACTERIAL)
DISEASE • *See p. 143*

Spots develop on the foliage (here of a calathea), often with a water-soaked appearance. A distinct yellow "halo" may develop around each spot (*see p. 13*).

**PLANTS AFFECTED** Depends on bacterium involved
**SEASON** Varies with host plant

## HOLES IN LEAVES

### IRIS SAWFLY
PEST • *See p. 139*

Brownish-grey, caterpillar-like larvae eat V-shaped notches from the leaf margins. Older larvae cause extensive defoliation, especially of the upper parts of the leaves.

**PLANTS AFFECTED** Waterside irises only
**SEASON** Early to mid-summer

### SLUGS
PEST • *See p. 176*

Slugs feed mainly at night and eat irregular holes in foliage or graze away the surface tissues. A silvery slime trail is sometimes left on plants. Snails (*see right*) cause similar damage.

**PLANTS AFFECTED** Many plants
**SEASON** All year

### SNAILS
PEST • *See p. 177*

Irregular holes are eaten in the foliage and a silvery slime trail may be found nearby. Slugs (*see left*) cause similar damage and a torchlight inspection may be required to identify the culprits.

**PLANTS AFFECTED** Many plants
**SEASON** Early spring to late autumn

### CAPSID BUG
PEST • *See p. 114*

Leaves at the shoot tip (here of a phygelius) develop many small, brown-edged holes. The leaves are often distorted and flower buds may be killed or damaged.

**PLANTS AFFECTED** Many plants
**SEASON** Late spring to late summer

### FLEA BEETLES
PEST • *See p. 127*

Small, rounded holes are scalloped out of the upper leaf surface (here of a wallflower). Small, blue-black or black, yellow-striped beetles, which jump off the foliage, may be seen.

**PLANTS AFFECTED** Brassicas and related plants
**SEASON** Mid-spring to late summer

### VINE WEEVIL (DAMAGE BY ADULT)
PEST • *See p. 186*

Adult vine weevils eat irregular notches from the leaf margins (here of a rhododendron). On shrubs the leaves closest to the ground are attacked.

**PLANTS AFFECTED** Many plants
**SEASON** Mid-spring to mid-autumn

# HOLES IN LEAVES CONTINUED

### ELEPHANT HAWK MOTH
PEST • *See p. 125*

Large holes are eaten in the leaves by blackish-brown or green caterpillars. The caterpillar's eye markings give them a snake-like appearance.

**PLANTS AFFECTED** Mainly fuchsias and busy Lizzies
**SEASON** Mid- to late summer

### LEAF-CUTTING BEES
PEST • *See p. 142*

Semi-circular or elliptical pieces are removed from the leaf margins (here of a rose). The smooth outline of the missing pieces distinguishes leaf-cutting bee damage from other leaf-eating pests.

**PLANTS AFFECTED** Many plants, especially roses
**SEASON** Early to late summer

### CABBAGE MOTH
PEST • *See p. 112*

Infested plants have many holes of various sizes in their foliage. On cabbages the caterpillars bore into the heart leaves and soil the edible parts with their droppings.

**PLANTS AFFECTED** Brassicas, onions, pelargoniums
**SEASON** Early summer to early autumn

### WATER LILY BEETLE
PEST • *See p. 188*

Both the adult beetles and their larvae feed on the upper leaf surface, grazing elongate strips which rot away to form holes. Damaged leaves turn yellow and decay prematurely.

**PLANTS AFFECTED** Water lilies
**SEASON** Late spring to late summer

### CHINA MARK MOTH
PEST • *See p. 116*

Elliptical pieces of various sizes are cut from the leaf margins of water lilies and used to form protective cases for the caterpillars. The larval cases can be found on the underside of the leaves.

**PLANTS AFFECTED** Water lilies
**SEASON** Early to late summer

### SHOTHOLE
DISEASE • *See p. 175*

Leaves develop spots (*see p. 18*) where the leaf tissue dies and falls away, leaving holes, usually with a slight browning around their edges, in the leaves (here of a *Prunus*).

**PLANTS AFFECTED** Many plants
**SEASON** Mainly summer and autumn

### ARUNCUS SAWFLY
PEST • *See p. 102*

Pale green caterpillars initially eat small holes between the leaf veins. However, by the time they are fully grown, they may have eaten everything except the larger veins.

**PLANTS AFFECTED** Goatsbeard (*Aruncus dioicus*)
**SEASON** Early to late summer

### PIGEONS
PEST • *See p. 160*

Pigeons rip pieces off leaves with their beaks, causing ragged tears on the foliage (here of a cabbage). In heavy attacks most of the soft parts of the leaves will be removed.

**PLANTS AFFECTED** Brassicas and peas
**SEASON** All year

### LACKEY MOTH
PEST • *See p. 141*

Hairy caterpillars, with red, blue and white stripes running along their bodies, cause extensive defoliation (here of a *Malus*). The feeding area is covered by silk webbing spun by the caterpillars.

**PLANTS AFFECTED** Many trees and shrubs
**SEASON** Mid-spring to early summer

# HOLES IN LEAVES CONTINUED

## SOLOMON'S SEAL SAWFLY
PEST • See p. 177

Whitish-grey caterpillars with black heads eat slots in the middle of the leaves. By the time they are fully grown, they will have caused extensive defoliation.

**PLANTS AFFECTED** Solomon's seal (*Polygonatum*)
**SEASON** Late spring to mid-summer

## VAPOURER MOTH
PEST • See p. 185

Irregular holes are eaten in the leaf margins by very hairy caterpillars. Plumes of black hairs project from their head and rear ends, with four tufts of buff hairs on the upper surface.

**PLANTS AFFECTED** Mainly trees and shrubs
**SEASON** Late spring to early summer

## MINT BEETLE
PEST • See p. 148

Irregular holes are eaten in the foliage by iridescent green beetles and their rotund, black larvae. Infestations are rarely heavy enough to cause serious damage to the plant.

**PLANTS AFFECTED** Wild and cultivated mints
**SEASON** Mid- to late summer

## GOOSEBERRY SAWFLY
PEST • See p. 133

The soft parts of the foliage are rapidly devoured, leaving just the stalks and larger veins. The larvae are pale green with many black spots or uniformly green in colour.

**PLANTS AFFECTED** Gooseberries, red/white currants
**SEASON** Mid-spring to late summer

## WINTER MOTH
PEST • See p. 191

Small holes are eaten in newly-emerged spring foliage; these holes increase in size as the leaves expand. The pale green caterpillars loosely bind leaves together with silk threads.

**PLANTS AFFECTED** Fruit and other trees
**SEASON** Early to late spring

## PEA AND BEAN WEEVIL
PEST • See p. 155

Uniform, U-shaped notches are eaten from the leaf margins (here of a broad bean leaf). The greyish-brown beetles do not usually cause extensive damage.

**PLANTS AFFECTED** Broad beans and peas
**SEASON** Late spring to late summer

# INSECTS ON LEAVES

### BEECH WOOLLY APHID
PEST • *See p.106*

Pale yellow aphids covered with a fluffy, white wax form dense colonies on the underside of beech leaves. Foliage at the shoot tips may be distorted.

**PLANTS AFFECTED** Beeches (*Fagus*)
**SEASON** Late spring to late summer

### CEDAR APHID
PEST • *See p.115*

Greyish-brown aphids feed at the base of the leaves (needles). Honeydew excreted by the aphids makes foliage sticky and allows the growth of sooty mould. Leaves may turn yellow and fall.

**PLANTS AFFECTED** Cedars (*Cedrus*)
**SEASON** Late spring to mid-summer

### BLACK BEAN APHID
PEST • *See p.107*

Dense colonies of black aphids with small, white markings on their upper surface infest the foliage, and sometimes the shoot tips and flowers. Ants often collect the honeydew they excrete.

**PLANTS AFFECTED** Beans, nasturtiums, dahlias
**SEASON** Late spring to late summer

### CABBAGE WHITEFLY
PEST • *See p.113*

Small, white-winged insects live on the underside of leaves and readily fly up when disturbed. Flat, oval, scale-like immature nymphs also occur on the lower leaf surface.

**PLANTS AFFECTED** Cabbage and other brassicas
**SEASON** All year

### MEALYBUGS
PEST • *See p.147*

Flattened, oval insects covered in a white, mealy or fluffy wax live on the underside of leaves and in the leaf axils. These sap-feeding insects excrete honey-dew which makes the foliage sticky.

**PLANTS AFFECTED** Many greenhouse plants
**SEASON** All year

### MEALY CABBAGE APHID
PEST • *See p.146*

Dense colonies of grey-white aphids live on the underside of brassica leaves and on the shoot tips of young plants. Mottled yellow patches develop on leaves.

**PLANTS AFFECTED** Brassicas and swedes
**SEASON** Late spring to early autumn

### ALDER SUCKER
PEST • *See p.98*

Flattened, greenish-black insects covered in a fluffy, white wax live on the stems and undersides of alder leaves during the spring. The fluffiness disappears in early summer when the nymphs mature.

**PLANTS AFFECTED** Alders (*Alnus*)
**SEASON** Mid-spring to mid-summer

### JUNIPER WEBBER MOTH
PEST • *See p.140*

Dead patches develop in juniper bushes as the foliage turns brown and dries up. The patches are bound together by silk webbing produced by the caterpillars.

**PLANTS AFFECTED** Junipers (*Juniperus*)
**SEASON** Late spring to early summer

### LARCH ADELGID
PEST • *See p.142*

Small, black, aphid-like insects concealed under fluffy, white wax feed by sucking sap from the foliage. Heavy attacks can make the foliage sticky with the pests' excrement.

**PLANTS AFFECTED** Larches (*Larix*)
**SEASON** Mid-spring to late summer

# STEMS AND BUDS

THE STEMS OF A PLANT can be compared to the skeleton of animals, their support giving the plant its characteristic shape and holding the leaves, flowers and buds in the position in which they function best. Since the stems contain the plant's vascular, or transport, system they are also fundamental to its function and when stems are damaged by a pest or by disease, the buds, flowers and leaves carried on them may deteriorate too.

Stems may be attacked in a variety of ways, by pests that gnaw the outer tissues (sometimes ringing the whole stem), feed from them by sucking sap or even tunnel through them. Pest damage may then allow pathogens to enter and so cause further infections, which in turn may spread down the stem and into the crown of the plant; if this occurs, the whole plant may be killed. The tissues of the stem may also be prevented from working by the presence of fungal wilts, which literally clog up the vascular system, so preventing it from functioning.

## STEM PROBLEMS

### CURRANT CLEARWING MOTH
PEST • *See p.121*

A white caterpillar with a brown head bores into the stems of currants, making them liable to snap. Emergence holes may be seen on stems in early summer.

**PLANTS AFFECTED** Currants
**SEASON** Late spring to late summer

### VERTICILLIUM WILT
DISEASE • *See p.185*

Soft stems and leaves wilt (*see p.29*). Conducting tissue is stained brown, causing longitudinal streaking which is only visible if the bark is removed.

**PLANTS AFFECTED** Many trees and shrubs
**SEASON** Symptoms most apparent in summer

### SHOT-HOLE BORERS
PEST • *See p.175*

Small holes in the bark, resembling damage by shotgun pellets, are the emergence holes of small, brownish-black beetles. Attacks mainly occur on trees growing under stressful conditions.

**PLANTS AFFECTED** Plums, cherries, almonds
**SEASON** Late spring to mid-summer

### PHLOX EELWORM
PEST • *See p.159*

Stems are stunted and abnormally swollen and there may be splitting at the base. Leaves at shoot tips are narrow and may consist of little more than the midribs (as on the top stem, above).

**PLANTS AFFECTED** Perennial and annual phlox
**SEASON** Mid-spring to mid-summer

### DIDYMELLA STEM ROT
DISEASE • *See p.123*

Blackish-brown, sunken blotches appear on the stem base (here of a tomato). Adventitious roots may form above soil level. Older leaves turn yellow.

**PLANTS AFFECTED** Tomatoes and aubergines
**SEASON** Mainly summer

### LILAC BLIGHT
DISEASE • *See p.145*

Young shoots wilt and die, with the older stems developing cankers. Angular brown spots appear on leaves which then die off. Grey mould growth is often present, masking the symptoms.

**PLANTS AFFECTED** Lilacs (*Syringa*)
**SEASON** Late spring and early summer

### BARK BEETLES
PEST • *See p.104*

The larvae of several species of beetle feed underneath the bark of recently dead or dying trees and shrubs. This network of tunnels was made by elm bark beetle grubs.

**PLANTS AFFECTED** Many trees and shrubs
**SEASON** All year

## STEM PROBLEMS CONTINUED

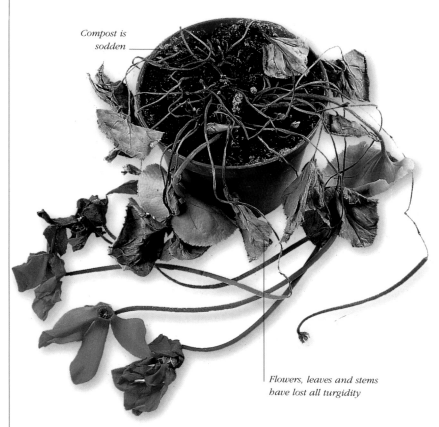

*Compost is sodden*

*Flowers, leaves and stems have lost all turgidity*

### WATERLOGGING DISORDER • *See p.188*

Foliage wilts and yellows, as if suffering from drought, but the condition does not improve if more water is applied. The entire top growth may flop over (as with the cyclamen above).

**PLANTS AFFECTED** Any plant
**SEASON** All year

### DAMPING OFF DISEASE • *See p.122*

Seedlings (here of lettuce) collapse. The stem base is discoloured and may shrink inwards, often appearing water-soaked initially. Fluffy white fungal growth may be present.

**PLANTS AFFECTED** All seedlings
**SEASON** Mainly late winter and spring

### WEEDKILLER DISORDER • *See p.188*
### DAMAGE – BROADLEAVED

Leaves are narrowed and may be cupped and leaf veins appear parallel. Here conifer foliage is distorted. Petioles and stems may show spiral twisting.

**PLANTS AFFECTED** Many plants
**SEASON** Mainly early to mid-summer

### GREY MOULD DISEASE • *See p.134*

Discoloured patches develop on stems (here of a rose) and may girdle them, causing dieback. Fluffy grey fungal growth and black fungal resting bodies can develop on affected areas.

**PLANTS AFFECTED** Soft and woody-stemmed plants
**SEASON** All year

### HONEY FUNGUS DISEASE • *See p.136*

A creamy white mycelium (fungal sheet), smelling strongly of mushrooms, develops beneath the bark on infected stem bases, trunks and roots. The woody material beneath discolours.

**PLANTS AFFECTED** Trees, shrubs, some other plants
**SEASON** All year

### RASPBERRY DISEASE • *See p.167*
### CANE SPOT

Purple spots with silvery-white centres develop on the canes. The spots enlarge rapidly and the canes may split and die off. Leaves and flowers may be affected.

**PLANTS AFFECTED** Raspberries, other cane fruits
**SEASON** Early summer to autumn

## STEM PROBLEMS CONTINUED

### BACTERIAL CANKER DISEASE • See p.104
Patches of bark flatten and sink inwards. Amber-coloured, resin-like ooze may appear close by. Foliage withers and dies off or buds fail to open. Branches may be killed by girdling cankers.

**PLANTS AFFECTED** Ornamental and fruiting *Prunus*
**SEASON** Mainly spring and autumn

### BEECH BARK DISEASE DISEASE • See p.106
Small areas of bark die off and sink inwards. Black, tar-like exudations appear and as the damage increases foliage may yellow and die.

**PLANTS AFFECTED** Beeches (*Fagus*)
**SEASON** All year

### SEIRIDIUM CANKER DISEASE • See p.174
Foliage (here on a cypress) yellows and dies. Bark becomes roughened as cankers develop. Resin oozes from the infected areas and tiny, black pycnidia appear. Cankers may ring stems.

**PLANTS AFFECTED** Several conifers
**SEASON** All year, mainly autumn and winter

### SMUT DISEASE • See p.176
Raised, blister-like areas develop on stems (here on *Trollius*). These erupt to produce black, powdery spore masses. Affected stems may be killed if the pustules are sufficiently large.

**PLANTS AFFECTED** Many plants
**SEASON** Mainly summer

### POOR PRUNING See p.294–295
This pruning cut – here on birch – has removed a branch too close to the trunk and the wound has failed to heal successfully. Infection has entered the wound, causing the wood to deteriorate.

**PLANTS AFFECTED** Any tree or shrub
**SEASON** All year

### ROSE ROOT APHID – EGGS PEST • See p.172
Large, dark brown aphids come up from the roots in autumn and lay clusters of eggs on the stems. The eggs are brown when freshly laid but turn shiny black.

**PLANTS AFFECTED** Roses
**SEASON** Eggs seen autumn to spring

### FIREBLIGHT DISEASE • See p.127
Flattened areas develop on the bark (here of a pear tree) and may exude bacterial ooze. The wood beneath is stained foxy-red brown. Flowers wither and die, followed by adjacent leaves.

**PLANTS AFFECTED** Pome fruits
**SEASON** Late spring to summer

### WOOLLY APHID PEST • See p.192
Fluffy, white patches develop on branches, especially on pruning wounds and in cracks in the bark. Infestations on young shoots cause knobbly swellings where the aphids have been feeding.

**PLANTS AFFECTED** Apple, pyracantha, cotoneasters
**SEASON** Mid-spring to mid-autumn

### BEECH BARK SCALE PEST • See p.106
An off-white, powdery wax accumulates in crevices in the bark of beech trees. It is secreted from tiny, sap-feeding scale insects. Heavy infestations are linked with beech bark disease (*see above*).

**PLANTS AFFECTED** Beeches (*Fagus*)
**SEASON** All year

## STEM PROBLEMS CONTINUED

### WILLOW BARK APHID PEST • See p. 190
Dense colonies of greyish-black aphids, up to 5mm long, develop on willow stems. The bark and ground under the tree becomes sticky with honeydew excreted by the aphids.

**PLANTS AFFECTED** Willows (*Salix*)
**SEASON** Mid-summer to mid-autumn

### BROWN SCALE PEST • See p. 110
Reddish-brown shells cover the bodies of sap-feeding insects or their eggs, especially on shrubs growing in sheltered places (here on a cotoneaster). Heavily infested plants lack vigour.

**PLANTS AFFECTED** Many woody plants
**SEASON** All year

### SLUGS PEST • See p. 176
The softer parts of leaves and leaf stalks are eaten (here of celery, causing elongate grooves between the veins). Damaged parts turn brown and may be infected by secondary rots.

**PLANTS AFFECTED** Soft tissue of many plants
**SEASON** Late summer to mid-autumn

### SCLEROTINIA DISEASE • See p. 174
Stems (here of a Jerusalem artichoke) become discoloured and covered in dense, white, fluffy fungal growth. Large black sclerotia (fungal resting bodies) are embedded in the fungal growth.

**PLANTS AFFECTED** Wide range of plants
**SEASON** Mainly early autumn, but all year

### APPLE AND PEAR CANKER DISEASE • See p. 100
Flattened areas of discoloured bark crack and split, forming concentric rings of flaky bark. White or red fungal bodies may develop on the cankered areas.

**PLANTS AFFECTED** Mainly apple and pear
**SEASON** All year

### SYCAMORE SOOTY BARK DISEASE DISEASE • See p. 180
Green or yellow staining develops under bark. Later, black spore masses or blister-like outgrowths develop on the stems. Large areas of bark die and break off.

**PLANTS AFFECTED** Sycamore (*Acer pseudoplatanus*)
**SEASON** All year

### IRREGULAR WATERING DISORDER • See p. 140
Longitudinal cracks or splits in bark may penetrate into the woody tissue. Older splits have swollen edges due to the formation of scar tissue.

**PLANTS AFFECTED** All woody plants
**SEASON** All year

### MUSSEL SCALE PEST • See p. 149
Bark may become heavily encrusted with greyish-dark brown scales, which may also spread onto fruits. These sap-feeding insects weaken the plant and cause shoots to die back.

**PLANTS AFFECTED** Apple trees and many shrubs
**SEASON** All year

### VOLES PEST • See p. 187
Branches or whole plants die as a result of these small rodents gnawing away the bark on the stems or roots. Damage often occurs in winter, but the results may not be apparent until summer.

**PLANTS AFFECTED** Many trees and shrubs
**SEASON** All year

## STEM PROBLEMS CONTINUED

### COTONEASTER WEBBER MOTH
PEST • *See p. 119*

The foliage turns brown and dries up where small, dark brown caterpillars have been feeding. They cover the infested branches with a silk webbing.

**PLANTS AFFECTED** Cotoneaster, hawthorn
**SEASON** Late spring to late summer

### JUNIPER SCALE
PEST • *See p. 140*

Flat, yellow-white, rounded scales encrust the stems, often causing yellowing of the green parts. Heavy infestations of this sap-feeder can cause stems to die back.

**PLANTS AFFECTED** Juniper, cypress, thuya
**SEASON** All year

### MEALYBUGS
PEST • *See p. 147*

Pinkish-grey or white insects infest the stems and leaf axils. They are often covered with a fluffy, white, waxy substance. Honeydew excreted by the mealybugs allows sooty mould to grow.

**PLANTS AFFECTED** Greenhouse and house plants
**SEASON** All year

### BRACKET FUNGUS
DISEASE • *See p. 110*

Bracket- or shelf-shaped, fungal fruiting bodies of variable size grow out from the tree (here a *Malus*), most commonly at the base of the trunk, but sometimes high in the crown.

**PLANTS AFFECTED** Most trees and some shrubs
**SEASON** All year, particularly autumn

### FOMES ROOT AND BUTT ROT
DISEASE • *See p. 128*

Dark red-brown, woody-textured bracket fungus grows on the trunk or butt of trees. The lower surface is creamy-white and covered in minute pores.

**PLANTS AFFECTED** Mainly conifers
**SEASON** All year (mainly autumn)

### DAEDALEOPSIS CONFRAGOSA
DISEASE • *See p. 122*

Semi-circular, rubbery, annual bracket fungus develops on trunks or branches, reddish-brown and smooth or slightly ridged with a paler undersurface.

**PLANTS AFFECTED** Deciduous trees
**SEASON** Mainly autumn

GALLERY OF SYMPTOMS

## STEM PROBLEMS CONTINUED

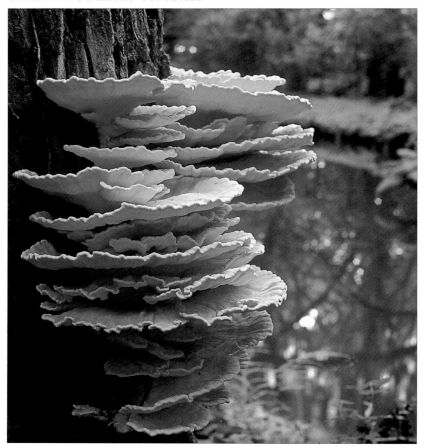

### LAETIPORUS SULPHUREUS
DISEASE • *See p.141*

Overlapping, annual, bright yellowish-orange fungal brackets are found on the trunk or large limbs of trees. The brackets have wavy edges.

**PLANTS AFFECTED** Trees

**SEASON** Mainly autumn

### CROWN GALL – AERIAL FORM
DISEASE • *See p.119*

Roughened, rounded growths break out from within the stem (here of a logan-berry), rupturing it. Galls become woody and if extensive, stems may die back.

**PLANTS AFFECTED** Many woody plants

**SEASON** All year, mainly summer

### PINEAPPLE GALL ADELGID
PEST • *See p.160*

Shoot tips are swollen and resemble green pineapples by mid-summer. The galls turn brown after the aphid-like insects emerge in late summer.

**PLANTS AFFECTED** Spruce (*Picea*)

**SEASON** All year

### BRACKET FUNGUS
DISEASE • *See p.110*

Single or overlapping, small bracket fungi grow on stems or trunks (here on dead wood). The upper surface may be concentrically zoned in shades of tan, brown, cream and grey.

**PLANTS AFFECTED** Deciduous trees and shrubs

**SEASON** All year, particularly autumn

### SMALL ERMINE MOTHS
PEST • *See p.176*

Black-spotted, creamy-yellow caterpillars feed together under white silk webbing that they spin over their feeding area. If numerous they can cause defoliation.

**PLANTS AFFECTED** A variety of hosts

**SEASON** Early to late summer

### MULBERRY CANKER
DISEASE • *See p.149*

These are small cankers which may girdle and so kill stems. Tiny red-brown pustules appear around the cankered area. Leaves and shoots may die back.

**PLANTS AFFECTED** Mulberry

**SEASON** All year (most obvious in summer)

## STEM PROBLEMS CONTINUED

### FOMES POMACEOUS
DISEASE • See p.128

Hoof-shaped or rounded, fungal fruiting bodies appear on the bark of dead or living stems (here of an apple tree). The upper surface is pale grey-brown.

**PLANTS AFFECTED** Various deciduous trees
**SEASON** All year, particularly autumn

### LEAFY GALL
DISEASE • See p.143

Numerous small, bunched, distorted, thickened or fasciated leaves grow, usually around the stem base (here of a pelargonium). Normal growth is also present.

**PLANTS AFFECTED** Annual or herbaceous plants
**SEASON** Mainly spring and summer

### WOOLLY VINE SCALE
PEST • See p.192

In early summer this scale insect deposits its eggs under pads of white wax (here on a pyracantha), which can be drawn out in long threads. Heavily infested plants lack vigour.

**PLANTS AFFECTED** Vines, currants and others
**SEASON** All year (eggs in early summer)

### MERIPILUS GIGANTEUS
DISEASE • See p.147

Large, fleshy, golden-brown fungal brackets are found around the base of trees or in the near vicinity of the roots. The undersurface is creamy-white.

**PLANTS AFFECTED** Oaks, beeches, robinias
**SEASON** Late summer and autumn

### BROOM GALL MITE
PEST • See p.110

Buds fail to grow into shoots and become very swollen. These galls gradually dry up and change from green to grey-brown. Heavily infested plants produce few flowers or new growth.

**PLANTS AFFECTED** Broom (Cytisus)
**SEASON** All year

### PINE ADELGID
PEST • See p.160

Small, black, aphid-like insects suck sap from the young shoots. The adelgids are hidden under a fluffy, white wax that is secreted from their bodies. This can resemble a mould.

**PLANTS AFFECTED** Pines (Pinus)
**SEASON** Mid-spring to late summer

### WEEDKILLER DAMAGE – BROADLEAVED
DISORDER • See p.188

Adventitious roots and other outgrowths form on stems (here of a brassica), causing them to become distorted. The foliage may not be affected.

**PLANTS AFFECTED** Plants without woody stems
**SEASON** Mainly late spring and summer

### ROBIN'S PIN CUSHION
PEST • See p.170

Roughly spherical swellings covered in reddish or yellowish-green, mossy leaves develop around the stems of certain roses and occasionally on foliage.

**PLANTS AFFECTED** Wild and species roses
**SEASON** Mid-summer to early autumn

### CUCKOO SPIT
PEST • See p.120

Globules of frothy white liquid appear on stems and sometimes on foliage. Inside the froth is a yellowish-green, sap-feeding insect, which is the nymphal stage of a froghopper.

**PLANTS AFFECTED** Many plants
**SEASON** Late spring to early summer

GALLERY OF SYMPTOMS

## STEM PROBLEMS *CONTINUED*

### PHYSICAL INJURY
*See pp. 70–71*

Winds, heavy snow, browsing animals or vandals may cause tearing of bark and injury to the wood tissues beneath. Jagged wounds are slow to heal and harmful pathogens may enter them.

**PLANTS AFFECTED** Woody plants

**SEASON** All year

### RASPBERRY RUST
DISEASE • *See p. 168*

Bright orange spore masses develop on the stems, often appearing from ruptures in stems. Large fissures remain on the stems (here of a blackberry) when all the spores have been dispersed.

**PLANTS AFFECTED** Raspberries, blackberries

**SEASON** Late spring to autumn

### HORSE CHESTNUT SCALE
PEST • *See p. 137*

Brownish-black scales secrete a white, waxy substance that covers their eggs. Later the scales die and fall away, leaving the bark covered in white patches.

**PLANTS AFFECTED** Many trees and shrubs

**SEASON** All year

### ASPARAGUS BEETLE
PEST • *See p. 103*

Both grubs (*above left*) and adults (*above right*) eat the foliage and also gnaw bark from the stems, causing them to dry up and turn a yellowish-brown colour.

**PLANTS AFFECTED** Asparagus

**SEASON** Late spring to early autumn

### FRUIT TREE RED SPIDER MITE
PEST • *See p. 129*

The spherical winter eggs are less than 1mm in diameter, but there may be so many that they give the bark a reddish colour, espcially on branch undersides.

**PLANTS AFFECTED** Apples, plums

**SEASON** Eggs early autumn to late spring

### FORSYTHIA GALL
DISEASE • *See p. 128*

Rough-surfaced, near-spherical woody galls, usually 10–15mm in diameter, develop on stems. Several galls may be fused together. The growth and development of the stem is not affected.

**PLANTS AFFECTED** Forsythia

**SEASON** All year

### ROSE CANKER AND DIEBACK
DISEASE • *See p. 171*

Stems die back and discoloration occurs, either in patches or all over the affected stem. Tissues may dry out. Growth above the affected areas dies off.

**PLANTS AFFECTED** Roses

**SEASON** All year

### BLIND SHOOTS
DISORDER • *See p. 108*

Shoots which should bear flower buds form, but no flower buds develop (here on a rose). Leaf and stem growth otherwise appear perfectly healthy, with no sign of dieback.

**PLANTS AFFECTED** Flowering shrubs

**SEASON** Mainly summer

### PAPERY BARK
DISORDER • *See p. 154*

Bark peels off as a thin, papery brown sheet. Small- to medium-sized shoots are most commonly affected, but any may show symptoms. Shoots can die back. Apple trees are particularly affected.

**PLANTS AFFECTED** Woody plants

**SEASON** All year, mainly spring and winter

## STEM PROBLEMS CONTINUED

### WITCHES' BROOMS
DISEASE • *See p.191*

Closely-packed, densely-branched clusters of twigs grow from normal stems. The stems do not die back, but bear numerous small leaves. Adjacent growth is perfectly normal.

**PLANTS AFFECTED** Trees

**SEASON** All year

### WILLOW SCAB
DISEASE • *See p.190*

Irregular, dark spots develop on the young stems and may cause girdling. Olive-brown pustules are found on the undersurface of leaves and on affected stems. Stems may die back.

**PLANTS AFFECTED** Willows (*Salix*)

**SEASON** All year, particularly spring

### CORAL SPOT
DISEASE • *See p.119*

Raised red, orange or coral-coloured, hard pustules develop on dead or dying stems. Affected stems die back further and if the infection spreads to the crown, the whole plant may be killed.

**PLANTS AFFECTED** Woody plants

**SEASON** All year

### LUPIN APHID
PEST • *See p.146*

Large, whitish-grey aphids cluster on the flower stems and underside of leaves. Heavy infestations cause the plant to wilt. Honeydew and sooty mould soil the foliage.

**PLANTS AFFECTED** Lupins

**SEASON** Late spring to late summer

### DEER
PEST • *See p.122*

Various species of deer come into gardens and eat foliage, flowers and stems. Nibbled stems invariably have ragged cuts caused by a deer's lack of incisor teeth in its upper jaw.

**PLANTS AFFECTED** Most plants

**SEASON** All year

### BLOSSOM WILT AND SPUR BLIGHT
DISEASE • *See p.108*

Raised, rounded, buff-coloured fungal pustules appear on twigs and stems (here of an apple tree), often associated with blossom wilt on adjacent flowers.

**PLANTS AFFECTED** Apple, pear, plum, cherry

**SEASON** Mainly late spring and summer

## BUD PROBLEMS

### BLINDNESS OF BULBS
DISEASE/DISORDER • *See p.108*

Flower buds fail to form, or do not develop fully (here on a daffodil) and open partially or not at all. Leaves are unaffected and appear healthy.

**PLANTS AFFECTED** Many bulbs

**SEASON** Spring

### BIG BUD MITES
PEST • *See p.107*

Infested buds (here on a blackcurrant stem) are abnormally swollen and rounded compared to normal buds. They usually dry up and fail to open in the spring.

**PLANTS AFFECTED** Blackcurrant, hazel, yew

**SEASON** Mid-winter to early summer

### RHODODENDRON BUD BLAST
DISEASE • *See p.168*

Buds form but turn dry and brown. Numerous tiny, black, bristle-like fungal growths develop on the surface. Affected buds remain on the plant.

**PLANTS AFFECTED** Rhododendrons

**SEASON** All year, evident late spring and summer

# FLOWERS

MOST GARDENS ARE packed full of plants grown largely because of their flowers, be they annuals, biennials, perennials, shrubs, climbers or trees. In fruit and vegetable gardens, flowers are often equally important, as without them the edible fruits would never develop. Keeping flowers free from pests and diseases is therefore vital.

Flowers are usually short-lived, yet during their brief lifespan there may well be several pests or pathogens which could attack and therefore spoil them. They may be discoloured (*p.42*) or disfigured by infestations of pests (*p.43*), deformed (*pp.44–45*) or damaged by pests (*p.46*). In many cases, however, even if a few blooms are spoilt, later flowers may be unscathed, having missed the attentions of the pest, or the infection period of the pathogen.

Flowering in general is also greatly influenced by cultural and weather conditions, in particular soil moisture and fertility and the level of sunshine. To a large extent, growing a plant in a suitable spot ensures a good display of flowers.

## DISCOLOURED FLOWERS

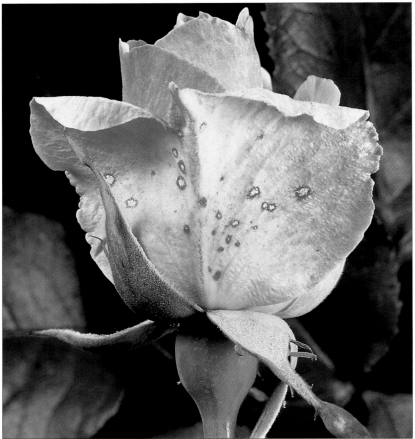

**GREY MOULD**  DISEASE • *See p. 134*
Spots of discoloration appear on the petals (here of a rose). Spots may be edged with a dark ring. No fungal growth is apparent. In extreme cases the whole flower deteriorates.

**PLANTS AFFECTED** Many flowers
**SEASON** Mainly spring and summer

**RHODODENDRON PETAL BLIGHT**  DISEASE • *See p. 169*
Spots of discoloration develop on the petals. These enlarge and appear water-soaked. The flower collapses and rapidly discolours all over.

**PLANTS AFFECTED** Rhododendrons and azaleas
**SEASON** Spring

**VIRUS**  DISEASE • *See p. 187*
Petals show "colour breaking", with streaks of discoloration (here on a tulip). The petals do not deteriorate. Flowers may be reduced in number and size and show distortion.

**PLANTS AFFECTED** Most plants
**SEASON** Mainly spring and summer

**WESTERN FLOWER THRIPS**  PEST • *See p. 189*
A pale flecking develops on petals (here of a gloxinia), and yellowish-brown insects are seen on flowers and foliage. Infested blooms have a shortened life.

**PLANTS AFFECTED** Mainly greenhouse plants
**SEASON** All year

# PESTS VISIBLE ON FLOWERS

### ROSE APHIDS
PEST • *See p.170*

Green, yellowish-green or pink aphids cluster on flower buds and stems. Heavy infestations reduce the quality of the blooms. White cast aphid skins litter the buds and foliage.

**PLANTS AFFECTED** Roses
**SEASON** Mid-spring to late summer

### POLLEN BEETLES
PEST • *See p.162*

Black or bronzy-green beetles, 2mm long, crawl around the centre of blooms, where they feed on pollen. No direct damage is caused, but they can be a nuisance on cut flowers in the house.

**PLANTS AFFECTED** Many flowers
**SEASON** Mid-spring to late summer

### MELON COTTON APHID
PEST • *See p.147*

Small black or dark green aphids infest the flowers (here chrysanthemum) and the undersides of leaves. Sooty mould grows on the aphids' excrement.

**PLANTS AFFECTED** Chrysanthemums and others
**SEASON** Mainly summer to autumn

### SLUGS AND SNAILS
PESTS • *See pp.176, 177*

Damage caused by slugs and snails is usually indistinguishable. On daffodils, above, foliage is ignored but petals can be reduced to brown-edged fragments. A silvery slime trail is usually left behind.

**PLANTS AFFECTED** Wide range of plants
**SEASON** Early to late spring

### APPLE SUCKER
PEST • *See p.102*

Flattened, pale green insects up to 2mm long suck sap from the blossom trusses. Heavy attacks make the petals turn brown and can be confused with frost damage.

**PLANTS AFFECTED** Apple trees
**SEASON** Mid- to late spring

### CUCKOO SPIT
PEST • *See p.120*

A frothy, white liquid covers yellowish-green insects (here on *Potentilla*), which are the immature nymphs of froghoppers. They also suck sap from foliage and stems (*see p.39*).

**PLANTS AFFECTED** Many plants
**SEASON** Late spring to mid-summer

### PEONY WILT
DISEASE • *See p.158*

Stem tissue is discoloured and shrinks inwards, commonly at the base or, as above, below flower buds. Fuzzy grey fungal growth develops and black sclerotia are found on or within stems.

**PLANTS AFFECTED** Peonies
**SEASON** Spring to autumn

### GLASSHOUSE RED SPIDER MITE
PEST • *See p.168*

Although mainly a foliage pest (*see pp.22, 23*), red spider mites also infest flowers, especially chrysanthemums. Yellowish-black mites crawl amongst the petals.

**PLANTS AFFECTED** Many greenhouse plants
**SEASON** Mid-spring to mid-autumn

### FASCIATION
DISORDER • *See p.126*

Stems become fused together, producing broad, ribbon-like stems (here on delphinium). The stems usually bear leaves and flowers, but become curled and distorted.

**PLANTS AFFECTED** Many plants
**SEASON** All year on perennial stems

# DEFORMED AND DISTORTED FLOWERS

### DROUGHT
DISORDER • *See p.124*

Leaves and flowers wilt and soft stems may do so also. If conditions do not improve, wilting may be permanent (as here on a lavatera) and flowers become shrivelled and brown.

**PLANTS AFFECTED** Most plants

**SEASON** Mainly summer

### SMUTS
DISEASE • *See p.176*

Anthers become swollen and distorted and packed full of masses of dark brown or black spores. Anthers burst to release spore masses. Flower stems may be distorted.

**PLANTS AFFECTED** Several plants

**SEASON** Mainly summer

### FROST DAMAGE (BUDS AND FLOWERS)
DISORDER • *See p.129*

Petals become discoloured (here of a camellia), usually brown and either soft or, less frequently, dry. Exposed flowers are most badly affected.

**PLANTS AFFECTED** Many flowers

**SEASON** Mainly spring and early winter

### ROSE BALLING
DISORDER • *See p.170*

Outer petals on an unopened or only partially opened bud are pale brown and dry. Initially, inner petals are unaffected, but they may die off when secondary organisms invade.

**PLANTS AFFECTED** Roses

**SEASON** Mainly summer

### BOLTING
DISORDER • *See p.109*

Plants grown for their leaf crop become elongated and produce flowers unusually early (here on a brassica). This limits the cropping period. Early cultivars are particularly susceptible.

**PLANTS AFFECTED** Beetroot, brassicas, lettuces

**SEASON** Mainly summer

### PROLIFERATION
DISORDER • *See p.166*

Flower buds form within the centre of the flower (here a rose). These remain as buds or develop into stems bearing flowers and occasionally leaves. Only the flowers or buds are affected.

**PLANTS AFFECTED** Roses, also pears and apples

**SEASON** Late spring and summer

# DEFORMED AND DISTORTED FLOWERS CONTINUED

## MICHAELMAS DAISY MITE
PEST • *See p.147*

The flowers are converted by the feeding activities of this tiny pest into rosettes of green leaves. Infested plants are often stunted, with scarring on the stems.

**PLANTS AFFECTED** *Aster novi-belgii*
**SEASON** Early summer to mid-autumn

## CHRYSANTHEMUM PETAL BLIGHT
DISEASE • *See p.117*

Small brown spots or larger water-soaked lesions develop on the petals. Flowers rot and die off rapidly. Leaf growth is not affected.

**PLANTS AFFECTED** Mainly chrysanthemums
**SEASON** Mainly late summer

## WIND DAMAGE
DISORDER • *See p.191*

Petals (here of a camellia) become scorched and discoloured, usually brown. Damaged areas are usually dry, not soggy. Flowers on exposed areas of the plant are most susceptible.

**PLANTS AFFECTED** Many plants
**SEASON** Mainly spring

## HEMEROCALLIS GALL MIDGE
PEST • *See p.135*

Infested buds are squat and abnormally swollen; they fail to open, go brown and dry up. Many small white maggots live in the buds at the base of the petals.

**PLANTS AFFECTED** Daylilies (*Hemerocallis*)
**SEASON** Late spring to mid-summer

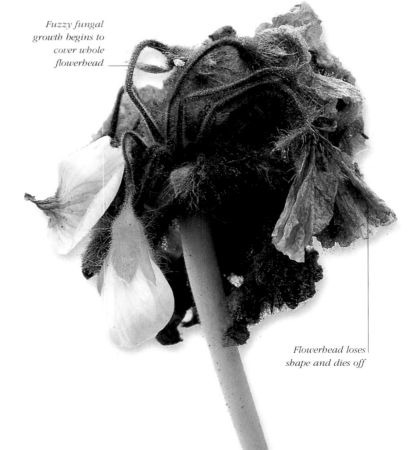

*Fuzzy fungal growth begins to cover whole flowerhead*

*Flowerhead loses shape and dies off*

## GLADIOLUS THRIPS
PEST • *See p.132*

Creamy-yellow or black, narrow-bodied insects, up to 2mm long, feed in the flower buds and on the foliage. Infested blooms have pale mottled petals, but buds may turn brown and fail to open.

**PLANTS AFFECTED** Gladioli
**SEASON** Mid- to late summer

## GREY MOULD
DISEASE • *See p.134*

At this advanced stage of grey mould, the flowers rapidly become covered in dense, grey fungal growth (here on a pelargonium). Dieback may spread. For earlier symptoms *see p.42*.

**PLANTS AFFECTED** Many flowering plants
**SEASON** All year

# DAMAGED FLOWERS

### NECTAR ROBBING
PEST • *See p.151*

Some short-tongued bumblebees cannot reach the nectar from the front of some flowers and so bite a hole at the base of the flower. Bees taking nectar in this way do not assist with pollination.

**PLANTS AFFECTED** Often, runner and broad beans
**SEASON** Summer

### SPARROWS
PEST • *See p.177*

House sparrows peck the petals of some flowers (here a crocus) to shreds, leaving fragments scattered nearby on the soil. They sometimes select certain colours, but are not consistent.

**PLANTS AFFECTED** Crocus, primroses and others
**SEASON** Early spring to summer

### BULLFINCHES
PEST • *See p.112*

The dormant flower buds (here of a plum) are pecked off and eaten during the winter. The outer bud scales are discarded and dropped to the ground. Bullfinches also eat seeds.

**PLANTS AFFECTED** Fruits and some ornamentals
**SEASON** Late autumn to mid-spring

### APPLE BLOSSOM WEEVIL
PEST • *See p.100*

Infested flowers fail to open fully and the petals turn brown. A white grub or the pupa will be found inside these "capped" blossoms.

**PLANTS AFFECTED** Apples
**SEASON** Mid- to late spring

*Male earwigs (top) have curved pincers, while females (bottom) have straight ones*

*Ragged holes are eaten in petals*

### CAPSID BUGS
PEST • *See p.114*

Flower buds (here of fuchsias) abort so that infested plants (*above left*) are largely flowerless. Foliage is distorted and has many small holes, due to capsid bugs damaging leaves at the shoot tips.

**PLANTS AFFECTED** Many plants
**SEASON** Late spring to late summer

### EARWIGS
PEST • *See p.125*

Earwigs emerge after dark, usually in mild weather, to feed on petals (here of a cosmos) and on foliage at shoot tips. This results in holes developing in the blooms and the expanded leaves.

**PLANTS AFFECTED** Clematis, dahlias and others
**SEASON** Late spring to early autumn

# FRUITS, BERRIES, NUTS AND VEGETABLES

IF THE EDIBLE PARTS of a plant are attacked by a pest or pathogen, the whole reason for growing the plant may well be lost. Fruits and berries are particularly prone to attack, as they usually have a high sugar and moisture content. Nuts, peas and beans are also valuable food sources for many creatures.

In addition, some ornamental trees and shrubs are very commonly grown mainly for their attractive show of berries, and when these are eaten, removed or spoiled by disease, the plant's appearance is ruined.

In many cases prolonged protection of the plant may be needed, perhaps by using barriers or traps to keep pests away from the flowers or the developing and ripening fruits. Chemicals can be used with great success, but again these may sometimes need to be applied before the first signs of damage are noticed.

## FRUIT PROBLEMS

### APPLE SAWFLY
PEST • *See p.101*

Most infested fruits drop off in early summer. Some do develop as ripe fruit but have distinctive ribbon scars where the young larvae fed under the skin earlier in the summer.

**PLANTS AFFECTED** Apples
**SEASON** Mid-spring to early autumn

### APPLE SCALD
DISORDER • *See p.101*

Reddish-brown patches of discoloration develop on the skin of the fruit. The flesh beneath is usually undamaged, but it may be discoloured. Exposed surfaces are most badly affected.

**PLANTS AFFECTED** Apples
**SEASON** Mainly summer

### BIRD DAMAGE
PEST • *See p.107*

Blackbirds, starlings and other birds will peck holes in ripening fruits. Wasps may enlarge this damage, and brown rot frequently develops once the skin has been broken.

**PLANTS AFFECTED** Apples, pears, plums
**SEASON** Mid-summer to early autumn

### APPLE CAPSID
PEST • *See p.100*

This pest sucks sap from the fruitlets and causes yellowish-brown, raised, corky swellings. The damage is only skin deep and does not affect the fruits' keeping qualities.

**PLANTS AFFECTED** Apples
**SEASON** Late spring to early autumn

### CODLING MOTH
PEST • *See p.118*

This is the cause of maggoty apples in late summer. The caterpillar feeds at the core and tunnels out through the side or the eye end of the ripening fruit when it is fully developed.

**PLANTS AFFECTED** Apples, pears
**SEASON** Mid-summer to early autumn

### WASPS
PEST • *See p.188*

These familiar insects enlarge damage started by birds, but can also initiate feeding on softer-skinned fruits (*see p.49*). Their presence can make fruit picking hazardous.

**PLANTS AFFECTED** Tree fruits, grapes
**SEASON** Mid-summer to early autumn

### BROWN ROT
DISEASE • *See p.111*

Soft, brown rot develops on fruit (here on apple) and soon becomes studded with raised, creamy-white pustules (*see also p.49*). These are often arranged in concentric rings.

**PLANTS AFFECTED** Tree fruits
**SEASON** Late summer and autumn

## FRUIT PROBLEMS *CONTINUED*

### ROSY APPLE APHID
PEST • *See p.173*

Small greyish-pink aphids infest the foliage and fruitlets in spring. They cause leaf curling and yellowing, and affected fruits remain small and distorted with a pinched appearance at the eye end.

**PLANTS AFFECTED** Apples

**SEASON** Early spring to late summer

### PEAR MIDGE
PEST • *See p.157*

Pear fruitlets become abnormally swollen, turn black and drop off the tree in early summer. These fruitlets contain many small, whitish-orange maggots. The whole crop may be lost.

**PLANTS AFFECTED** Pears

**SEASON** Late spring to early summer

### PEAR STONY PIT VIRUS
DISEASE • *See p.157*

Fruits become knobbly and pitted, with hard patches throughout the flesh. Affected fruits usually only appear on isolated branches.

**PLANTS AFFECTED** Pears

**SEASON** Late summer to early autumn

### PEAR SCAB
DISEASE • *See p.157*

Blackish-brown, scabby patches appear on the skin of the fruits. The fruits are often small and misshapen and may be spoiled by splits, since secondary organisms rapidly cause rotting.

**PLANTS AFFECTED** Pears

**SEASON** Late summer to early autumn

### QUINCE LEAF BLIGHT
DISEASE • *See p.166*

Numerous small, dark red or blackish-brown spots are found on the skin. Fruits may be distorted and under-sized, and occasionally split.

**PLANTS AFFECTED** Quinces

**SEASON** Late summer or early autumn

### APPLE BITTER PIT
DISEASE • *See p.100*

Pale brown spots of discoloration appear within the flesh, or sometimes on the fruit's skin. Those on the skin are usually slightly sunken. The fruit is otherwise perfectly normal in shape and size.

**PLANTS AFFECTED** Apples

**SEASON** Mainly late summer and in store

### APPLE SAWFLY
PEST • *See p.101*

After initially tunnelling under the skin, apple sawfly larvae bore into the core of fruitlets. Wet, blackish-brown excrement comes out of the entry hole. For apple sawfly damage to mature fruit *see p.47.*

**PLANTS AFFECTED** Apples

**SEASON** Mid-spring to mid-summer

### APPLE FRUIT SPLIT
DISORDER • *See p.101*

Straight or branched splits develop in the skin. The splits are usually single, but several may occasionally appear on one fruit. Secondary organisms such as brown rot may invade and cause rotting.

**PLANTS AFFECTED** Most fruits, especially apples

**SEASON** Mid- to late summer

### PLUM MOTH
PEST • *See p.161*

Infested plum fruits tend to ripen prematurely. They contain a pinkish caterpillar, which eats the flesh around the stone, leaving behind a trail of brown excrement pellets.

**PLANTS AFFECTED** Plums, damsons, gages

**SEASON** Late summer to early autumn

# FRUIT AND BERRY PROBLEMS

## WASPS
PEST • *See p.188*

These familiar yellow and black stinging insects are attracted to ripe fruits. They can eat out extensive cavities in some softer-skinned fruits. For wasp damage to apples and pears *see p.47*.

**PLANTS AFFECTED** Tree fruits, grapes

**SEASON** Mid-summer to mid-autumn

## RED BERRY MITE
PEST • *See p.168*

Blackberries fail to ripen completely, with parts or all of the fruit remaining red. Ripening becomes progressively unsatisfactory as the season develops. The cause is microscopic mites.

**PLANTS AFFECTED** Blackberries

**SEASON** Late summer to mid-autumn

## BROWN ROT
DISEASE • *See p.111*

Fruits (here a cherry) turn brown and rot and rapidly develop numerous raised, creamy-white pustules (*see also p.47*). The fruits then dry out, but remain hanging on the tree.

**PLANTS AFFECTED** Tree fruits

**SEASON** Late summer to autumn

## POWDERY MILDEW
DISEASE • *See p.165*

Fruits (here grapes) develop a white, off-white or greyish-beige fungal coating in patches or all over their surface. They fail to swell fully, may be small or distorted and may crack or split.

**PLANTS AFFECTED** Many fruits

**SEASON** Summer

## AMERICAN GOOSEBERRY MILDEW
DISEASE • *See p.98*

Powdery or felty white fungal growth develops on the skin, soon turning buff or pale brown. The fungus may be rubbed off. Fruits may be under-sized.

**PLANTS AFFECTED** Gooseberries, blackcurrants

**SEASON** Summer

## PYRACANTHA SCAB
DISEASE • *See p.166*

Greenish-black, scabby fungal growth develops on the berries, which may cause them to split and die off. In extreme cases the whole surface of the berry may be obscured.

**PLANTS AFFECTED** Pyracanthas

**SEASON** Summer to autumn

# FRUIT AND VEGETABLE PROBLEMS

## RASPBERRY BEETLE
PEST • *See p.167*

The stalk end of the fruit dries up and becomes greyish-brown. An elongate, whitish-brown beetle grub, up to 8mm long, may be found feeding in the core or plug of the fruit.

**PLANTS AFFECTED** Cane fruits

**SEASON** Mid-summer to early autumn

## STRAWBERRY GREY MOULD
DISEASE • *See p.179*

Soft, pale brown areas develop on the fruit and rapidly increase in size. Fuzzy grey fungal growth develops on the surface and the fruit rapidly rots.

**PLANTS AFFECTED** Strawberries

**SEASON** Late spring to early autumn

## TOMATO GREENBACK
DISEASE • *See p.182*

A partial or complete ring of unripened tissue appears around the stalk end of a ripening fruit. The flesh and skin remains either green or yellow.

**PLANTS AFFECTED** Tomatoes

**SEASON** Summer

## BLOSSOM END ROT
DISEASE • *See p.108*

A sunken, leathery, dark brown to black patch develops at the bottom end of the developing fruit (here tomatoes). The remainder of the fruit ripens normally, but remains distorted.

**PLANTS AFFECTED** Tomatoes, peppers

**SEASON** Summer

## TOMATO FRUIT SPLITTING
DISORDER • *See p.182*

Fruits develop normally and then split shortly before they are ready to pick. The split may dry out or may become infected with secondary fungi.

**PLANTS AFFECTED** Tomatoes

**SEASON** Summer or early autumn

## GREY MOULD
DISEASE • *See p.134*

Fruits (here a tomato) turn soft and discolour and become covered in fuzzy grey fungal growth. On tomatoes, ghost spotting symptoms (yellow circles) on otherwise normal fruits may appear.

**PLANTS AFFECTED** Many fruits

**SEASON** Summer and early autumn

# VEGETABLE PROBLEMS

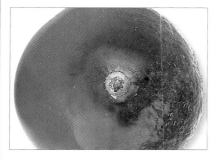

### TOMATO BLOTCHY RIPENING
DISORDER • See p. 181

Randomly distributed, hard, yellow or green patches of flesh remain unripened. The rest of the fruit develops and ripens normally.

**PLANTS AFFECTED** Tomatoes
**SEASON** Summer and early autumn

### TOMATO BLIGHT
DISEASE • See p. 181

Brown, discoloured areas develop within the fruits and are visible on the surface of both ripe and unripe fruits. Discoloured areas may shrink inwards and suffer secondary rotting.

**PLANTS AFFECTED** Tomatoes and potatoes
**SEASON** Summer and early autumn

### BEAN SEED BEETLES
PEST • See p. 105

Beans, especially broad beans, kept for seed may have circular holes where adult seed beetles have emerged. The grubs feed inside on the cotyledons, but damaged seeds will usually germinate.

**PLANTS AFFECTED** Beans and peas
**SEASON** Mid-summer to mid-autumn

### BEAN SEED FLY
PEST • See p. 105

White fly maggots feed on the germinating seeds, making holes in the cotyledons and stems. Seedlings may be killed or emerge with the growing point killed.

**PLANTS AFFECTED** Runner and French beans
**SEASON** Mid-spring to early summer

### PEA MOTH
PEST • See p. 155

Small cream-coloured caterpillars with dark spots feed inside pea pods on the developing seeds. Piles of excrement pellets indicate where the caterpillars have been feeding.

**PLANTS AFFECTED** Peas
**SEASON** Early to late summer

### PEA LEAF AND POD SPOT
DISEASE • See p. 155

Yellow or brown, often sunken, patches develop on the pods. Dark brown or black pycnidia (fungal fruiting bodies) may be visible on affected areas.

**PLANTS AFFECTED** Peas
**SEASON** Summer

### SWEETCORN SMUT
DISEASE • See p. 180

Individual kernels on sweetcorn cobs become enlarged and distorted. Initially pale grey in colour, they rupture to release masses of powdery black spores. The rest of the plant develops normally.

**PLANTS AFFECTED** Sweetcorn
**SEASON** Summer

### MICE
PEST • See p. 147

Rows of germinating peas, beans and other seeds are disturbed by burrowing field mice. Only seeds and corms are eaten; the shoots are left lying on the soil surface.

**PLANTS AFFECTED** Peas, beans, sweetcorn
**SEASON** Early spring to mid-summer

### CARROT FLY
PEST • See p. 114

Elongate, yellow-white fly maggots tunnel in certain tap roots, causing orange-brown discoloured lines on the outside of the roots. Secondary rots may extend the damage.

**PLANTS AFFECTED** Carrots, parsnips, parsley
**SEASON** Early summer to late autumn

## VEGETABLE PROBLEMS *CONTINUED*

### ONION NECK ROT  DISEASE • *See p.153*
The neck area of the onion turns pale brown and soft and appears semi-transparent. Fuzzy grey fungal growth and black sclerotia may develop. The onion can become mummified.

**PLANTS AFFECTED** Onions, shallots
**SEASON** Late summer and while in store

### ONION EELWORM  PEST • *See p.153*
Microscopic, worm-like animals live within the bulb and leaves. Infested plants are swollen and distorted, with softened tissues that are susceptible to secondary rots.

**PLANTS AFFECTED** Onions, leeks
**SEASON** Late spring to early autumn

### ONION WHITE ROT  DISEASE • *See p.153*
Dense, fluffy white fungal growth appears around the basal plate. Small black sclerotia may be visible in amongst this and the bulb starts to rot.

**PLANTS AFFECTED** Onions, leeks, shallots, garlic
**SEASON** Summer

# NUT PROBLEMS

### NUT WEEVIL  PEST • *See p.151*
The kernel is eaten by a white grub with a brown head. The fully-fed grub bores a circular hole in the ripe nut shell (here of a hazelnut) when it leaves to pupate.

**PLANTS AFFECTED** Hazelnuts and cob nuts
**SEASON** Mid-summer to early autumn

### SQUIRRELS  PEST • *See p.178*
Grey squirrels often strip trees before the nuts are fully ripe and the ground becomes littered with broken nut shells. Nuts are buried, sometimes in lawns, for consumption during the winter.

**PLANTS AFFECTED** Hazelnuts, cob nuts, walnuts
**SEASON** Late summer to mid-autumn

### ACORN GALL WASP  PEST • *See p.98*
Oak acorns are converted into ridged, yellow-brown structures, which are sometimes called knopper galls. The wasp grub feeds in a seed-like capsule inside the gall.

**PLANTS AFFECTED** Oaks (*Quercus*)
**SEASON** Late summer to mid-autumn

# SOIL, ROOTS, TUBERS AND BULBS

PESTS AND PATHOGENS that attack below ground level are common, but because the symptoms they cause are usually first seen above ground, they may well be misidentified. Often the only way to identify a soil-borne problem conclusively is to dig up all or part of the plant, a process which in itself may obviously cause considerable and long-term damage.

Most soil-borne problems, or pests or pathogens which attack roots, tubers, bulbs or corms directly, are potentially very damaging, simply because the root system or the base of the plant is so fundamental to the survival of the plant as a whole. Early investigation, identification and, where possible, control is therefore essential. Generally, control is difficult, however, since locating and targeting an underground pest or pathogen with a pesticide is not easy.

## SOIL PROBLEMS

### VINE WEEVIL (DAMAGE BY LARVAE)
PEST • See p.186

Plump, white, legless larvae with pale brown heads feed on roots, corms and tubers, particularly on plants grown in containers. The plant lacks vigour.

**PLANTS AFFECTED** Many plants
**SEASON** Late summer to mid-spring

### LEATHERJACKETS
PEST • See p.143

Leatherjackets are the larvae of daddy-long-legs. They are greyish-brown, legless maggots with no obvious head. They eat roots and seedling plants and also cause problems in lawns (see p.58).

**PLANTS AFFECTED** Many garden plants
**SEASON** Mid-spring to late summer

### WOODLICE
PEST • See p.192

Woodlice may be grey or pinkish-brown and up to 15mm long. They live mainly in the surface layers of the soil, where they feed on decaying vegetation and occasionally seedlings.

**PLANTS AFFECTED** Seedlings
**SEASON** All year

### FUNGUS GNAT (LARVAE)
PEST • See p.129

The larvae of fungus gnats, or sciarid flies, are white, legless maggots with black heads. They feed on decaying vegetation and seedlings.

**PLANTS AFFECTED** Greenhouse and house plants
**SEASON** All year

### CHAFER GRUBS
PEST • See p.116

Chafer grubs have plump, white bodies, curved like the letter C, with brown heads and three pairs of legs. They eat roots and make cavities in root vegetables and potato tubers.

**PLANTS AFFECTED** Many garden plants
**SEASON** Mostly mid-spring to mid-autumn

### CUTWORMS
PEST • See p.121

Cutworms are the earth-coloured caterpillars of several species of moths. They kill small plants by severing the stems or by gnawing away the outer bark at soil level.

**PLANTS AFFECTED** Mostly annuals and vegetables
**SEASON** Mainly mid-spring to mid-autumn

### SWIFT MOTHS (LARVAE)
PEST • See p.180

These are slender, white-bodied caterpillars with brown heads which feed on roots, corms and rhizomes, especially in herbaceous borders.

**PLANTS AFFECTED** Many garden plants
**SEASON** Mainly mid-spring to mid-autumn

## SOIL PROBLEMS CONTINUED

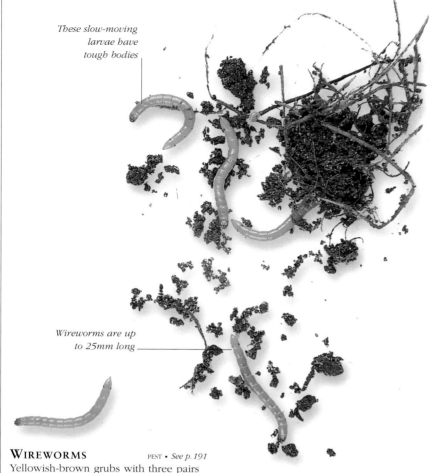

*These slow-moving
larvae have
tough bodies*

*Wireworms are up
to 25mm long*

### MILLIPEDES
PEST • *See p.148*

These spotted snake millipedes are creamy-white spotted with red, with legs on every body segment. They feed on rotting vegetation and enlarge damage initiated by other pests and diseases.

**PLANTS AFFECTED** Seedlings, potatoes
**SEASON** Spring to autumn

### WIREWORMS
PEST • *See p.191*

Yellowish-brown grubs with three pairs of short legs at their front eat roots and bore into root vegetables. They cause most damage where grass has been dug to create flower or vegetable gardens.

**PLANTS AFFECTED** Seedlings, potatoes and others
**SEASON** All year

### SPRINGTAILS
PEST • *See p.178*

Small, white insects come to the surface of pot plants or are washed out of the pot when the plant is watered. They cause no harm to the plant as they feed on dead plant material.

**PLANTS AFFECTED** Pot plants
**SEASON** All year

## BULB PROBLEMS

### BLUE MOULD ON BULBS
DISEASE • *See p.109*

Reddish-brown lesions develop on the flesh of the bulb and may be sunken. Fluffy, white or blue-green fungal growth may appear on top of lesions.

**PLANTS AFFECTED** Most bulbs in store
**SEASON** Mainly late summer to autumn

### NARCISSUS EELWORM
PEST • *See p.150*

Plants are stunted and distorted. If the bulb is cut in half transversely, brown concentric rings can be seen where the microscopic eelworms are feeding. Infested bulbs rot and die.

**PLANTS AFFECTED** Daffodils
**SEASON** All year

### NARCISSUS BULB FLY
PEST • *See p.150*

Infested bulbs rot in the soil or produce a few distorted leaves in the spring. One large maggot feeds in the centre of the bulb and fills it with its muddy-brown excrement.

**PLANTS AFFECTED** Daffodils, amaryllis and others
**SEASON** Late summer to early spring

## BULB PROBLEMS CONTINUED

### TULIP FIRE
DISEASE • See p. 183

Tiny, irregularly shaped, black sclerotia (fungal resting bodies) develop on the bulb, clustered around the neck. The bulb may rot or appear firm. Foliage that appears above ground may be withered.

**PLANTS AFFECTED** Tulips

**SEASON** Late summer

### GLADIOLUS CORM ROT
DISEASE • See p. 131

Brown, ridged areas of discoloration appear on the corm. The corm dries out and becomes mummified. No fungal growths or sclerotia are visible.

**PLANTS AFFECTED** Gladioli, crocus, bulbous iris

**SEASON** In storage

### GLADIOLUS CORM ROT – FOLIAGE SYMPTOMS
DISEASE • See p. 131

Yellow flecking and, later, striping develops towards the tips of the outer-most leaves. The discoloration spreads and the leaves turn brown and die.

**PLANTS AFFECTED** Gladioli, crocus, bulbous iris

**SEASON** Late spring and summer

# ROOT AND TUBER PROBLEMS

### FOOT AND ROOT ROTS
DISEASE • See p. 128

Roots discolour (here on daphne), often turning black or brown and shrinking inwards. Deterioration usually starts at the root tips and roots may disintegrate.

**PLANTS AFFECTED** Most plants

**SEASON** All year

### IRIS RHIZOME ROT
DISEASE • See p. 139

Roots and rhizomes deteriorate rapidly, turning soft and slimy. The infection usually starts just beneath the leaf, but spreads quickly. Foliage may fail to develop, or yellows and dies back.

**PLANTS AFFECTED** Rhizomatous irises

**SEASON** Mainly spring and summer

### CLUBROOT
DISEASE • See p. 118

Numerous irregularly shaped swellings develop on roots of cabbage and related plants. Swellings do not contain any insect larvae or grubs. The entire root system may be distorted and swollen.

**PLANTS AFFECTED** Any in Cruciferae family

**SEASON** All year

### ROOT APHIDS
PEST • See p. 170

Infested plants grow slowly and tend to wilt in sunny weather. Aphids, usually globular and creamy-brown, feed on the roots. They often secrete a white, waxy powder from their bodies.

**PLANTS AFFECTED** Lettuce, beans and others

**SEASON** Mid- to late summer

### POTATO CYST EELWORMS
PEST • See p. 163

Pinhead-sized, spherical cysts, which may be white, pale yellow or brown, form on the roots. These are the swollen bodies of the female eelworms.

**PLANTS AFFECTED** Potatoes, tomatoes

**SEASON** Mid- to late summer

### ROOT MEALYBUGS
PEST • See p. 170

These sap-feeding insects have white, elongate bodies up to 2mm long. They secrete a white, waxy powder from their bodies, which coats roots and soil particles.

**PLANTS AFFECTED** Pot plants

**SEASON** All year

## ROOT AND TUBER PROBLEMS *CONTINUED*

### POTATO SPRAING
DISEASE • *See p. 165*

Tan-coloured arcs of discoloration appear in the flesh. Corky tissue may be present around the discoloured areas. The tubers are occasionally distorted. Foliage and stems may be mottled yellow.

**PLANTS AFFECTED** Potatoes

**SEASON** Late spring onwards

### POTATO COMMON SCAB
DISEASE • *See p. 163*

Raised, roughened, scabby patches develop on the skin of the tuber. Scabs have ragged edges. The damage is usually superficial.

**PLANTS AFFECTED** Potatoes

**SEASON** Late spring onwards

### POTATO SILVER SCURF
DISEASE • *See p. 165*

Pale and inconspicuous silvery-grey markings develop on the tuber skin. The size, shape, flesh and eating quality of the potato are not affected.

**PLANTS AFFECTED** Potatoes

**SEASON** Mainly in storage

### POTATO BLIGHT
DISEASE • *See p. 163*

Slightly sunken, darkened areas develop on the skin. Internally, the flesh is discoloured by a dry reddish-brown rot. Secondary organisms cause soft rotting and an unpleasant smell.

**PLANTS AFFECTED** Potatoes and tomatoes

**SEASON** Late summer onwards

### POTATO POWDERY SCAB
DISEASE • *See p. 164*

Small, near-circular scabby areas appear on the tuber skin. The scabs have raised margins and burst to release masses of brown spores.

**PLANTS AFFECTED** Potatoes

**SEASON** Summer onwards

### WIREWORMS
PEST • *See p. 191*

Elongate, yellowish-brown grubs with three pairs of legs at the head end bore into the tubers. They are mainly a problem where grass has recently been dug over to make a vegetable garden.

**PLANTS AFFECTED** Root vegetables, potatoes

**SEASON** Early summer to mid-autumn

## ROOT AND TUBER PROBLEMS CONTINUED

### VIOLET ROOT ROT

DISEASE • *See p. 186*

The surface of the root (here of a carrot) or tuber is covered in a network of dark purple fungal strands. Sclerotia may be present. Roots and tubers may rot.

**PLANTS AFFECTED** Various plants

**SEASON** Mainly spring to autumn

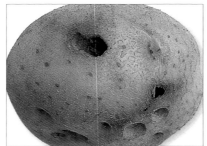

### SLUGS

PEST • *See p. 176*

Slugs make small, round holes in the skin of potato tubers but hollow out extensive cavities once they are inside. Some potato cultivars are particularly susceptible to slug damage.

**PLANTS AFFECTED** Potatoes, root vegetables

**SEASON** Mid-summer to mid-autumn

### POTATO GANGRENE

DISEASE • *See p. 164*

Slightly sunken, round-edged areas develop on the tuber shortly after lifting. Lesions may remain defined or enlarge. The tuber rots, turning pale pink.

**PLANTS AFFECTED** Potatoes

**SEASON** Summer and autumn

### POTATO DRY ROT

DISEASE • *See p. 164*

One end of the tuber shows wrinkled skin and then shrinks inwards rapidly, forming concentric rings of wrinkles and discoloration. Pink, white or blue-green fungal pustules may develop.

**PLANTS AFFECTED** Potatoes in store

**SEASON** Any time of year

*Secondary infection causes rotting of shoulder area*

*Deep fissure running length of root*

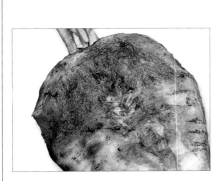

### PARSNIP CANKER

DISEASE • *See p. 154*

Roughened, reddish-brown, orange or black cankered areas develop on the root, most commonly around the shoulders. Affected tissues may develop secondary rotting.

**PLANTS AFFECTED** Parsnips

**SEASON** Mainly autumn and winter

### IRREGULAR WATERING

DISORDER • *See p. 140*

Longitudinal cracks appear in roots (here of a carrot). Top growth appears perfectly normal. Secondary organisms may cause deterioration of root.

**PLANTS AFFECTED** Many root crops

**SEASON** Mainly summer and autumn

# LAWNS

LAWNS ARE VERY POPULAR and one of the most widely used areas of a garden. Even if you do not want a perfect lawn, but prefer an area which is practical, serviceable and a good shade of green, there will still be times when problems need to be dealt with.

There are several fungal infections and pests which are common. The majority of problems, whether caused by a pest, a pathogen or poor cultivation, result in areas of grass turning yellow or brown. At first sight, differentiating between the various causes may not be easy. Most fungal problems do involve the development of visible (if small) fungal growths and a few, such as fairy rings or other toadstools, are very distinct. Lawn pest problems are generally easier to identify and culprits more readily visible.

In all cases, prompt action is advisable, as once turf grasses weaken or their growth becomes thinner, coarse grasses, mosses and weeds colonize rapidly. Among these, dog lichens (*see facing page*) are commonly mistaken for a disease, but in fact are half fungus and half algae.

## LAWN PROBLEMS

**LEATHERJACKETS**    PEST • *See p.143*
These daddy-long-legs larvae generally cause damage when they are becoming fully grown. Patches of turf turn yellow-brown and holes may be seen where birds have been searching for the grubs.

**PLANTS AFFECTED** Turf, roots of many plants
**SEASON** Mid-summer

**SLIME MOULD**    DISEASE • *See p.176*
Numerous grey, yellow or off-white spheres are clustered together on individual blades of grass. These may appear to "dissolve" when wet. Turf growth and health are not affected.

**PLANTS AFFECTED** Many plants, including turf
**SEASON** Mainly summer and autumn

**BURROWING BEES**    PEST • *See p.112*
Small, conical heaps of powdery soil form on the lawn. Each has an entrance hole at the top where a solitary bee leaves and enters. No significant damage is caused to the lawn.

**PLANTS AFFECTED** Turf
**SEASON** Mid-spring to mid-summer

**CHAFER GRUBS**    PEST • *See p.116*
Plump, creamy-white grubs with brown heads sever roots, making it easy for foxes, badgers and birds to rip up turf to feed on them. The grubs are curved, with three pairs of legs (*see p.53*).

**PLANTS AFFECTED** Turf, other plant roots
**SEASON** Early autumn to early spring

**TURF THATCH FUNGAL MYCELIUM**    DISEASE • *See p.184*
Bleached yellow or reddish patches of grass develop. Dense white or off-white fungal growth appears around the grass roots.

**PLANTS AFFECTED** Turf
**SEASON** All year

**TURF DOLLAR SPOT**    DISEASE • *See p.183*
Patches of grass die off and turn straw-yellow. The individual patches are usually 10cm or less in diameter. The fungus causing this disease is most prevalent during warm, damp weather.

**PLANTS AFFECTED** Turf
**SEASON** Mainly early autumn

**EARTHWORMS**    PEST • *See p.125*
Some earthworms deposit heaps of muddy excrement on the surface of lawns. These worm casts become smeared by feet or mowers and spoil the lawn's appearance.

**PLANTS AFFECTED** Turf
**SEASON** Mainly autumn to spring

## LAWN PROBLEMS *CONTINUED*

### ANTS
PEST • *See p. 99*

Heaps of fine soil appear on the lawn's surface above ant nests; if they are disturbed, black or reddish-yellow ants will be seen. Ant heaps get in the way of mowers and make the lawn uneven.

**PLANTS AFFECTED** Turf

**SEASON** Late spring to early autumn

### TURF RED THREAD
DISEASE • *See p. 184*

Patches of grass begin to deteriorate. Numerous pale red or pink, gelatinous, horn-like fungal growths are attached to individual grass blades. Later, fluffy, pale pink fungal growth develops.

**PLANTS AFFECTED** Turf

**SEASON** Mainly late spring to early autumn

### DOG LICHENS
*See facing page*

Small, unevenly shaped, grey, leaf-like growths, off-white beneath, develop amongst grass, particularly when turf is sparse and weak. Lichen is not a pest or disease; it should be treated as a weed.

**PLANTS AFFECTED** Turf

**SEASON** All year

### TURF SNOW MOULD
DISEASE • *See p. 184*

Patches of dying grass develop and then enlarge and turn brown. Deteriorating grass may be covered in pale pink, fluffy fungal growth.

**PLANTS AFFECTED** Turf

**SEASON** Mainly autumn and winter

### MOLES
PEST • *See p. 148*

Moles create a network of tunnels in the soil in which they live, and leave the excavated soil in heaps on the surface. Collapsed mole tunnels give the lawn an uneven surface.

**PLANTS AFFECTED** Turf

**SEASON** All year

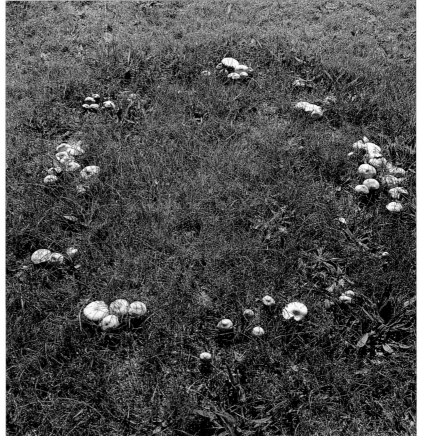

### TURF FAIRY RINGS
DISEASE • *See p. 183*

Rings of dead or dying grass, or sometimes other patterns of discoloration, develop on a lawn. One or more rings of lush grass grow adjacent to these. Brown toadstools grow out of the ring.

**PLANTS AFFECTED** Turf

**SEASON** Mainly late summer and autumn

# BENEFICIAL GARDEN CREATURES

GARDENS ARE INHABITED BY a wide range of insects and other animals, but only a small proportion are plant pests. Most are neither particularly helpful nor harmful to the gardener. There are, however, some which are very beneficial, as they feed on pests, pollinate flowers or improve soil.

Predatory and parasitic insects may not be sufficient to prevent pests causing damage, but they do stop pests from having things all their own way. Pollinating insects are essential for good crops of most tree, bush and cane fruits, and some vegetables. Any insects that move from one flower to another can effect the transfer of pollen.

Most insecticides do not discriminate between pests and beneficial insects. It is therefore important to avoid the unnecessary use of chemicals, saving them for plants which would be seriously damaged by pests or diseases if control measures were not taken.

### LADYBIRDS
Most ladybirds (*see p.141*) prey on aphids, but some feed on scale insects, spider mites or mealybugs, and others on powdery mildew fungal spores. Not all are red with black spots; some are yellow with black spots, others are orange or brown with white spots.

### FLOWER BUG
Flower bugs, also known as anthocorid bugs, are 3–4mm long and are general predators of aphids, thrips, other small insects, mites and their eggs. They have needle-like mouthparts with which they pierce their prey and suck out the contents.

### PARASITIZED APHIDS
Several species of tiny parasitic wasp lay eggs in young aphid nymphs. The wasp grub develops as an internal parasite that kills the aphid, causing it to become very swollen and straw brown in colour. The adult wasp emerges through a hole it makes in the dead aphid's abdomen.

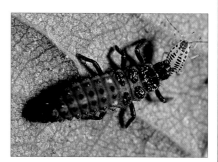

### LADYBIRD LARVA
Although less familiar than the adult beetles, ladybird larvae are equally voracious predators of aphids and other pests. The larvae, up to 11mm long, are black with orange or white markings. A larva will eat 200–400 aphids before it is fully fed and ready to pupate.

## HOVERFLY (LARVA)
Both the larval and adult stages (*see p.138*) of hoverflies feed on aphids. The larvae have flattened, semi-transparent bodies, up to 16mm long. The adult flies have yellow and black striped abdomens, giving them a wasp-like appearance, although their bodies usually look smoother.

## TACHINID FLIES
There are many species of tachinid flies, between 4 and 18mm long. Most have larvae that are parasites of insects, especially caterpillars. The maggots feed inside the body cavity of their host insect, causing its death by the time the tachinid larvae are ready to pupate.

## EARTHWORMS
Earthworms (*see p.125*), up to 20cm long, burrow tunnels in the soil through which water can drain and air can reach plant roots. Worms feed on dead plant material and their presence in compost heaps speeds up the conversion of debris to compost.

## BROWN CENTIPEDE
This centipede, up to 30mm long, lives in the surface layers of the soil and also shelters under stones, logs and pots. It is active at night, preying on a wide range of small soil animals and their eggs. The front pair of legs curve forward and are modified to act as jaws.

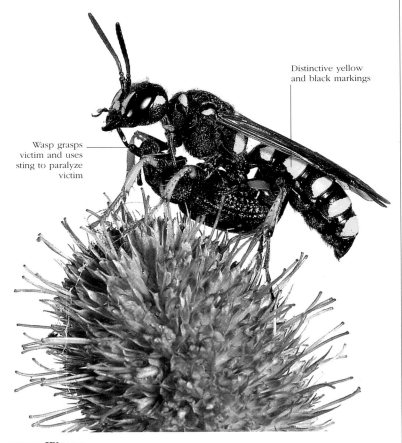

Distinctive yellow and black markings

Wasp grasps victim and uses sting to paralyze victim

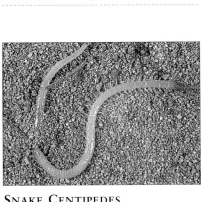

## SNAKE CENTIPEDES
Snake centipedes live below soil level and are often found when soil is dug. They have slender, yellowish-brown bodies, up to 60mm long, adapted for moving through the soil where they hunt for small insects. If disturbed they writhe in a snake-like manner.

## SOLITARY WASPS
Unlike social wasps, these tunnel out individual nests in rotten wood and sandy soil, or make use of beetle tunnels in dead wood. They provision their nests with aphids, flies, weevils or other insects to feed their larvae. There are many species of various sizes.

## BUMBLEBEES

Bumblebees make their nests in the ground, often down abandoned mouse holes. They are valuable pollinating insects as they will visit flowers in poor weather when honeybees stay in their hives. At full strength in mid-summer a bumblebee nest houses 200–500 bees.

## HONEYBEES

Many plants, especially most fruits, require insects to carry pollen from one flower to another to fertilize the flowers and allow seeds and fruits to set. The hairy bodies of honeybees, 12mm long, become heavily dusted with pollen when they visit flowers to collect nectar.

## WOLF SPIDERS

Most spiders spin sticky webs to capture flying or crawling insects, but wolf spiders, 5–8mm long, run rapidly over the soil or vegetation to hunt down their prey. The females are frequently seen in early summer carrying their silken egg sacs under their abdomens.

## SHREW

These small, mouse-like predators, which have elongate snouts, are rarely seen but may be numerous in hedgerows and rough grass. They are active during the day and at night, when they feed on a wide range of insects, slugs, woodlice and earthworms.

## PARASITIC WASPS

Many species of parasitic wasps lay their eggs in the larvae, pupae or eggs of other insects. The larvae develop inside the host insect, which is killed when the parasite is ready to pupate. Some wasps have long egg-laying organs to enable them to reach concealed insects.

## DEVIL'S COACH HORSE

This beetle, 25mm long, is often found sheltering under stones, logs or pots. It is one of the staphylinid or rove beetles, which are characterized by short wing cases that leave most of the abdomen uncovered. It eats slugs and soil pests such as cutworms and leatherjackets.

### SONG THRUSH

The song thrush specializes in feeding on snails. It breaks the snail's shell by smashing it against a stone. They often make regular use of certain stones and these "anvils" become littered with pieces of snail shell. Their diet also includes worms, insects and berries.

### GREEN LACEWING

Lacewings, 12–15mm long, derive their name from the many veins that give the wings a lace-like appearance. Other types of adult lacewings (see p.141) have black or brown bodies. The larval stages are voracious predators of aphids and sometimes other small insects.

### CARABID BEETLES

Also known as ground beetles (see p.134), these insects live mainly on the soil surface, but some climb up plants in search of insects to eat. They are from 2 to 35mm long and mostly black, brown or metallic green. The larvae prey on a range of insects, slugs and pest eggs.

### COMMON TOAD

Toads, like frogs, need ponds in which to breed in the spring, but for the rest of the summer they may be found well away from water. They are mainly active after dark, when they feed on a variety of insects, woodlice, worms and spiders.

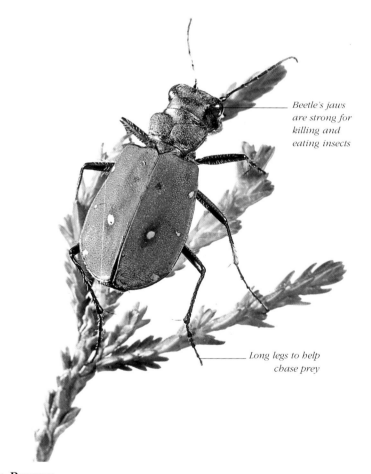

*Beetle's jaws are strong for killing and eating insects*

*Long legs to help chase prey*

### HEDGEHOGS

Although sometimes active in daylight, hedgehogs are largely nocturnal. Their diet includes slugs, worms and insects, particularly beetles. Hedgehogs need somewhere sheltered where they can hibernate undisturbed during the winter months.

### TIGER BEETLE

This colourful predator, 11–13mm long, occurs mainly in sandy areas. It is a type of ground, or carabid, beetle. Its larvae live in pits in the soil from which they emerge to seize passing insects. The adults run over the soil surface and readily fly up in sunny weather.

# GARDEN HEALTH AND PROBLEMS

GARDEN PLANTS CO-EXIST with many other living things – some beneficial, many benign, but some harmful. Favourable growing conditions and good gardening practice make plants strong and more able to tolerate potentially harmful organisms, but when serious attack by pests or diseases occurs, the gardener must choose an appropriate method with which to bring the problem under control.

# THE BALANCE OF LIFE

EVERY GARDEN FORMS a miniature ecological system that is unique, with its own combination of plants, soil conditions, topography and climate. These factors, together with the interaction of the various creatures – including predators, parasites, pathogens, saprophytes and herbivores – that live within the garden, determine what thrives, what survives and what deteriorates or dies. The aim of the gardener is to encourage desirable growth – usually, that of ornamental and cropping plants – but this must be achieved within a healthy ecological balance that involves other life forms; a garden purged of insects or fungi, for example, would soon become unproductive. The activities and life processes of other forms of plant and animal life, together with their death and decay, are crucial to the growth and health of garden plants. Since all the organisms in a garden will influence the overall "working" and functioning of the plants and the garden as a whole, any action taken by the gardener may have an effect on plants or animals other than that being treated. When considering pest or disease control it is therefore important to consider any such effect.

## THE ENERGY FOR GROWTH

Plants manufacture energy by the process of photosynthesis, in which the action of sunlight on green leaf pigments enables plants to combine

## INTERACTION IN THE GARDEN

Even in the smallest corner of the garden, plants, creatures and environmental factors interact to form a rich and complex ecological system. Any action taken by the gardener may affect numerous links in the varied and interlocking cycles and webs of life.

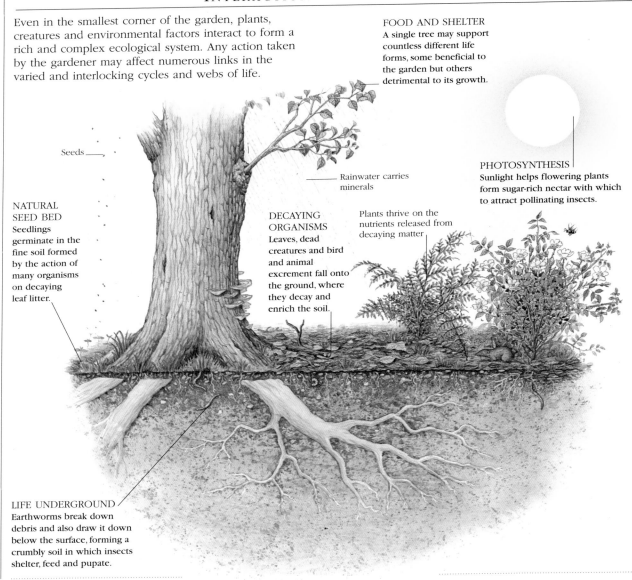

**FOOD AND SHELTER**
A single tree may support countless different life forms, some beneficial to the garden but others detrimental to its growth.

Seeds

Rainwater carries minerals

**PHOTOSYNTHESIS**
Sunlight helps flowering plants form sugar-rich nectar with which to attract pollinating insects.

**NATURAL SEED BED**
Seedlings germinate in the fine soil formed by the action of many organisms on decaying leaf litter.

**DECAYING ORGANISMS**
Leaves, dead creatures and bird and animal excrement fall onto the ground, where they decay and enrich the soil.

Plants thrive on the nutrients released from decaying matter

**LIFE UNDERGROUND**
Earthworms break down debris and also draw it down below the surface, forming a crumbly soil in which insects shelter, feed and pupate.

carbon dioxide from the air with water to make sugars and starch. Sugars may be used for other purposes, such as aiding pollination and seed dispersal. Many plants are insect-pollinated, and the secretion of sugar-rich nectar by flowers is an effective means of encouraging bees and other insects to visit blooms. While doing so insects pick up pollen when they come into contact with the anthers, and this is transferred to the stamens of flowers that are subsequently visited.

Sugars also make fruits and berries sweet and attractive to birds and mammals. The seeds are dispersed away from the parent plant when they are spat out or have passed through the animal's gut.

## ESSENTIAL NUTRIENTS

The starches and sugars used by plants for growth must be augmented by essential nutrients (*see pp.68–69*), drawn from the soil or, in the case of parasitic plants, from their hosts. For the cycle of life to continue, these nutrients must be returned to the soil. Dead or waste organic material, whether decomposing plant or animal matter, is broken down by the actions of fungi, bacteria, earthworms and others that feed on decaying plant material.

Certain larger fungi and many micro-fungi, described as saprophytic, live on dead organic material and, by breaking it down, may release components which are then used by other organisms in the vicinity. Their primary role, however, is one of clearing up debris which would otherwise accumulate. In a similar way numerous garden creatures live on debris and are to be encouraged because of their recycling function.

### RECYCLING ACTIVITY
*Woodlice occasionally cause plant problems, but have a valuable role in processing garden debris.*

As any organic matter deteriorates it releases and uses certain nutrients and so alters the range and quantity of materials available to plants and other organisms in the garden. Although deteriorating plant material may harbour plant pathogens, it may also provide a useful overwintering or breeding site for beneficial insects, so it is important to reach a compromise between extremes of garden tidiness.

## FOOD AND FEEDERS
Plants form the basis of all food webs. There is a huge range of invertebrate animals, including nematodes, millipedes, insects, mites, woodlice, slugs and snails, that feed directly on plants. Many

### PLANT ASSOCIATIONS
*Gardeners' time frames often differ from those of nature. While mistletoe on tree branches (usually, of apple) will weaken the host over many years, most gardeners prefer the short-term attraction and novelty value of its presence.*

of them specialize in feeding on particular parts of a plant; some may feed on a variety of plants, while others may be specific in their choice of food plant.

Most pests and disease-causing organisms have a relatively restricted host range, attacking only one type of plant, or perhaps a few closely related ones. An outbreak of one pest or disease could therefore remove the main food source of another potential pest or pathogen, so indirectly minimizing the damage it causes. Plant pests in turn are under attack from predators and parasites which help to prevent pests becoming so abundant that plants are killed.

## OTHER BENEFICIAL ASSOCIATIONS
Fungi, bacteria and viruses are equally active in attacking plants, but there are also some which have a beneficial effect. Many plant roots have developed a mutually beneficial relationship with certain fungi or bacteria that help host plants to absorb nutrients from the soil (*see overleaf*).

### PEST PREDATOR
*A hedgehog is a valuable garden ally, naturally culling slugs and other plant pests, although it also feeds on earthworms.*

# SOIL AND NUTRIENTS

THE STRUCTURE AND texture of soil, and the range of nutrients it contains, has a great impact on plant growth and development. Texture – basically whether a soil is light and sandy, heavy and clayey, or something in between – influences plants' ability to develop strong root systems through which to take up water and nourishment, and affects the . soil's ability to retain moisture and, to a certain extent, nutrients. These nutrients comprise a range of elements and minerals. Although some are present in only small ("trace") quantities, deficiencies, either due to natural factors or caused by heavy demands by plants that go unreplenished, will cause symptoms in plants that are often taken to be the result of attack by pests and diseases (*see table, facing page*). Sometimes, pest and disease problems follow on from deficiency, as the plant is weakened and more vulnerable to attack.

Soil pH or acidity also influences the range of nutrients that are in a form available to plants. For example, soils of a high pH (alkaline, limey or chalky) may "lock up" iron and manganese so that plants become chlorotic. In fact, in an alkaline soil most trace elements are less readily available than in an acid or neutral soil. Similarly a low pH (acid) soil is more likely to produce plants which suffer a deficiency of magnesium. An unusually high level of one nutrient may also influence the levels or

**CROSS-SECTION, GARDEN SOIL**
*Garden soil has a distinct structure, with three layers – the top-soil, the subsoil and the bedrock (not shown here). The topsoil is the uppermost layer and is usually a dark brown colour. It contains a high level of organic matter and soil organisms and is the layer in which you dig and plant. The subsoil has a much lower level of organic matter, nutrients and organisms.*

availability of another. Very commonly, tomatoes fed with a high potash feed to encourage a heavy crop of fruits may also show yellowing between the veins and around the edges of older leaves due to magnesium deficiency.

## BALANCING NUTRIENT LEVELS

Checking the nutrient status and soil pH is a worthwhile job, especially when taking on a new garden with unfamiliar soil. Various kits are available (*see facing page*) and need not be expensive. These enable you to determine the soil pH and give an outline of the levels of major nutrients. Serious deficiencies are quite rare in gardens and feeding with a general fertilizer or regular use of plenty of organic material and good husbandry is usually enough to keep most plants adequately fed. Occasional use of specific materials to, for example, encourage flowering and fruiting or lush growth, may also be beneficial. It is worth bearing in mind, however, that overfeeding can be damaging both to plants and other soil organisms.

## OTHER SOIL LIFE

Many garden creatures, including a number of pests, spend their lives, or a stage in their lives, within the soil. However, in general gardeners are far better occupied in improving soil texture and composition than in attempting to control its inhabitants. Although it is easy to become preoccupied with potentially

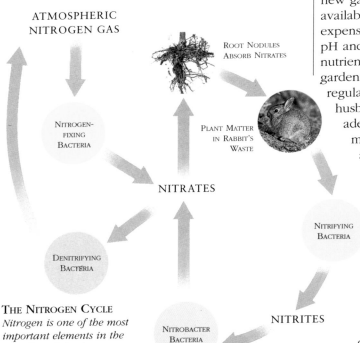

ATMOSPHERIC NITROGEN GAS

ROOT NODULES ABSORB NITRATES

NITROGEN-FIXING BACTERIA

PLANT MATTER IN RABBIT'S WASTE

NITRATES

DENITRIFYING BACTERIA

NITRIFYING BACTERIA

NITROBACTER BACTERIA

NITRITES

**THE NITROGEN CYCLE**
*Nitrogen is one of the most important elements in the growth and development of all living organisms.*

# USING A KIT TO TEST SOIL NUTRIENTS

**1 Preparing.** This test for phosphorus is part of a kit to test the pH and major nutrient levels of soil. Mix 1 part soil and 5 parts water and allow to stand until the liquid is fairly clear. Use the pipette to transfer liquid to the testing chambers.

**2 Testing.** Make sure the water extract reaches the level marks on the colour chart. Open the correct capsule (blue in this test) and pour the powder into the chamber. Fit the cap onto the tester and shake until the powder has dissolved.

**3 Results.** Allow the colour to develop in the test chamber for a few minutes. Then compare this with the colour chart against a white background and in natural daylight. Read off the results and take further action if required.

damaging garden organisms, the vast majority of creatures in soil are harmless and indeed some are actually beneficial (*see pp.60–63*). The activity of each type of organism affects the population levels of others living in the vicinity and so any action taken by the gardener may well have a knock-on effect on something else.

Some animals and fungi can enhance the productivity or the performance of plants; examples include pest predators and parasites, and mycorrhizae – fungal associations formed on the root of a plant which allow the plant to become more efficient at nutrient uptake. Mycorrhizae may sometimes be visible in the soil as tiny fungal strands attached to the roots – these are often presumed to be harmful. Some also produce fungal fruiting bodies or toadstools around the base of the plant; again it may be presumed, incorrectly, that all toadstools around plants are harmful.

## NITROGEN FIXERS

Certain bacteria that "fix" nitrogen, making it available to the plant, inhabit nodules on plant roots. If these roots are left in the soil once the crop has been harvested, they will also improve the nutrient status of the soil for any plants grown there subsequently. The nitrogen-fixing bacteria are invisible to the naked eye but the nodules are seen as distinct swellings on the roots (*see facing page*). The presence of these root nodules on peas and beans can be used to great benefit when following a rotation system for growing vegetables (*see p.96*).

## NUTRIENT REQUIREMENTS AND ASSOCIATED PROBLEMS

| Element | Specifically aids | Symptoms of deficiency | Possible causes | Remedies |
|---|---|---|---|---|
| NITROGEN (N) | Leaf and shoot growth. | Pale foliage, poor growth and may develop reddish or yellow discoloration. Older leaves affected first, as the nitrogen is moved from these into the newer growth when in short supply. | Most common on soils with low organic matter or where un-decomposed organic matter is added to soil in large quantities. | See p.151 |
| PHOSPHORUS (P) | Root growth. | The plant's overall growth, including flowering and fruiting and the growth and development of seeds, is affected. | Most common on acid soils or following heavy watering or rain. | See p.159 |
| POTASSIUM (K) | Flowering and fruiting; wood ripening; tissue strengthening. | Leaves may show scorching of the tips. Fruiting and flowering may be reduced and on tomatoes it causes blotchy ripening. | Most common on sandy soils and on chalk. | See p.162 |
| MAGNESIUM (MG) | Chlorophyll formation (the green pigment in plants). | Leaves, particularly the older ones, show yellowing between the veins and around the edges. When it is in short supply, magnesium is moved from the older tissues into the new ones, so young foliage rarely shows any symptoms. | Most common on sandy soils and following heavy watering or rain, or heavy applications of potassium. | See p.146 |
| CALCIUM (CA) | Formation of plant cell walls. | The most commonly seen symptoms are bitter pit on apples and blossom end rot of tomatoes. Foliage may also be reduced in size and distorted. | Most common on acid soils, and on plants growing in any soil which is not adequately watered to allow the plant to take up calcium. | See p.113 |
| IRON (FE) | Chlorophyll formation. | Leaf yellowing, mainly between the veins and most apparent on the young leaf growth. | Most common on alkaline soils. | See p.139 |
| MANGANESE (MN) | Chlorophyll formation. | Yellowing between the veins on young leaves is the most common symptom. | Most common on alkaline soils and waterlogged or poorly draining soils. | See p.146 |

# NATURAL UPSETS

IN ANY NATURAL environment there are factors that are not ideal for plant growth. These naturally occurring problems are often the result of adverse weather conditions that either cause physical wounds through which infections may enter, or, more commonly, simply weaken the entire plant so that it is more likely to succumb to pests and diseases.

Often the gardener can compound problems, either through "contributory negligence" – leaving drought-stricken plants unwatered, for example – or by unintentionally inflicting damage such as that caused by poor pruning or by overtight stem or tree ties. By far the most common contributory factor, however, is the poor choice of plants for a given site. If a plant is not suited to the prevailing conditions and climate, or microclimate, it will rarely thrive and hence will be more prone to problems. A less than fully hardy plant may be so set back in a cold, exposed site each year that it never makes a good specimen. A moisture-loving plant will fare badly on dry soil; a Mediterranean plant may rot in waterlogged ground. Always check before buying that the plant stands a good chance of survival in the conditions your garden offers.

## WATER AND DROUGHT

Water is a fundamental plant requirement, used in photosynthesis (*see pp.66–67*), to maintain turgidity (the ability of stems to stand upright) and to transport nutrients into and around the plant tissues.

Plants vary in their requirement for water and some are better than others at surviving dry conditions. Wilting is the most common symptom of drought; if soil moisture levels are restored fairly

## DAMAGE CAUSED BY ENVIRONMENTAL FACTORS

### FROST DAMAGE
*A scorched appearance is the most common symptom of frost damage. Soft new growth is most vulnerable to damage, with blackening and withering of shoot tips; pruning should not be left so late in the season that regrowth cannot ripen.*

Frost-damaged
*Pieris japonica*
leaves

### LIGHTNING DAMAGE
*The larger and more ragged wounds are, the more slowly they will heal. Until a wound has completely callused over with scar tissue, it is vulnerable to invasion by harmful pathogens.*

### ANIMAL VANDALISM
*Few plants can withstand repeated injury caused by browsing deer and rabbits, which may cause considerable problems in gardens in rural areas or close to parkland.*

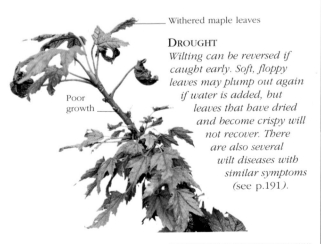

Withered maple leaves

### DROUGHT
*Wilting can be reversed if caught early. Soft, floppy leaves may plump out again if water is added, but leaves that have dried and become crispy will not recover. There are also several wilt diseases with similar symptoms (see p.191).*

Poor
growth

rapidly it can be reversed. Prolonged drought can, however, cause parts or all of the plant to die. If water levels are consistently low, but never in drastically short supply, the plant may survive, but be small and fail to perform well. Poor flowering, buds that fail to develop, and small fruits with tough skins are common symptoms. Dry soil is most obviously associated with dry weather, but light soil texture and drying winds may also be factors.

## WATERLOGGING AND IRREGULAR WATERING

Excess water can be just as harmful as drought. If soil is waterlogged, plant roots may deteriorate rapidly, largely due to the lack of oxygen. Water-logged soil is also more likely to contain harmful pathogens, such as damping-off or foot- and root-rotting fungi. In the early stages, however, chlorosis or yellowing, often associated with the veins of leaves, is a common symptom.

**POOR PRUNING**
*Stubs left after pruning are readily colonized, as here, by rotting fungi that may spread.*

Plants also find it difficult to adapt to an erratic water supply. Plants that are growing in containers are very susceptible, as they tend to dry out readily and may then be overwatered by an over-enthusiastic gardener. Symptoms include poor or slow growth, bud drop and cracked or split fruits.

## EXTREMES OF TEMPERATURE

In addition to drought symptoms, high temperatures often cause scorching of leaves, particularly on plants growing under glass. Low temperature injury is most commonly seen as frost damage (*see facing page*), causing withering of shoots, foliage or flowers or even death of the whole plant. Plants growing in very cold areas or in containers may also suffer root injury due to freezing of the root ball. This may then be followed by poor growth, dieback or even death. Fruit set is frequently affected by low temperatures as the activity of pollinating insects may be restricted, and flowers themselves damaged.

Snow and hail too can injure leaves or, in the case of snow, weigh down and break branches. This problem is particularly common when a heavy fall of snow has partially melted and re-frozen. In some cases, however, snow may act as an insulating "blanket" and prevent cold wind or frost injury of the plants which it covers.

Water-filled flesh

CROSS-SECTION

**NATURAL PROTECTION**
*Cacti (above and right) are very well adapted to drought conditions. Evaporation from the plant's surface is minimized by the thick, tough outer skin, and the spines (reduced leaves) protect the water-rich, fleshy core from browsing animals.*

## STRONG WINDS

Wind may cause physical injury and scorching of foliage and a prevailing strong wind may also prevent a plant growing to its full potential size, or even alter its shape. Wind blows away moisture shells around a transpiring plant leaf and so will, in turn, increase the plant's water requirements and make it more likely to suffer from drought if a plentiful supply of water is not available. In very windy seasons or sites, pollination may also be poor, again due to pollinating insects being restricted in their activities.

Healthy *Primula obconica* flowers

Flowers dying off

Leaves turn yellow and flop over

**WATERLOGGING**
*The water in saturated soil displaces air between soil particles, creating an anaerobic environment in which plants cannot take up water. The symptoms therefore resemble those of drought, and plants may mistakenly be given yet more water.*

# TYPES OF PLANT DISEASE

T HE DISEASE-CAUSING organisms or pathogens most prevalent in gardens are fungi and viruses. In addition, several bacterial infections and a few diseases caused by mycoplasmas occur.

## FUNGAL INFECTIONS

This major group of infections may be caused by microfungi that are extremely small but usually still visible to the naked eye. Examples include downy mildews, powdery mildews, fungal leaf spots and rusts. A few fungal infections are systemic (found throughout the plant), but the majority are restricted to certain plant parts. The larger fungi – those which produce easily visible fungal structures such as fungal brackets or toadstools – are less common but may prove extremely damaging.

## VIRUS INFECTIONS

Virus particles are sub-microscopic, that is to say invisible even with the aid of a normal microscope, but can be seen with an electron microscope. The symptoms they cause are, however, generally easily visible, most commonly as yellow leaf markings. Virus particles are systemic so are found throughout an infected plant, but their symptoms may only be apparent on one area, or their appearance may vary from one part of the plant to another. Virus infections may occur alone or in combination with other virus infections. It is impossible to identify precisely which are involved from the symptoms alone. Mycoplasmas are sub-microscopic and similar to viruses but much less common.

## BACTERIAL INFECTIONS

Problems caused by bacterial infections are usually localized, but as many cause rapid deterioration of plant tissues, their effect may be quite dramatic. Sometimes the bacteria are apparent as bacterial ooze, for example on a tree infected with bacterial canker or fireblight. Many bacterial infections which result in softening of plant tissues are accompanied by a distinct, and often unpleasant, smell.

## HOW DISEASES ARE SPREAD

Most fungal infections are spread from plant to plant or within a plant by spores. These are occasionally large enough to be visible with the naked eye. Their type, size and quantity depends largely on the specific fungus. Most spores are spread either on air currents or by rain splash. Bacteria are spread by similar means. Some infections may also be carried by animals, including insects as they move from plant to plant. Soil-borne infections are usually spread either in association

## MICROSCOPIC AND SUB-MICROSCOPIC ORGANISMS

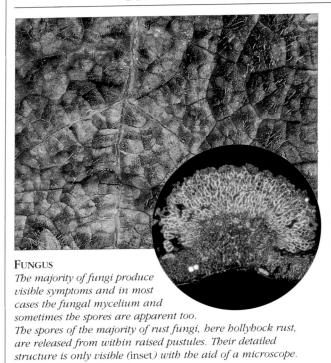

**FUNGUS**
*The majority of fungi produce visible symptoms and in most cases the fungal mycelium and sometimes the spores are apparent too.*
*The spores of the majority of rust fungi, here hollyhock rust, are released from within raised pustules. Their detailed structure is only visible (inset) with the aid of a microscope.*

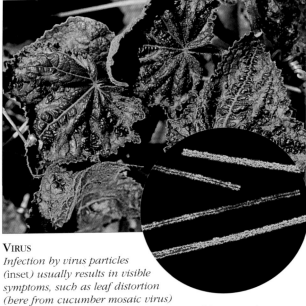

**VIRUS**
*Infection by virus particles (inset) usually results in visible symptoms, such as leaf distortion (here from cucumber mosaic virus) and yellow patterning, often accompanied by general poor growth and stunting. Virus infected plants may also be symptomless, or the extent of the symptoms may vary greatly.*

**SPREADING DISEASE**
*Using cutting tools such as secateurs on an infected plant and then on a healthy one without disinfecting them can spread disease.*

**COLOUR BREAKING**
*A characteristic symptom of virus attack is slashing or striping of flower colour. Usually unwanted, it has in some cases, particularly in tulips, been developed through breeding to produce novel petal patterns in otherwise healthy plants.*

with soil particles or in soil moisture. Some, such as *Phytophthora*, have spores that can propel themselves, called zoospores, and these spread in water. Weather conditions may affect disease spread and development, and in most cases moisture on the plant surface encourages infection and spread. Humidity and the resultant film of moisture on the plants may also cause a marked increase in disease spread and development.

Viruses are most commonly spread when sap from an infected plant is transferred to another potential host. Pruning tools, physical injury, grafting and the taking of cuttings are all common routes. Even handling – for example, dividing – an infected plant may be sufficient to transfer virus particles to a healthy host. Some viruses are also transmitted by garden creatures which, because of this role, are known as vectors. The most common are sap-feeding pests such as aphids, leafhoppers and nematodes. Viruses may also be seed-borne.

## INCREASING VULNERABILITY

A plant which has suffered a severe attack of, say, powdery mildew, with consequent loss of leaves, may be severely stressed, particularly if infection occurs in several consecutive years. The plant may well then be more prone to attack by another pathogen, such as a leaf spot or honey fungus, both of which more readily attack stressed plants. In a similar way an apple which has been attacked by codling moth is very prone to attack by brown rot fungus, which gains entry via the wounded skin of the fruit.

It cannot be reiterated too often that stresses brought about by unsuitable growing conditions or extremes of weather (*see pp. 70–71*) and problems associated with nutrient deficiencies and phytotoxicity (damage to the plant caused by excessive nutrient levels or use of unsuitable chemicals) are often responsible for lowering a plant's ability to fight off infection and disease development.

## MOST COMMON PLANT DISEASES

| Disease | Type | Method of spread | Conditions making plant more susceptible | Season |
|---|---|---|---|---|
| POWDERY MILDEWS | Fungi | Water/rain splash; air currents. | Dry soil/root conditions; moist or humid environment around leaves. | Spring, if dry, and summer to autumn. |
| DOWNY MILDEWS | Fungi | Water/rain splash; air currents. | Humid or wet air around foliage; damp spring and autumn weather; stagnant conditions. Plant debris in vicinity; dead/damaged plant parts. | All year. |
| RUSTS | Fungi | Water/rain splash; animals; wind. | Damp, warm, stagnant conditions; moisture film on leaves. | Mainly damp summers and autumn. |
| FUNGAL CANKERS | Fungi | Water/rain splash; air currents. | Damage to bark from pruning wounds; pest attack; frost crack. | All year. |
| FUNGAL WILTS | Fungi | Pruning tools; infected root material; infested soil. | Damage creating open wounds when spores in vicinity. | All year. |
| BACTERIAL SOFT ROTS | Bacteria | Water splash, water films; pruning tools; insects. | Damage to the plant's surface allowing entry of bacteria; poor storage conditions. | All year. |
| FUNGAL LEAF SPOTS | Fungi | Mainly water/rain splash; some air currents. | Poor growing conditions; weather extremes; other diseases. | Mainly spring and summer. |
| HONEY FUNGUS | Fungus | Underground rhizomorphs (fungal strands); root contact. | Any other predisposing stresses such as other diseases, poor growing conditions, weather extremes or excessive pruning. | All year. |
| CUCUMBER MOSAIC VIRUS | Virus | Sap-feeding pests; handling; pruning and taking cuttings; occasionally seed-borne. | Presence of any factors which encourage spread (*see left*). | Mainly late spring and summer. |
| PHYTOPHTHORA | Fungus | Soil- and water-borne. | Injury to plant; over-wet soil conditions. | All year. |

# TYPES OF GARDEN PEST

ANIMALS BECOME PESTS when their activities begin to have an adverse effect on plants. They may do this by reducing plant vigour so that yields of fruits and vegetables, or the decorative qualities of ornamentals, are impaired. In addition to the direct damage caused by their feeding activities, pests can have indirect effects on plants. Those feeding on the edible parts of fruits and vegetables may not be numerous enough to reduce yields, but nevertheless make them unpalatable due to the presence of insects and their excrement. Many sap-feeding insects soil their host plants with a sugary excrement called honeydew that encourages the growth of sooty moulds.

Pests can also transmit some virus and fungal diseases of plants on their mouthparts or bodies. Some plant diseases are dependent on the presence of pest-damaged tissues and cannot become established without this initial help.

While some garden pest problems are local, others are the result of migrations from elsewhere. Pests may be imported in purchased plants or gifts. Pollen beetles and flea beetles can fly in from agricultural crops, such as oil seed rape, while cabbage white butterflies, which are capable of longer flights, may come to Britain from elsewhere in Europe.

Snail uses its rough "tongue" or radula to break up plant tissues

**SOLITARY FEEDER**
*While snails can occur in great numbers in gardens, they are essentially lone operators when browsing on plants. Each individual is, however, capable of causing extensive damage.*

Pests affect plants in a variety of ways. Some attack specific plant parts, while others feed on or damage several different types of plant tissue.

### SAP-FEEDING PESTS
Sap-feeding pests such as aphids and scale insects have needle-like mouthparts that are inserted into plants in order to suck sap. No visible holes are made in the leaves, but many sap-feeders have a toxic saliva that results in leaf distortion and/or discoloration. Some sap-feeding pests – including aphids, leafhoppers and thrips – can transmit virus diseases. Most sap-feeders infest the foliage, stems and flowers, but there are some species of aphids and mealybugs that also attack roots.

### FLOWER PESTS
Relatively few pests specialize in feeding on flowers or flower buds, but these are attacked by many general plant pests such as earwigs and caterpillars. Pests specifically associated with flowers are pollen beetles, hemerocallis gall midge, apple blossom weevil and Michaelmas daisy mite. Some birds, such as bullfinches and house sparrows, also destroy flowers and flower buds.

### GALL-FORMING PESTS
Some pests secrete chemicals into their host plants when they deposit eggs or as they feed. This induces a variety of abnormal growths known as galls that enclose the causal animals. The gall-forming habit is found among many types of insects, mites and nematodes. Plant galls can also be caused by fungi and bacteria.

### ROOT-FEEDING PESTS
The larvae of a wide range of insects bore into or destroy roots, corms, bulbs and tubers. Their activities are often slow to be discovered, and can be difficult to eliminate. Some controls involve

**PESTS EN MASSE**
*A single aphid or even a small colony presents no threat to a healthy plant. The speed at which they multiply, however, quickly results in heavy infestations that soon sap plants' vigour.*

**FEEDING DAMAGE**
*The depredation caused by feeding is a major factor in pest damage. However, they also create open wounds on plants that may act as entry points for secondary infections.*

disposing not only of the affected plant but of the soil in the vicinity of the rootball. Many root pests arrive as eggs laid by winged insects; others may be imported into gardens via container-grown plants. Plants purchased or received as gifts should always be tipped out of their pots and the rootball examined for signs of pest activity before planting.

## FRUIT PESTS

Fruits, berries and seeds are attacked by the larvae of many insects, including a variety of moths, sawflies and beetles. Wasps, birds and squirrels can also be a problem. Some pests, such as capsid bugs, reduce the quality of fruits by causing skin blemishes; others render them inedible. Damage to ripening fruits renders them liable to infection by fungal rots, so they cannot be stored.

## STEM PESTS

Most pests attacking plant stems feed from the outside and include most of the sap-feeding pests. Bark beetles feed as larvae under the bark of woody plants while the larvae of leopard moth and clear-wing moths bore into the centre of stems. Mammals, such as rabbits, deer, voles and squirrels, can damage bark, especially during the winter and spring.

## LEAF PESTS

Plant foliage can bear a greater variety of pests than any other part of a plant. In addition to the sap-feeding pests listed above, there are pests that feed on the leaf tissue itself, such as slugs, snails, earwigs, caterpillars of sawflies, moths and butterflies, and adult vine weevils. Some insects, such as the larvae of certain flies, moths, sawflies and weevils, feed within the leaves as leaf miners.

## HOW PESTS ATTACK

WHITEFLIES

**SAP-FEEDERS**
*Adelgids, aphids, capsid bugs, froghoppers, leafhoppers, mealybugs, red spider mites, scale insects, suckers, thrips, whiteflies.*

EARWIG

**FLOWER PESTS**
*Aphids, apple blossom weevil, bullfinches, caterpillars, earwigs, hemerocallis gall midge, house sparrows, Michaelmas daisy mite, pollen beetles, slugs, snails, thrips.*

VINE WEEVIL GRUBS

**ROOT-FEEDERS**
*Cabbage root fly larvae, carrot fly larvae, chafer grubs, cutworms, eelworms, leatherjackets, onion fly larvae, root aphids, root mealy-bugs, slugs, swift moth caterpillars, vine weevil grubs, wireworms.*

GALL WASP LARVAE

**GALL-FORMERS**
*Adelgids, aphids, eelworms, gall midges, gall mites, gall wasp larvae, sawflies, suckers.*

WASPS

**FRUIT PESTS**
*Apple sawfly, bean seed beetles, birds, codling moth larvae, nut weevil, pea moth larvae, plum moth larvae, plum sawfly, raspberry beetle, squirrels, strawberry seed beetle, wasps.*

SCALE INSECTS

**STEM PESTS**
*Many sap-feeding pests (see above), plus clearwing moth larvae, deer, leopard moth larvae, bark beetles, rabbits, squirrels, voles.*

CATERPILLAR

**LEAF PESTS**
*All sap-feeding pests (see above), plus adult vine weevils, earwigs, eelworms, flea beetles, leaf-cutter bees, moth and butterfly caterpillars, sawfly larvae, slugs, snails.*

# PEST LIFE CYCLES

THE LIFE CYCLES of pests are as diverse as the pests themselves, but they can be divided into two broad categories: complete and incomplete metamorphosis. The latter involves little more than a gradual increase in size as the pest reaches maturity, with the immature stages usually resembling the adult animal. This form of growth is shown by eelworms, slugs, snails, millipedes, woodlice, mites and some insects, including earwigs, aphids, leafhoppers, froghoppers, whiteflies, mealybugs, scale insects, capsid bugs and thrips. The feeding habits of the immature stages are often the same as the adult and so all active stages in the life cycle are capable of causing damage.

## COMPLETE METAMORPHOSIS

More advanced insects, such as butterflies, moths, beetles, flies, ants, bees, sawflies and wasps, undergo complete metamorphosis. Here, the larval stages are very different in appearance from the adult and there is a pupal stage in which the larva, usually enclosed within a protective casing, changes into the adult insect. With these insects the feeding habits of the larvae are often different to those of the adult and it is usually the larvae that cause

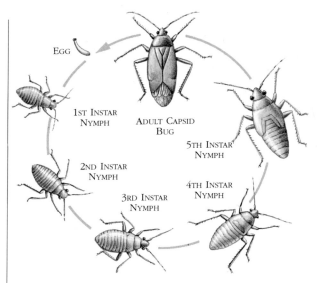

**INCOMPLETE METAMORPHOSIS, CAPSID BUG**
*Pests that undergo incomplete metamorphosis look broadly similar at each stage of their lives and gradually acquire the adult features.*

damage to plants. For example, caterpillars of butterflies and moths feed on leaves while the adults feed on nectar. Some beetle pests, however, cause damage at both the larval and adult stages.

## HOW PESTS DEVELOP

Pests generally start life as eggs, although aphids give birth to live young for most of the year, only resorting to producing overwintering eggs in the autumn. Invertebrate animals, such as insects, mites, woodlice and millipedes, do not have any bones or internal skeletons. Instead they have a hard outer covering to their bodies. Like a suit of armour, this restricts growth, so must be shed from time to time to allow the animal to increase in size. This process is known as ecdysis. After the old outer skin has been cast off, the soft new skin underneath is stretched by the animal before it hardens, thus creating more space for growth. The number of stages, or instars, between the newly-hatched nymph or larva and the adult animal is usually between four and twelve, depending on the species.

## HOW PESTS REPRODUCE

Most adult pests are either male or female and mate in a conventional manner. There are, however, exceptions. Slugs and snails are hermaphrodites with both male and female genitalia. They mate with others of their species and both partners will subsequently lay eggs. Some insects are capable of

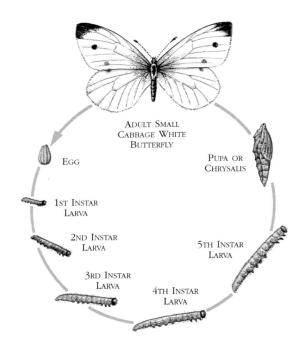

**COMPLETE METAMORPHOSIS, CABBAGE WHITE BUTTERFLY**
*The more advanced types of insect undergo a complete transformation in appearance and often in feeding habits during their lives.*

**APHID GIVING BIRTH**
*Aphids give birth to live young in spring and summer, but as day length shortens and temperatures fall in autumn, they lay eggs instead, which will overwinter.*

breeding without mating (parthenogenesis). Some, like vine weevil and aruncus sawfly, only occur as females when adult. Aphids occur as females only during spring and summer; males are produced in late summer when mating is necessary to produce overwintering eggs. Some gall wasps have alternating asexual all-female generations and sexual generations with males and females.

## PEST LIFESPANS

The length of time required to complete the life cycle depends on factors such as temperature, the availability of food and the nature of the pest. Most pests are able to complete their development within a year, but some require much less, while others take longer. Wireworms, for example, feed on roots in the soil for up to five years before they pupate and become adult click beetles. Aphids, by cutting out the egg stage, are able to go from young nymph to adult insect in as little as seven days.

Pests that have more than one generation in a year complete their development more rapidly in warmer weather. Glasshouse red spider mite takes 55 days to complete its life cycle at 10°C (55°F), but only 12 days at 21°C (70°F). Most outdoor pests have one to three generations a year. Glasshouse pests like whitefly and red spider mite can go through as many as 12 generations a year due to the warm, sheltered conditions in which they live.

## SEASONAL CHANGES

Pests have evolved alongside the plants they feed on and so they begin to attack their hosts at the time of year when the appropriate plant tissues will be available. This is strongly related to temperature and day length, which not only control plant growth processes such as bud burst, flowering, fruiting and aging, but also bring pests out of winter dormancy and allow egg hatching. Similarly, the life cycles of predatory insects, such as ladybirds, are related to that of their prey, aphids.

## SURVIVING THE WINTER

The means by which pests overwinter varies, although each species will usually have one particular stage in its life cycle in which it survives the colder months of the year. This may be the egg, one of the nymphal or larval stages, the pupa or adult. The overwintering stage generally occurs in some sheltered place, such as in the soil, in the crevices of bark or inside dense shrubs such as conifers. Many insects and mites go into a form of hibernation called diapause from which they will not emerge until they have experienced a sufficient period of cold. This prevents them emerging prematurely whenever the winter weather turns mild. Some glasshouse pests, such as whitefly, continue feeding and breeding throughout the year, albeit more slowly in winter than in summer.

## ANNUAL POPULATIONS

While some pests are a problem every year, others are only occasional problems. Serious outbreaks may be due to a combination of favourable weather conditions and a relative absence of predators, parasites and diseases of the pest. Minor pests often need a run of several good reproductive cycles before they are numerous enough to become a significant problem. Hot, dry summers favour certain pests such as spider mites and thrips, while slugs and snails thrive under damp conditions. Cold winters do not kill pests as all those native to northern Europe have a cold-tolerant overwintering stage. Mild winters can allow some pests, such as aphids, that would normally go dormant, to continue feeding and breeding, resulting in higher populations in spring.

LADYBIRD LARVA EATING APHID          ADULT LADYBIRD EATING APHID

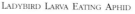

**THE BENEFICIAL LADYBIRD**
*Both the larval and adult stages of ladybirds feed avidly on aphids, making them doubly valuable to gardeners. Some are also predators of scale insects, thrips, mealybugs or mites.*

# APPROACHES TO CONTROL

WHEN SOMETHING goes wrong in a garden and pests or diseases are to blame the first consideration is whether control measures are necessary, and if so what action should best be taken. Not all infections or infestations merit control. A powdery mildew infection at the end of the season on a herbaceous plant may look unsightly, but if the leaves are going to die back shortly anyway, the infection has little if any detrimental effect on the plant and may not be worth treating. Similarly, it may be easier and more sensible to pick off one or two caterpillars on a cabbage than attempt to control them with an insecticide.

Some problems cause rapid and severe damage, while others may cause some damage but are probably more worrying to the gardener than they are to the plant. It is therefore important to balance the need to spray with the damage the pest or disease actually causes. Farmers and growers have an "economic threshold" – the point at which a pest or disease is causing or is likely to cause sufficient loss to merit the time and money spent spraying. This is something the gardener should consider, modifying the equation to include the extent to which a problem spoils a plant's appearance, before reaching for the sprayer.

## USING GARDEN CHEMICALS

There is no doubt that chemical controls (*see pp.82–83*) offer rapid, safe and usually complete solutions to pest or disease problems. If correctly used, garden products today have been so carefully formulated and tested (*see pp.80–81*) that possible adverse side-effects are minimal. Most gardeners

## INTEGRATED CONTROL IN THE GARDEN

The integrated approach to controlling problems combines appropriate plant choice and good gardening practice, preventive measures to avoid problems, and sensible use of chemicals only when warranted by serious pest or disease attack.

CHEMICAL SPRAYS
Spraying may be necessary to combat serious crop pests and diseases.

AVOIDING PROBLEMS
Traps and barriers (*see pp.86–89*) prevent many pest problems from occurring.

SPECIAL CARE
Roses rarely escape problems and should be inspected regularly.

Grease band

COMPOST HEAP
Composting healthy plant material returns nutrients to the soil.

TIDY FLOWER-BEDS
Tidy beds with well-spaced plants minimize conditions that encourage disease.

Barrier thwarts flying pests

USING A TRELLIS
Climber held away from wall by trellis allows good air circulation.

Crop rotation prevents a buildup of problems

OUT OF REACH
Training keeps crops off the ground where they may be eaten by insects and mice.

LAWN PROBLEMS
The perfection or otherwise of a lawn, and thus the amount of chemicals that need be used, is generally entirely a matter of personal taste.

COMPANION PLANTING
Flowering plants attract beneficial insects that help to control vegetable pests.

EDIBLE CROPS
All chemicals formulated for garden crops are safe to use on fruits and vegetables, provided that harvest intervals are observed.

who do not follow organic gardening methods are also happy to use inorganic or chemical fertilizers. The inorganic approach to gardening is still more widespread than the organic, but increasing numbers of people who would not call themselves organic gardeners use many organic methods.

## "GREEN" GARDENING

In recent years organic or so-called "green" gardening has become increasingly popular. Many gardeners have started to question the use of too many garden chemicals and some use purely organic methods. Organic gardening avoids the use of manufactured chemicals, but some organic gardeners will happily use products such as derris or Bordeaux mixture as these are both made from "natural" ingredients.

The organic approach also advocates the use of only naturally occurring materials, such as manure, for improving soil fertility, and materials such as leaf mould for mulching or improving soil texture.

It embraces other environmental concerns, too, such as the use of coir or bark-based growing media in preference to peat-based products, and of recycled materials wherever possible.

Organic gardening also makes full use of what are, essentially, good gardening techniques, such as the pruning out of infected material and good garden hygiene (see pp.92–96). Resistant or tolerant plant cultivars and trap and barrier methods (see pp.86–89) are also important to the organic gardener.

To succeed at organic gardening it is also important to take all possible steps to encourage populations of beneficial creatures such as naturally occurring pest predators and parasites (see pp.90–91). The avoidance of many chemicals will make this easier, but it should be remembered that some of the products accepted by many organic gardeners (for example, derris) are not selective in their activity and so may damage the beneficial organisms which the organic gardener is trying so hard to encourage.

## INTEGRATED CONTROL IN THE GREENHOUSE

The enclosed environment of a greenhouse can increase problems, but may also assist control. While pests flourish in the warm, sheltered atmosphere, they may also be targeted much more successfully by traps, sprays, smokes and the release of the predators or parasites known as biological controls.

Shading

**IRRIGATION SYSTEM**
A simple automatic system ensures that plants receive an even and adequate water supply, avoiding drought stress. Watering plants from below, rather than above, also minimizes disease problems by preventing a film of water from resting on leaves and stems.

Sticky traps are very successful within enclosed spaces

**TEMPERATURE CONTROL**
A greenhouse heater, shading and a thermometer are aids in preventing high- and low-temperature injury.

**HEALTHY PLANTS**
Plants will be much more resistant to problems if they are well cared for.

Heater with automatic switch

**OUT OF REACH**
Gap behind staging foils pests crawling up the walls to reach plants.

**SPECIAL CONDITIONS**
Plants can be given more favourable conditions and special growing mediums in a controlled environment.

**BENEFICIAL CREATURES**
Frogs and toads, which are insectivorous, are welcomed in the greenhouse by most gardeners.

Floor kept well swept, clean and dry

Saucers of water prevent insects from crawling up legs of staging

**RODENT CONTROL**
Humane traps to protect crops enable mice to be evicted without injury from the greenhouse.

# CHEMICAL CONTROLS

PLANT PESTS AND diseases have always been a problem for gardeners and farmers alike. It is, however, only fairly recently that an industry has existed to develop and market products for keeping plants in good health. In the nineteenth century these included some chemicals based on arsenic, cyanide, mercury and other chemicals far too dangerous to be used today. A few chemicals, such as the plant-derived insecticides derris and pyrethrum, sulphur, and the copper fungicide Bordeaux mixture, survive from that time and are still in use in gardens today.

## SAFETY STANDARDS

Most of the chemicals currently in use have, however, been developed since 1945. Some developed at that time, such as the insecticides DDT and aldrin, have subsequently been withdrawn, when it was discovered that they persist in the environment and accumulate in the bodies of birds of prey and other predators. The manufacturers of pesticides and the government departments that regulate them have learned from mistakes such as this and environmental safety is now a high priority.

Before any new product can be marketed it has to be thoroughly researched to test factors such as its effectiveness, its persistence, any effects on

**PATENT FUMIGATOR**
*Chemicals used in the past by gardeners could be toxic; some methods of application were equally intimidating.*

non-target plants and animals, and its long- and short-term effects on humans. These tests start in the laboratory and build up to field trials. The data required to support an application to market a new product takes about ten to twelve years to compile and will cost tens of millions of pounds. No other chemicals, apart from pharmaceuticals, are as thoroughly tested before they can be sold to the public.

Insecticides and fungicides are developed for the agricultural market and only a small proportion of the chemicals available to professional growers are marketed for garden use. These are chemicals at the safer end of the toxicity spectrum and are supplied as formulations which can be applied without the need for special protective clothing. The manufacturers and the government regulators have carried out their parts in producing safe pesticides; the final responsibility for safe use lies with the gardener, who must read and follow the instructions before applying a chemical.

## PESTICIDE RESISTANCE

It is possible for some pests and fungal diseases to build up an immunity or tolerance to chemicals, so that a pesticide that was once an effective treatment

**APPLYING CHEMICALS ON FARMLAND**
*Garden chemicals are largely developed from those used in agriculture and in the professional horticultural trades. However, the formulations in which they are supplied for use in the garden are much safer, eliminating the need for protective clothing and other special precautions, although sensible safety rules should be observed and the manufacturer's instructions strictly followed.*

no longer works. Continual use of the same chemical can exert a powerful selection pressure by killing the more susceptible individuals, but leaving those with some tolerance to pass this characteristic on to the next generation. Pests can develop full resistance in about 40 generations and, since some pests have five or more generations a year, a chemical can fail within a comparatively short time.

Resistance problems are less frequent in gardens than in agriculture or commercial horticulture. This is because most gardeners use chemicals over small areas, and susceptible strains of pests and diseases continually come in from other gardens where different or no chemicals are used. Resistant pests are mainly found in greenhouses, where the enclosed environment and a pest's rapid reproduction rate can make chemicals ineffective against whitefly, red spider mite and aphids. Some diseases, such as grey mould, rose black spot, powdery mildews and apple scab, can become tolerant of fungicides, especially the systemic types.

### REDUCING RESISTANCE

Gardeners can reduce the likelihood of resistance occurring by using chemicals sensibly. Where possible, non-chemical alternatives, such as biological controls (*see pp.90–91*) or the removal of badly diseased plants, should be considered. It should be noted that some different active ingredients are chemically related and it is likely that a pest or disease resistant to one will also tolerate other chemicals in the same family. When pesticides are used, the continual use of the same product should be avoided.

### PHYTOTOXICITY

Plants are sometimes damaged by insecticides or fungicides; this is known as phytotoxicity. The manufacturer's instructions should be read and

**GLASSHOUSE TRIALS**
*Before potential pesticides are used outdoors, extensive glasshouse trials are conducted. These are designed to test the chemical's efficacy and lack of harmful effects on plants.*

followed carefully to ensure that the correct dilution is being applied. Plants known to be harmed by a chemical will be listed by the manufacturer, but this information is not known for all potential plant and chemical interactions.

The risk of such damage occurring can be minimized by not treating plants already under stress as a result of exposure to bright sunlight, extremes of temperature or dryness at the roots. Seedlings are more susceptible than mature plants, and petals are more likely to be scorched than leaves. Where there is doubt and there are a number of plants to be sprayed, it is sensible to treat just a part of one plant and watch for adverse reactions over the next few days. Plants that have been damaged may suffer a significant check in growth but will usually make a recovery.

**USING CONTROL DOMES TO TEST CHEMICALS**
*Small-scale tests in enclosed environments, such as the control domes illustrated here, are used to detect possible environmental problems before field trials are held.*

WITHIN A CONTROL DOME

SYSTEM OF CONTROL DOMES

# CHEMICALS AND THEIR ACTIONS

MOST INSECTICIDES and fungicides are contact in action. This means that insecticides must cover the plant surface thoroughly so that the pest is either hit directly or picks up the pesticide from the treated surface. Similarly, contact fungicides kill fungal spores when they germinate if coverage of the plant surface has been adequate.

Systemic insecticides and fungicides are absorbed into the plant and may be transferred to parts which have not been directly treated. This movement may be within the leaves or up to the shoot tips, but systemic insecticides and fungicides will not move from the foliage down to the roots. Systemic fungicides kill fungal growth within plant tissues and have a curative action, rather than the preventive role of contact types. Systemic insecticides are particularly useful for controlling sap-feeding pests (see pp. 74–75), including those concealed in distorted leaves that are difficult to reach with contact sprays.

## THE WAYS CHEMICALS WORK

Different problems require chemicals to be applied in different ways. Sometimes, a combination of methods of application is needed to achieve complete elimination.

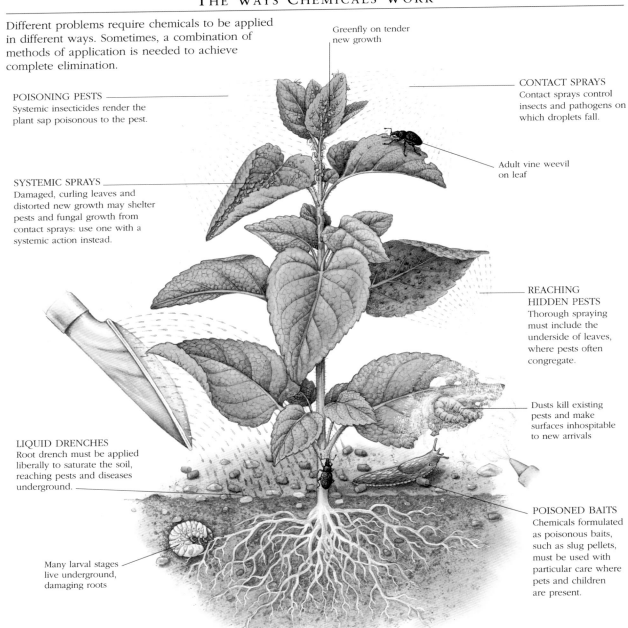

Greenfly on tender new growth

POISONING PESTS
Systemic insecticides render the plant sap poisonous to the pest.

SYSTEMIC SPRAYS
Damaged, curling leaves and distorted new growth may shelter pests and fungal growth from contact sprays: use one with a systemic action instead.

LIQUID DRENCHES
Root drench must be applied liberally to saturate the soil, reaching pests and diseases underground.

Many larval stages live underground, damaging roots

CONTACT SPRAYS
Contact sprays control insects and pathogens on which droplets fall.

Adult vine weevil on leaf

REACHING HIDDEN PESTS
Thorough spraying must include the underside of leaves, where pests often congregate.

Dusts kill existing pests and make surfaces inhospitable to new arrivals

POISONED BAITS
Chemicals formulated as poisonous baits, such as slug pellets, must be used with particular care where pets and children are present.

## APPLICATION METHODS

| Chemical type | Advantages | Disadvantages |
| --- | --- | --- |
| LIQUID SPRAY | Contact and systemic types available. Most have a wide range of applications. Inexpensive. | Handling and accurate measuring of concentrate needed. Requires special equipment for application. |
| LIQUID DRENCH | Contact and systemic types available. Most have a wide range of applications. Inexpensive. | Handling and accurate measuring of concentrate needed. Difficult to work out precise amounts. |
| DUST | Used direct from pack, so little is wasted. It is easy to apply direct to the target organism. | Only contact types available. May leave unsightly deposit on plant. |
| READY-TO-USE SPRAYS AND PUMP-GUNS | No need to handle concentrate as contents already correctly diluted. Quick and simple to use. | Expensive and really only suitable for treatment of relatively small areas of infection/infestation. |
| PELLETS | In measured doses, so easy to apply correct quantity. Repellents in them to deter animals. | May be eaten by pets, children or wildlife. Can become less active after rain. |
| AEROSOL | Quick, simple and easy to use. Required dose applied, so no wastage. No equipment needed. | Propellant may be potentially flammable and may prove phytotoxic to plants. Expensive. |
| SMOKE | Allows chemical to permeate to all areas within closed environment in which it is applied. | Very limited range available. Only suited to enclosed spaces or chemical will diffuse. |

A few pests can be controlled by chemicals that are mixed with something palatable to form a poisoned bait. These include rat poisons, slug pellets and ant baits.

Pesticides generally control a wide range of insects, including beneficial types if they are on the plants when they are sprayed. Such chemicals are said to have a broad spectrum of activity. There are others which are selective, such as pirimicarb, which controls aphids but leaves most other pests and beneficial insects unharmed.

Other chemicals may be relatively selective by having a short persistence. Organic insecticides, such as derris, pyrethrum and insecticidal soaps, are

### PROTECTING PONDS
*Fish are extremely susceptible to harm from garden chemicals, which should not be used on plants in or near ponds. Avoid chemical fertilizers also where they may leach into pond water.*

### CHEMICALS IN THE FOOD CHAIN
*There is a danger that birds may pick up insects and other pests that have been treated with pesticides, either to eat themselves or to feed to their young. It is difficult to prevent this happening, but it is rare for them to eat sufficient amounts for harmful effects to be observed.*

broad-spectrum chemicals, but remain effective on the plant for less than 24 hours, so their impact on non-target insects is limited.

### PERSISTENCE OF CHEMICALS
The rate at which insecticides and fungicides lose their effectiveness and break down into other less active compounds varies. This is largely due to the degree of stability of the chemical's molecular structure, but is also affected by factors such as temperature and exposure to sunlight. Some organochlorine insecticides, such as DDT and aldrin, are virtually indestructible and this is why they have been withdrawn from use.

No chemical currently available to gardeners remains active against pests or diseases for more than two or three weeks after application, and the period of protection is often much less. It is, therefore, important to treat plants thoroughly and apply the chemical at the correct time or times in order to get the best results. Insecticides and fungicides used on fruits, vegetables and culinary herbs may require an interval between treatment and harvesting. For the chemicals currently available to amateur gardeners, this harvest interval varies from nothing to three weeks.

### HARM TO OTHER CREATURES
Many chemicals, in particular some insecticides, have the potential to be harmful to creatures other than garden pests and other insects. If used with care, and precisely according to the manufacturer's instructions, there should be no danger. Product labels should state clearly where risks exist and it is important to take all possible efforts to keep any creatures which could be harmed out of the area where chemicals are being used. In most cases, once a spray has dried on the plant which has been treated, it no longer poses a threat to pets, children or wildlife.

# APPLYING CHEMICALS

GARDEN CHEMICALS are a great help in controlling pest and disease problems. If used with care and only when needed they pose little or no threat to the environment. Changes in legislation and the increasing concerns about safety to the environment and the user mean that the range of chemicals available is constantly being reviewed. All chemicals are also rigorously tested before being marketed.

When considering chemical control always ascertain whether the situation really necessitates their use. Slight pest infestations or those which occur towards the end of the season may not pose a sufficient threat to the plant.

Selecting the most suitable chemical is also important and wherever possible one which controls only the pest organism should be used. Formulations which contain both a fungicide and an insecticide are only appropriate when both the specified pest and disease are present at a level that warrants control.

## READING THE LABEL

Always consult the product label to determine precisely what the chemical can be used for and whether it is formulated for use against the problem and on the plant you have in mind.

Products which carry a label recommendation for use on edible crops may also specify a minimum interval which must elapse between the application of the chemical, and the consumption of the fruit or vegetable. Provided that this interval is observed, it is perfectly safe to eat the crop.

The timing, application rate and frequency of application are also important and again information about all of these is available on the label. Applying

**HEALTHY VEGETABLES**
*To make sure that vegetables and fruit are safe to eat after chemical treatment (cabbage being treated for whitefly, above), keep to the interval recommended on the product label.*

a chemical at the incorrect rate, frequency or time may prove ineffective, damaging or simply wasteful and so it is essential to follow the instructions.

## RULES FOR SAFETY

When applying chemicals there are a number of general rules which should be followed.

Chemicals should be used sensibly and only when necessary, and must always be applied according to the instructions. Avoid contact with your eyes and skin, and avoid inhaling the chemical. Although there is no stated need to wear any protective clothing for garden chemicals, you could wear rubber gloves when handling concentrates. Different products should not be mixed together unless this is recommended on the label or product leaflet.

Keep children and animals away and never eat, drink or smoke while applying chemicals. Do not spray or dust in windy, hot, sunny or completely calm weather, to minimize the threat of chemical drift affecting adjacent plants. Spray in the late afternoon, early evening or early morning.

## EQUIPMENT FOR APPLICATION

You may need to invest in some special equipment to apply chemicals. This should all be labelled clearly and kept separately, so that it is used for

## HAZARD SYMBOLS

OXIDIZING AGENT

HARMFUL OR IRRITANT

Garden chemicals have been formulated so that their contents will not cause any harm, as long as they are used correctly. The symbols shown here are standard throughout the European Union and will be found on some garden products. The only garden chemical with the oxidising agent symbol is the weedkiller sodium chlorate. It indicates that there can be fire risk in its use.

# MIXING AND APPLYING CHEMICAL SPRAYS

**1 Measuring.** Measure out the required amount of concentrate accurately. Avoid making up more than necessary and always read the instructions carefully. Wear rubber gloves and take care not to spill any of the chemical.

**2 Mixing.** Part-fill the sprayer with water and then carefully pour the concentrate in. Rinse out the measurer with water and add this to the sprayer. Top up to the required level with water and close all containers.

**3 Applying.** If possible spray both surfaces of the leaves. This is particularly important if contact action sprays are used. Spray at dusk to minimize risk to pollinating insects and to help prevent scorching.

no other purpose (*see below*). Use completely different equipment again for the application of weedkillers, as even trace amounts of these can cause serious damage to garden plants. A small sprayer is suitable for most chemicals unless you have a very large garden. One that produces a fine spray is needed for controlling most pests and diseases. Insecticide or fungicide drenches or weedkillers can be applied using a watering can fitted with a dribble bar.

Clean all equipment regularly and when rinsing it out after use, spray all washings onto empty land or suitable plants. Never dispose of any dilute or concentrated chemicals by pouring them down a lavatory or drain, into watercourses, or onto land where they may contaminate underground water.

## APPLYING CHEMICALS IN ENCLOSED SPACES

When chemicals such as sprays or dusts are used in enclosed spaces such as greenhouses or conservatories, particular care is needed to avoid contact or inhalation. Temperatures are often higher and ventilation much lower than outside in the garden. Greenhouses or conservatories which open directly into the house pose further difficulties and wherever possible, doors or windows into the house should be tightly closed.

The risk of scorching the plants being treated is also greater, due both to the raised temperatures and to the magnifying effect that glass may have on the sun's rays. To minimize problems, timing of application, in order to avoid the hotter, brighter parts of the day, is particularly important.

## STORING CHEMICALS

It is very important to store garden chemicals safely, since even small amounts can sometimes be harmful to people, animals and garden plants. Below are tips to help you store chemicals correctly. Rules for safe storage should also be applied to the equipment with which you apply chemicals. It is worth marking items such as watering cans clearly so that they are not confused with equipment for general garden use.

**STORAGE TIPS**
• Store chemicals in a safe place, preferably a locked cupboard inaccessible to children or animals.
• Keep chemicals in their original containers, with their tops tightly closed.
• Make sure all containers are clearly labelled and kept separately.
• Clear out cupboards annually, so that you do not accumulate old and obsolete products. To dispose of them, wrap them well and contact your local council. Check for and get rid of leaking products similarly.

**SAFETY HAZARDS**
• Do not transfer any chemical out of its original container into another one with an incorrect label or no label at all.
• Only buy sufficient chemicals for one to two years' use and do not store them for a long period of time. Chemicals become less effective once they have been opened and may go out of date.
• Do not keep surplus quantities of made-up solutions and sprays. Dispose of them and rinse out equipment directly you have finished treating plants.
• If you need to get rid of old chemicals, do not simply add them to household rubbish nor pour them down household drains. Dispose of them safely (*see left*) without removing them from their original containers.

# CHEMICAL-FREE CONTROLS

R ELYING SOLELY on chemicals to control pests and
diseases can bring long-term problems in the
form of pesticide resistance, harm to non-pest
animals and consequent damage to the environment. Integrated control is a more environmentally
friendly approach to tackling pests and diseases. It
includes recognizing the role that natural enemies
play in reducing pest numbers and devising
complementary control measures that enable pest
predators and parasites to survive. This may involve
more careful monitoring of pest and disease
infestations to determine whether chemical
treatments are needed and, if so, when they should
be applied to achieve the best effect. This approach
can significantly reduce the number of spray
applications given compared with a standard
spray programme.

Good garden practices (*see pp.92–96*), such as
removing badly infested or diseased plants,
ensuring plants have adequate water and nutrients,
growing plants and cultivars suitable for the local
soil and climate conditions and selecting cultivars
with pest or disease resistance where available, are
all part of an integrated control approach to limiting
pest and disease problems, as is the use of
biological controls (*see pp.90–91*). Picking off pests

## A BEER TRAP FOR WASPS
*Half-fill an old jar with beer and seal it with some paper and
an elastic band. Make a small hole in the paper and, in their
attempts to reach the beer, the wasps should fall through.*

by hand at an early stage can prevent many
problems from escalating. Pests that are most active
at night, such as slugs and snails, are best gathered
up with the aid of a torch on damp evenings.
On plants grown in or around garden ponds, you
should always pick off any pests by hand, since it
is dangerous to fish and other pond life to apply
chemicals in or near water.

There are also a number of traps and barriers
that either prevent pests reaching their intended
goal, or draw them to a more tempting destination.
Some, such as beer traps for slugs and wasps,
eliminate the pests; others are monitors, giving an
indication of when controls will be most effective.

## PHEROMONE TRAPS
Pheromones are volatile chemicals produced by
some insects as a means of communicating with
others of the same species. They play an important
role in maintaining the organization of social insects
such as ants, bees and wasps. Pheromones are also
used by some insects, usually females, as a means
of attracting males for mating purposes.

The pheromones of some pest species, such as
codling moth, plum moth and carnation tortrix,
have been identified and the chemicals synthesized
for use in pheromone traps, which are sold in kit
form (*see facing page*). Once the trap is in position

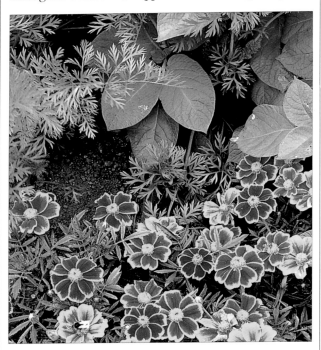

## COMPANION PLANTING
*Companion plants were once thought actually to deter pests from
crops. It is more likely, however, that their bright flowers attract
pest predators, initiating a form of biological control (see p.90).*

# HOW TO ASSEMBLE A PHEROMONE TRAP

**1 Kit contents.** A self-assembly pheromone trap kit should include a fold-together, open-sided box with hanging wire, some sticky sheets and some synthetic pheromone lures.

**2 Assembling.** Make up the box and place a sticky sheet in the bottom of it. Put a pheromone lure on top of this sheet. The sheet and pheromone should be replaced after six weeks.

**3 Hanging.** Hang the trap by its wire on the tree, at about head height and within the central canopy. It should be put in position at the expected beginning of the moth's flight period.

male moths fly into it, expecting to find a virgin female, but instead get caught on the sticky sheet. The main function of these traps is to record the moths' main flight and egg-laying period accurately so that, if necessary, insecticides can be applied at the right time to control the newly-hatched larvae. On isolated apple or plum trees pheromone traps may catch sufficient male moths to reduce the mating success of females, and so limit the number of fertile eggs that are laid.

**A STICKY TRAP**
*This trap is used to catch winged insects, who are attracted by the yellow colour of the strips. The strips are covered in non-drying glue and so trap the insects when they fly onto them.*

## GREASE BANDS AND STICKY TRAPS

Some insects cannot fly and it is possible to prevent them from gaining access to their host plants by placing a sticky barrier in their path. Grease bands are mainly applied to the trunks of fruit trees to trap winter moth and other related species that have wingless females. These emerge from pupae in the soil during autumn to spring and crawl up the trunk to lay eggs on the twigs. Grease bands applied to the trunk and tree stake in mid- to late autumn will reduce the number of eggs laid.

Other crawling insects that can be deterred by sticky or greasy barriers are earwigs, ants and adult vine weevils. A band around the base or the rim of containers can be very effective. Smearing grease around stems stops earwigs climbing up into flowers such as chrysanthemums and dahlias. Flea beetles can be made to jump on to a sticky trap: grease a piece of wood or card, and draw it, grease side down, along the tops of plants, being careful not to touch them with it, but agitating the foliage with your hand or a stick. Flea beetles jump when disturbed, and so will be caught on the trap.

Sticky traps are also useful in greenhouses for catching winged insects, such as whiteflies, thrips and fungus gnats. The best traps are made up of yellow plastic strips that are covered in non-drying glue. They are suspended above the plants and when insects alight they are unable to escape. Such traps may not give adequate control on their own, but they can give an early warning of the presence of pests. It is often easier to see whitefly on a yellow trap than to find them by searching on the underside of leaves. Early detection of an infestation allows the pest to be controlled, either by introducing biological controls or by using insecticides, before the plants are damaged.

## OTHER TRAPS

Other chemical-free traps can be used to protect flowers from earwigs, and roots from wireworms. Grapefruit and beer traps can be used against slugs and snails. The earwig trap uses an inverted flower pot filled with dry grass and placed over a cane amongst susceptible plants. Earwigs feed at night and will crawl into the trap to shelter from the daylight. You can then remove the pot and pick out the earwigs. To limit wireworm damage, fix a potato on a stick and bury it: the wireworms will move in, preferring it to plant roots. The potato with its pest inhabitants can be regularly removed and renewed.

Hollowed-out grapefruit half skins can be propped up on stones on the ground in flower-beds; slugs and snails are attracted to the smell, crawl in and remain there. The grapefruit skin can be picked up in the morning, complete with the slugs, which can be disposed of by drowning them in a strong salt solution. A jar half-full of beer can be almost buried in the ground, where the smell will attract slugs and snails, which fall in and drown. (For a beer trap for wasps, *see p.86.*)

## BARRIERS

There are various barriers that can be constructed in the garden to deter pests and provide a non-chemical alternative to control. They range in size from fruit cages (*see below*) and large glass cloches to individual plastic cloches (*see facing page*).

To protect smaller fruit bushes, such as strawberries or currants, against bird and animal damage, use mesh netting to construct a low frame, making sure that the base is secured with bricks to prevent the pests getting in underneath. Apple and pear trusses can be individually enclosed in bags made from muslin or old tights. Although time-

**EARWIG TRAP**
*Earwigs are a particular nuisance on chrsysanthemums, because they crawl to the top of stems and eat the buds and petals. This earwig trap is an easy and effective control.*

consuming this can be effective and should be done when the fruit is almost ripe – the time it is most likely to be attacked by wasps or birds.

In a greenhouse there are a variety of traps that can be used (*see p.79*); these include putting legs of tables or any other staging in saucers or pots of water. This will prevent insects climbing up onto the plants.

Some animals, such as hedgehogs and frogs, are highly beneficial in a garden (*see pp.60–63*) and should be encouraged. However, others, such as deer, rabbits and moles, can cause extensive damage. Fences can be used against rabbits and deer; wire netting buried at least 15cm below

FRUIT CAGE

PROTECTIVE FLEECE

**PROTECTIVE COVERS**
*The fruit cage (far left) is made from wire or plastic netting supported by metal posts. This type of cage can help protect small fruit trees or bushes from bird and animal attack. The fleece (left) protects and insulates plants and seedlings, while allowing light and air to reach them. This tunnel is constructed by using strong wire hoops to support a sheet of the woven polypropylene material.*

ground with 75cm above should deter rabbits. Deer will need a much taller barrier or electric fence. Mice (*see p.147*) and moles (*see p.148*) need traps.

## HORTICULTURAL FLEECES

In recent years several types of horticultural fleece have become available. These are finely spun light-weight materials, usually white or green in colour and made of polypropylene or other plastics, that can be laid over developing vegetable crops. These fleeces give the young plants valuable protection against wind and cold, and also exclude pests such as pigeons, cabbage caterpillars, carrot fly, cabbage root fly and flea beetles.

When using fleece to give protection against pests it is essential also to adopt a crop rotation system (*see pp.92–96*). If, for example, carrots are sown where carrots or parsnips were grown the previous year and then covered with a fleece, adult carrot flies may emerge from pupae buried in the soil and will then be trapped under the covering with the crop.

## BRASSICA COLLARS

Seedlings of brassicas and other related vegetables, such as swedes, are prone to attack by cabbage root fly. A ready-made collar, or one made using a piece of carpet, underlay or cardboard, is placed on the ground around the stem of each seedling. This prevents the female flies from laying their eggs in the soil; instead, they will lay them on the collar, where they will dry up before hatching.

## HOT WATER TREATMENT

This form of pest control is mainly carried out by commercial growers, but can be attempted by gardeners. It is used against certain bulb pests, such

**AN INDIVIDUAL CLOCHE**
*A cut-off plastic bottle placed over young plants in early spring makes an ideal slug deterrent and also acts as a mini-cloche, enabling soil around the roots to warm up more quickly.*

as narcissus eelworm and bulb fly, and also against eelworms on some herbaceous plants such as chrysanthemum. The technique involves immersing dormant bulbs or plant stools in water held at a temperature that is high enough to kill the pest without harming the plant. For details of hot water treatments, see chrysanthemum eelworm (*p.117*) and narcissus eelworm (*p.150*).

Care must be taken to ensure that the temperature is kept constant throughout the period of treatment, since too much heat damages the plants while insufficient heat allows the pest to survive. After treatment, bulbs are planted out in fresh soil, while herbaceous plants are potted up to produce cuttings.

## REPELLENTS

Repellents are used principally against birds and mammals, where the aim is to deter feeding or encourage them to move elsewhere, rather than to cause them any harm.

Humming tapes are used to protect plants against pigeons and other birds, while devices that produce ultrasonic sounds are available for frightening away moles and cats. Other means of keeping pets and pest animals away from plants rely on using irritant substances, such as pepper dust, or compounds which give plants unpleasant tastes or smells. Repellents may need frequent replenishment to maintain the effect. It is also advisable to change the type of repellent in use to avoid the target animals becoming familiar with the sound, smell or taste and losing their fear of it.

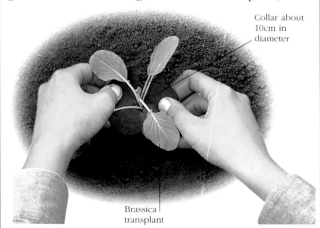

Collar about 10cm in diameter

Brassica transplant

**BRASSICA COLLAR**
*Prevent female cabbage root flies from laying their eggs in the soil around the stems of vulnerable young brassicas by placing a disc or collar around the stem base.*

# BIOLOGICAL CONTROLS

THIS APPROACH to control involves enlisting the help of pests' natural enemies – predators, parasites and fungal, bacterial or viral diseases. These natural enemies limit pest populations, but are not always successful in stopping them from damaging plants. This is often because pests breed more rapidly in the spring and early summer so that heavy infestations are already present by the time beneficial insects become numerous.

Biological control occurs naturally in gardens; ladybirds eat aphids, for example, and thrushes attack snails. Today, however, the term "biological control" is more commonly taken to refer to the use of natural enemies that are introduced under suitable conditions to give effective control. This approach has more success in greenhouses where specific predators and parasites can be purchased for use against most of the major pests. There are few successful biological controls for plant diseases, but in some parts of the world insects, mites and fungal diseases are being used to control alien plants that have become weeds.

## BENEFITS OF BIOLOGICAL CONTROL

Biological controls may often be more suitable to use than pesticides. When used on food plants, such as tomato and cucumber, for example, there is

Thrush needs to find a stone to break snail's shell

**USEFUL BIRD**
*While a few birds are a nuisance in the garden, the song thrush has a useful role, feeding on snails. Conveniently placed flat stones in or near beds and borders provide them with anvils on which to smash the shells.*

no problem with pesticide residues on the edible parts. Similarly, the risk of chemicals causing spray damage is avoided.

Predators and parasites can seek out pests living in places that are difficult to reach with pesticides. Once released, natural enemies can be left to get on with controlling the pests; chemical controls often require repeat applications, particularly during the summer when high temperatures allow pests to breed rapidly. Biological controls are selective, controlling particular pests, with no harmful effects on non-target species. Some pests, such as glasshouse whitefly and red spider mite, may acquire resistance to pesticides, making biological control the only effective treatment.

## ENCOURAGING PEST PREDATORS

The mixture of flowers and other plants found in a typical garden is helpful in providing, directly and indirectly, food and shelter for a wide range of beneficial insects, birds and pest-eating mammals such as bats, shrews and hedgehogs. Installing a pond can also help as it provides a breeding site for frogs and toads. In the autumn many of these pest predators will seek out sheltered places where they can survive the winter. A thorough tidying-up of the garden at the end of the summer may be pleasing to the eye, but could result in the loss of many potential allies in the battle against pests.

The choice of chemicals used on garden plants can also have a significant effect on natural populations of pest predators. Some pesticides are more selective than others for pests and their use limits the harm caused to non-pest animals. Pirimicarb, for example, gives good control of aphids, but has little effect against other pests or against most beneficial insects and mites.

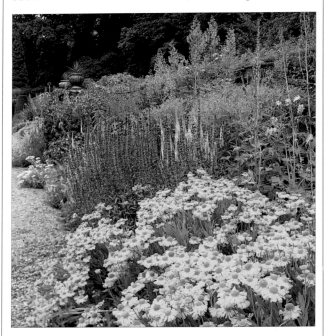

**A COLOURFUL GARDEN**
*Bright flowers, particularly in shades of purple and yellow, will attract many types of beneficial insects, such as hoverflies, whose larvae feed on greenfly and other aphids.*

## BUYING AND USING BIOLOGICAL CONTROLS

The predators, parasites and pathogenic agents that are commercially available are intended for use against specific pests. It is therefore essential that pests are accurately identified. The various biological controls are compatible with each other and can be used in combination; however, they must not be used together with any chemical control, or immediately afterwards. Predators and parasites are very susceptible to most insecticides and must be regarded as an alternative to chemical control rather than an additional treatment. The instructions for use should give a safe interval which must elapse between the application of chemicals and of the biological control.

With the exception of the caterpillar control *Bacillus thuringiensis*, which is used like a conventional spray, all the biological controls are living animals which should be released onto the infested plants as soon as they have been obtained. They can sometimes be bought through garden centres, but more usually are sold by mail-order firms who advertise in gardening magazines.

## APPLYING BIOLOGICAL CONTROLS

Pathogenic nematodes are watered into the soil or potting compost, while the caterpillar bacterium *Bacillus thuringiensis* is sprayed onto foliage. The nematodes must have moist, well-drained soil to enable them to survive and move between the soil particles. The vine weevil nematodes need relatively warm soils, in the range 14–21°C (57–70°F), though the slug nematode can work in a soil temperature

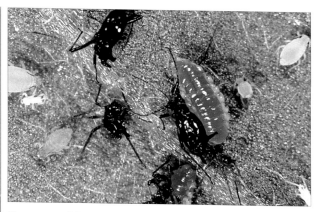

**PREDATORY MIDGE**
*Aphids can be among the most damaging and common of pests. The larvae of a tiny fly,* Aphidoletes aphidomyza, *though only up to 3mm long, will kill and eat up to 80 aphids each.*

from 5°C (41°F) upwards. The caterpillar bacterium is killed by bright sunshine and so should be applied in the evening, on the plants where caterpillars are present. The other predators and parasites that are available to gardeners (*see the table below*) need time to breed before they are numerous enough to overcome pests. It is therefore necessary to introduce these biological controls before pest infestations become heavy. Although they can survive lower temperatures, these predators and parasites need warm daytime temperatures, usually at least 21°C (70°F), and good light, if they are to be effective. Because of this their period of use is likely to be between early spring and mid-autumn.

## TYPES OF BIOLOGICAL CONTROL

| Pest | Name of biological control | Type of predator or parasite | How it works |
|------|---------------------------|------------------------------|--------------|
| APHIDS | Aphidoletes aphidomyza | Predatory midge | Larvae eat aphids. |
| | Aphidius species | Parasitic wasps | Larvae develop inside the aphids' bodies. |
| CATERPILLARS | Bacillus thuringiensis | Bacterium | Causes a fatal disease in moth and butterfly caterpillars that have eaten treated leaves. |
| FUNGUS GNATS (SCIARID FLIES) | Hypoaspis miles | Predatory mite | Feeds on the flies' larvae. |
| GLASSHOUSE RED SPIDER MITE | Phytoseiulus persimilis | Predatory mite | Eats the eggs, nymphs and adult spider mites. |
| GLASSHOUSE WHITEFLY | Encarsia formosa | Parasitic wasp | Attacks the pest's scale-like nymphal stages. |
| MEALYBUGS | Cryptolaemus montrouzieri | Predatory ladybird beetle | Both the adult and larval ladybirds eat the eggs and active stages in the mealybug's life-cycle. |
| SLUGS | Phasmarhabditis hermaphrodita | Pathogenic nematode | Infects slugs with a fatal disease. |
| SOFT SCALE | Metaphycus helvolus | Parasitic wasp | Develops in the immature scale nymphs. |
| VINE WEEVIL | Heterorhabditis megidis and Steinernema carpocapsae | Pathogenic nematodes | Infect vine weevil grubs with a fatal disease. |
| WESTERN FLOWER THRIPS | Amblyseius species | Predatory mites | Eat the immature thrip nymphs. |

# PREVENTIVE MEASURES

WELL GROWN, vigorous plants are not immune to pests and diseases, but they are less likely to be badly damaged. Attention should be paid to selecting plants and cultivars that are likely to do well under the local soil and climatic conditions. It is worthwhile paying more for good quality seeds, bulbs and plants from reputable suppliers. Cheap lots from dubious sources may be in poor condition, or carrying pests and diseases that may be difficult to eliminate once introduced into a garden or greenhouse.

### RESISTANT CULTIVARS

Many cultivars of plants have been developed in recent years to withstand common strains of fungal or bacterial disease, or susceptibility to pest attack. These include some fruits and vegetables that virtually eliminate the risk of several serious problems in the kitchen garden. Resistant cultivars suggested in the *A–Z of Pests, Diseases and Disorders, pp.97–192*, are to be recommended for general garden use.

### CARE OF YOUNG PLANTS

Take care to ensure that planting or sowing is done properly; sowing too early or late (when soil may be cold and wet), or planting pot-grown plants too deeply or with a pot-bound root system, will result in plants that will always struggle and fail to make good growth. When growing your own plants,

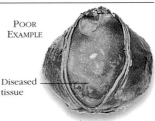

**RESISTANT CULTIVAR**
*This calabrese, 'Trixie', has been bred for its resistance to clubroot, a devastating soil-borne disease.*

ensure that you raise strong, healthy specimens by following procedures correctly. It is important, for example, to thin, prick out and pot on seedlings (*see facing page*) as soon as necessary, rather than neglecting them until the task is well overdue; crowded seedlings, competing for limited resources of nutrients, water and light, will be weak and undersized, and may never make vigorous plants. Also, the younger a plant is, the more swiftly it is able to recover from being transplanted.

### MAINTAINING HEALTHY PLANTS

Plant problems can arise at any time of the year but are more prevalent during the spring to autumn season, during their peak growing period. It is a good idea to make regular tours of inspection in order to detect problems before they become serious. It is important to look for plants which are under-performing; this may show as slow growth, often with symptoms of discoloration, distortion or damage that should be investigated as soon as possible.

Plants showing signs of attack by pests or diseases should not be ignored, since it is very likely that other plants may also become affected. Treatment with a suitable chemical is most often completely effective when the problem is detected and dealt with early on. In a group of plants, removal of the affected one is a sensible precaution to help stop the problem spreading.

## CHOOSING HEALTHY PLANTS AND BULBS

Vigorous growth

**SELECTING HEALTHY BEDDING PLANTS**
*Choose plants that are free of any sign of pests or disease. Look for compact growth, with a good leaf colour and a well-developed root system. Avoid those with yellowing leaves, weak growth or premature flowering.*

**BUYING BULBS**
*Since there is a good deal of variety in the quality and size of bulbs available to the gardener, they should be closely examined before buying. Bulbs should be as fresh and firm as possible (see bottom right) and without signs of mould or discoloured patches. The bulb shown top right has both diseased tissue and damaged outer scales, so it would not be a good choice. Undersized bulbs may not flower in their first year, so always try to choose the largest bulbs available of the chosen cultivar.*

POOR EXAMPLE

Diseased tissue

GOOD EXAMPLE

# GOOD PLANTING

**1 Sowing.** Sowing into a seed tray is a good way of raising large numbers of seedlings, but they will soon need more space to allow them to develop. Overcrowding will restrict their growth.

**2 Pricking out.** Select the stronger seedlings and carefully remove them from the tray. Handle them by gently holding the seed leaves rather than the stem, which is easily bruised.

**3 Transplanting.** Space the seedlings out and plant out or put into pots once they are well established. Avoid damping off by using fresh compost, rather than garden soil, and tap water.

As plants grow and develop, thin out any which have become overcrowded. This improves their vigour and allows air to circulate around stems and leaves, thus reducing damp air conditions that favour germination and infection by fungal spores.

## FEEDING PLANTS

Most plants will benefit from supplementary feeding, to increase yields of fruits and vegetables and to increase the quality and quantity of flowers. Organic fertilizers include garden compost, farmyard manure, bone meal, dried blood and seaweed extracts. These all provide a variety of nutrients, although not necessarily in the right balance required for optimum plant growth.

Man-made fertilizers can be essentially single nutrient compounds, such as ammonium sulphate (supplying nitrogen) or super phosphate (phosphorus), but more widely used by gardeners are general fertilizers which provide all the major nutrients. These are available as granules for adding to the soil, or in soluble form for watering in or spraying on the leaves (foliar feeding). The slow-release fertilizer pellets now available will give container-grown plants the nutrients they need throughout the growing season.

Feeding is generally done during the spring and early summer, when plants are making rapid growth. Fertilizers applied in autumn or winter may be largely wasted; dormant plants do not need feeding, and nutrients added then may be washed away by rain before spring.

## WATERING

Watering is always crucial in a garden, particularly at critical stages in a plant's development, such as the seedling stage and when flowers and fruits are being produced. In dry periods, it is important to water before plants show signs of drought stresses, and also to water at regular intervals.

Watering is best done in the evening so that less water is lost by evaporation, and should be into the soil as well as on the plant itself. Do not water plants in sun as the plant can be scorched. Covering the soil surface with a mulch will help retain moisture in the soil. Rain butts can be used to collect water from the roof. This source of water is suitable for most plants but is best avoided for seedlings and cuttings, as it can harbour fungal diseases that cause damping off and stem rots.

**SUPPORTING CLIMBING PLANTS**
*Plants that cannot support their stems need tying in to keep them off the ground, where growth will be vulnerable to insects and slugs. Even self-clinging climbers need help when young.*

### DRAINAGE

Waterlogged soil can be just as damaging as dry earth to plants. A heavy soil, such as clay, has only small spaces between the soil particles, making it moisture-retentive and difficult for water to drain through. Improve heavy soil texture by digging in plenty of coarse organic matter, such as compost, leaf mould or manure, to improve aeration.

Thin soils over chalk are very alkaline. Water drains through easily, as do soil nutrients, so that acidic types of organic matter such as manures are needed to give bulk and counteract the alkalinity. Sandy soil, too, is very light, needing plenty of mulching and feeding, but it is easy to work, and warms up quickly in spring.

### WEEDING

Good weed control is important. Weeds compete with cultivated plants and if they gain dominance, make them weak and more vulnerable to problems. Most weeds can be removed by hand or by hoeing, but in lawns and garden paths using a suitable chemical weedkiller may be more convenient. Also, like cultivated plants, weeds may attract common pests, viruses and fungal diseases, and are a source of reinfestation after the cultivated plants have been sprayed. They also help pests and diseases survive in the absence of cultivated host plants and thus diminish the value of crop rotation.

**REMOVING DEAD PLANT MATERIAL**
*Unless features such as winter seed heads are required, dead or dying flowers and foliage, together with their stalks or stems, should be removed, as they can become sites for rots and fungal growth, which may spread to living plant material.*

**REMOVING SUCKERS TO KEEP PLANTS STRONG**
*Rose suckers grow from the roots of the plant, since roses are grafted onto a rootstock. Gently clear away soil round the sucker to its union with the root, and cut or pull it off.*

### REMOVING OTHER UNWANTED GROWTH

To keep plants strong, suckering growth from the roots needs removing, particularly on grafted plants such as roses and most fruit trees. These are usually grafted onto rootstocks which will produce sucker growth of inferior quality to, but usually of greater vigour than, the named cultivar. If possible, the suckers should be traced back to the roots from which they arise and cut off at source. Cutting them off at soil level simply encourages more suckers. Avoid digging the soil around roses, cherries and other plants prone to suckering as damage to the roots can stimulate the growth of suckers.

### DEAD, DISEASED AND DAMAGED GROWTH

Although the routine pruning of plants that require it is usually best carried out at a particular time of year, it is, with very few exceptions, safer to remove any dead, diseased or damaged growth from a plant as soon as it is noticed, rather than waiting for the usual pruning time. Branches that have been broken by storms or other causes should be tidied up with a clean cut to speed callusing. Branches infected with fungal diseases, such as coral spot, canker or silver leaf, should be cut out well below the signs of infection, to prevent the further spread of these diseases.

### ROUTINE PRUNING

Not all plants require pruning but many woody plants do benefit from some form of thinning or reduction in size. This may be done on an annual basis – as with apple trees, in order to promote the development of spurs that will produce flower buds and hence fruits. Other plants may need infrequent pruning, with the intention of doing no more than

**BAD CUT**

Ragged tissue invites infection

**BAD CUT**

Budless stub will die back

**BAD CUT**

Too-sharply angled cut endangers selected bud

**CORRECT CUT**

### GOOD AND BAD PRUNING CUTS
*Make sharp pruning cuts close to a strong bud, with the cut sloping away from the bud. Avoid making ragged cuts or leaving stubs above the bud. Use a sharp knife or secateurs.*

removing the occasional branch or shoot that is growing too far in the wrong direction. Always prune to correct situations where branches cross each other, since in chafing together they may create wounds that are open to infection.

Pruning also provides an opportunity to thin shrubs and trees with a dense system of branches to provide a more open centre. This improves growth and the development of flowers and fruits by allowing sunshine in. There is also a better flow of air amongst the foliage and branches which will

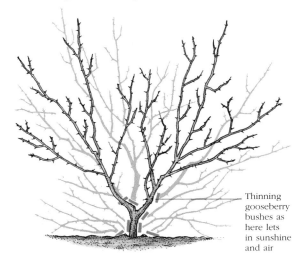

Thinning gooseberry bushes as here lets in sunshine and air

### PRUNING TO AVOID CROWDED GROWTH
*Cut out some of the older stems in order to encourage new growth. Thin the stems if overcrowded to create a more open centre. This will help reduce the incidence of diseases.*

help deter the germination of fungal spores. With flowering shrubs the removal of about one-third of the older stems each year will revitalize the plants by stimulating the development of young shoots.

## MAKING CORRECT CUTS
Bad pruning can cause problems, rather than curing or preventing them, so it is important to make cuts correctly with clean, sharp tools. Invest in good quality secateurs to prune stems up to the thickness of a pencil; anything larger is better tackled with loppers or a pruning saw. Shoots should be cut cleanly back to a strong bud with a slanting cut (*see left*). Never leave a stump of stem above the bud as this will encourage dieback.

Do not cut a branch off absolutely flush with the trunk or branch from which it arises. There will be a slight thickening or collar at the base of the branch and the cut should be made just outside this point. If the collar is also removed, the growth of scar tissue over the pruning wound is impaired and can result in a large exposed area of inner wood.

It is sometimes necessary to remove large branches from trees. Before doing so, it is generally wise to seek the advice of a professional arboriculturist; major work such as this can result in an unstable or unbalanced tree.

Large branches are heavy, and it is often advisable for safety reasons to remove them in smaller pieces instead of making a single cut. This also reduces the danger of the branch dropping as it is being cut and tearing a strip of bark from the trunk. This

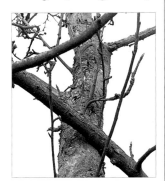

### CROSSING BRANCHES
*Crossing branches rub against each other, wearing away the bark and creating entry points for disease.*

risk can be further reduced by initially making a cut on the underside of the branch before completing the task by sawing down from the top.

## WOUND PAINTS
The use of proprietary wound paints, usually bitumen-based, on large pruning cuts is less fashionable today than in the past. There is some evidence to show that they can restrict the natural healing process that occurs around correctly-made pruning cuts. However, where trees and shrubs are known to be susceptible to fresh wound-infecting diseases, such as silver leaf or coral spot, it may be helpful to use a wound paint.

95

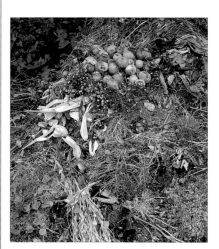

**COMPOST HEAP**
*Some types of infested or diseased material must never be composted, but many pathogens may be destroyed by the heat generated in the centre of the heap. Never leave rotten apples on top of compost like this, but bury them deep within the heap.*

**RAKING UP OLD LEAVES**
*Dead leaves harbour slugs and encourage grey mould (*Botrytis*) and other diseases. Keep the garden tidy by collecting plant debris and putting it on the compost heap.*

## DISPOSAL OF DEBRIS AND PRUNINGS

At the end of the growing or cropping season, tidy up plants and rake up fallen leaves to remove hiding-places for pests. Composting garden debris is an excellent way of returning nutrients and organic matter to the soil, but there are some basic rules that must be followed when composting unhealthy plant material. Most pests and diseases cannot survive in the heat generated by compost heaps, but it is advisable to cover infested plants with compost or with a layer of grass clippings or weeds. Plants with root pests or diseases should never be put onto compost heaps but burned or bagged up for removal with the domestic rubbish, as should prunings that come from plants affected by canker, silver leaf and coral spot.

## REPLANTING AND CROP ROTATION

Growing the same or closely-related plants in the same piece of ground in successive years can lead to a build-up of problems. Vegetables, bedding plants and bulbs are, however, particularly prone to some pests and diseases, and changing the growing site is not guaranteed to protect them. Many insects have winged adults which will have little difficulty in locating their host plant's new position. Similarly, many diseases are spread by wind-blown spores. There are, however, some less mobile problems, such as eelworms and root rots, which can be avoided by adopting a crop rotation system.

There are other good reasons for changing the growing sites of plants. Roses and apples suffer from replant disorders. If replacement plants are put in the same place from which an apple or rose has recently been removed, it is likely that the new plant will fail to thrive. Micro-organisms associated with the roots of the old plant seem to have an antagonistic effect on the new plant. Crop rotation in the vegetable garden also helps make better use of the available nutrients, as plants differ in their nutritional requirements. When devising a crop rotation system it is advisable to group related plants together in order to get maximum advantage when next year the planting sites are changed round. A simple rotation for the vegetable garden involves a three-year cycle; at the end of this time each of the three groups will have been grown for one year in a different area.

### SIMPLE CROP ROTATION

| GROUP 1 | GROUP 2 | GROUP 3 |
|---|---|---|
| Cabbage | Potato | Beans |
| Brussels sprouts | Tomato | Peas |
| Cauliflower | Carrot | Onions/Shallots |
| Kale | Parsnip | Leeks |
| Broccoli | Celery | Lettuce |
| Turnip | Parsley | Sweetcorn |
| Swede | Beetroot | Marrow |
| Radish | Spinach beet | Cucumber |

| FIRST YEAR | SECOND YEAR | THIRD YEAR |
|---|---|---|
| GROUP 1 | GROUP 2 | GROUP 3 |
| GROUP 2 | GROUP 3 | GROUP 1 |
| GROUP 3 | GROUP 1 | GROUP 2 |

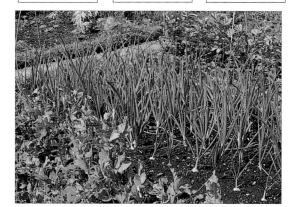

A BED OF GROUP 3 CROPS, INCLUDING PEAS AND BEANS

# A–Z of Pests, Diseases and Disorders

This dictionary section contains entries for all the most common pests, diseases and disorders likely to be found in the garden. Some occur in varying forms on many plants, and have a general entry with cross-references. Many others are largely host-specific, and will be found under plant names. Suggested controls and preventive measures give a choice between chemical and non-chemical methods wherever appropriate.

## ACER GALL MITES

**SYMPTOMS** In early summer, small, raised, reddish pimples appear on the upper leaf surfaces of maples and sycamores (see **p.17**), or a felt-like growth of yellowish-white or purplish-red hairs develops on leaf undersides.
**CAUSE** Microscopic mites that suck sap and secrete chemicals into the foliage that induce the abnormal growths. *Artacris macro-rhynchus* causes pimple galls and *Eriophyes eriobius* is responsible for felt galls.
**CONTROL** None of the chemicals available to gardeners is effective, but no control is needed, because apart from creating the galls, these mites have no harmful effect on their hosts. On small plants, light infestations may be controlled by removing affected leaves.

## ACORN GALL WASP

**SYMPTOMS** In late summer, oak acorns develop as sticky, ridged, yellowish-green structures (see **p.52**) instead of the familiar nut and cup. Later, the acorn (or knopper) galls become brown and woody and drop from the tree. In some years nearly all acorns on an oak tree are affected.
**CAUSE** A gall wasp, *Andricus quercuscalicis*, lays a single egg in a developing acorn in early summer. The egg hatches into a small, white, legless grub which secretes chemicals that cause the developing acorn to be converted into a gall. The grub pupates inside the gall, and the adult wasp emerges the following year.

**ACORN GALL**

**CONTROL** No control is available; the tree itself is unharmed by gall wasps.

## ADELGIDS

Only conifers, particularly pines (*Pinus*), spruce (*Picea*), larch (*Larix*), Douglas fir (*Pseudotsuga menziesii*) and silver fir (*Abies*), are attacked by adelgids. These black aphid-like insects feed by sucking sap from the foliage, stems and bark. Some species, particularly on spruce, cause galls at the shoot tips which they inhabit, but on other conifers they move about, although invariably covered by a fluffy, white waxy substance secreted from their bodies. Heavy infestations may cause some yellowing of the foliage, but generally the conifers will continue to make satisfactory growth. Control measures are limited, but are detailed in individual entries.
• *see also* DOUGLAS FIR ADELGID *p.124;* LARCH ADELGID *p.142;* PINE ADELGID *p.160;* PINEAPPLE GALL ADELGID *p.160.*

## ALDER SUCKER

**SYMPTOMS** Between mid-spring and early summer a fluffy, white waxy substance covering small, greenish-black insects appears on the foliage and shoot tips of alder trees (see **p.32**).
**CAUSE** The woolly wax is secreted from the bodies of the immature nymphs of *Psylla alni*. The adult stage is a winged aphid-like insect that does not produce this

substance. Both nymphs and adults are sap-feeders but cause little harm.
**CONTROL** Usually unnecessary. However, if the infestation is causing concern small trees could be sprayed with dimethoate, heptenophos with permethrin, malathion or pirimiphos-methyl.

## ALGAE

**SYMPTOMS** Powdery deposits develop on the plant's surface; these are commonly green or grey-green and can be rubbed off, leaving no discoloration on the plant tissue beneath. The plant's growth and development are not affected.
**CAUSE** Harmless microscopic, non-flowering plants which develop in films of moisture on plant surfaces. They are particularly common on plants growing in damp situations. Leaves and trunks or stems are most commonly affected, especially those on the north-facing side of the plant.
**CONTROL** No control is needed as the plant is not damaged. If deemed to be unsightly, algal deposits can usually be washed off with a sponge and soapy water. Improve air circulation around the plant and reduce damp-ness to prevent regrowth.

## ALTERNARIA LEAF SPOT

**SYMPTOMS** Grey or brown spots appear on affected foliage. These frequently show concentric rings of tissue dieback with central, raised pycnidia, or fruiting bodies, of the fungus. In extreme conditions spots may join together, causing extensive areas of leaf to be killed. In most cases,

however, the damage remains as distinct near-circular spots. Areas of pale or variegated foliage are most readily affected, when the leaf may wither and die back. Plants commonly affected include potatoes, carrots and brassicas, and also cinerarias.
**CAUSE** Various species of the fungus *Alternaria*. The spores are air-borne or carried by rain or water splash.
**CONTROL** Remove and dispose of affected leaves as soon as they appear. This should help to limit spread. Spray affected plants with a fungicide containing mancozeb.
• *see also* BRASSICA DARK LEAF SPOT *p.110;* POTATO EARLY BLIGHT *p.164.*

## AMERICAN GOOSEBERRY MILDEW

**SYMPTOMS** Whitish-grey powdery fungal growth appears on the upper surface of gooseberry foliage (see **p.25**), stems and fruits (see **p.49**). Severely affected leaves, especially the young growth towards the shoot tips, may become distorted and die off. Shoot tips may also die back. The mildew on the skin of affected berries turns pale brown or buff in colour as it ages, and can be scraped off. Infected fruit is edible, but unsightly, turning brown when cooked. Blackcurrants may also be affected.
**CAUSE** The fungus *Sphaerotheca mors-uvae*, which is encouraged by stagnant air around the branches and by excessive use of high-nitrogen fertilizers. It overwinters on the branches and within the buds.

**CONTROL** Prune out affected branches as soon as they are seen, and prune to thin growth and improve air circulation within the crown of the bush. Avoid excessive use of nitrogenous fertilizers; a general fertilizer is usually preferable. Spray with a fungicide containing carbendazim, bupirimate with triforine, or copper sulphate. Consider growing cultivars showing resistance to this infection, such as 'Invicta' or 'Greenfinch'.

## ANEMONE RUST

*See* PLUM RUST *p.161.*

## ANEMONE SMUT

**SYMPTOMS** Anemone stems and leaves show dark, raised blisters, which burst to release dark brown to black spores. The plant may show some dieback, but is rarely killed. Anemone smut may also affect globe flowers (*Trollius*).
**CAUSE** The fungus *Urocystis anemones*, which can overwinter in infected plant debris and on other plant material on the soil surface. It is easily spread by rain splash.
**CONTROL** Remove affected parts promptly, before the blisters erupt. Destroy severely affected plants and choose a fresh site for growing new anemones or globe flowers.

## ANTHRACNOSE

Various fungi cause patches or spots of discoloration to appear on infected plant tissue. Leaf, stem or, in beans, pod tissue may be killed, and in severe cases the whole plant may die back. Precise symptoms vary with the host plant and the

particular fungus involved.
• *see also* BEAN ANTHRACNOSE *p.105;* PLANE ANTHRACNOSE *p.160;* WILLOW ANTHRACNOSE *p.190.*

## ANTIRRHINUM RUST

**SYMPTOMS** Concentric rings or circles of pustules full of dark brown spores appear on the underside of snapdragon (*Antirrhinum*) leaves (*see* **p.15**). A chlorotic (yellow) "halo" develops around each patch of spores, and the upper leaf surface may also become yellowed. Affected leaves die, and if left unchecked, the infection may spread to stems and flower buds. Severely infected plants fail to flower and may be killed.
**CAUSE** The fungus *Puccinia antirrhini*, which may be seed-borne and which can overwinter on snapdragons if provided with sufficient warmth to survive the winter.
**CONTROL** Remove and dispose of severely infected plants promptly. Otherwise, spray with a fungicide containing penconazole or mancozeb. Remove all snapdragons at the end of the season and do not grow them on that site for at least two years. Try growing cultivars which show a degree of resistance to the infection, such as the Tahiti series, 'Monarch' and 'Royal Carpet Mixed'.

## ANTS

Several species of ants occur in gardens, including the black ant, *Lasius niger*, the yellow meadow ant, *Lasius flavus,* and red ants of various *Myrmica* species. They are often seen on plants infested with aphids and other sap-feeding insects. Ants collect the

sugary honeydew excreted by these insects and may protect aphids by driving off predators. Fine heaps of soil appear on lawns (*see* **p.59**) and in flower-beds, where soil excavated from the underground nests the worker ants build for the queen ant is deposited on the surface. This can interfere with mowing and partly bury low-growing plants.

ACTUAL LENGTH: 3–10MM

**BLACK ANT**

**CONTROL** Ants can be reduced in numbers but they cannot be eliminated from gardens. They are generally a nuisance rather than a significant pest. Many queen ants that fly into gardens are in fact killed and eaten by ants from existing nests before they can establish new colonies. Rigorous attempts to eliminate nests may increase the survival rate of queens flying in from nearby gardens, and lead to even more nests in the near future. Control measures should be confined to nests which are causing real problems. In gardens, treat nest sites with dusts containing bendiocarb, lindane, permethrin or pirimiphos-methyl, or spray with deltamethrin, permethrin, cypermethrin or chlorpyrifos. Ants in buildings can be killed with baits containing borax, bendiocarb or trichlorphon, or nest entrances and ant pathways can be sprayed with aerosol lacquers that leave persistent deposits of diazinon, cypermethrin or permethrin.

## APHIDS

Many species of aphid occur in gardens and few plants are never attacked. The more important aphid pests are described in more detail in individual entries. Generally, affected plants show reduced growth with varying degrees of leaf distortion. The upper leaf surface becomes sticky with the sugary honeydew excreted by aphids; a black sooty mould will grow on the honeydew, especially under damp conditions. Aphids may be green, yellow, pink, black, greyish-white or brown, and are usually about 2mm in length, although some conifer aphids can be up to 6mm long. All nymphs shed their outer skins several times as they grow larger. These whitish cast skins accumulate on the upper surface of leaves below those where the aphids are feeding. Some virus diseases, especially of strawberries and raspberries, are spread on the mouth-parts of aphids as they move from one plant to another.
**CONTROL** Control on specific plants may be detailed in individual entries (*see below*). Control overwintering eggs on fruit trees and bushes by spraying thoroughly with tar oil wash in winter. Use dimethoate, malathion, heptenophos with permethrin, pirimiphos-methyl, permethrin, bifenthrin or pirimicarb on plants in active growth. Pirimicarb is a selective insecticide that controls aphids but not other pests; it also leaves bees, ladybirds and most other beneficial insects unharmed. Effective organic insecticides include pyrethrum, derris and insecticidal soaps.
• *see also* BEECH WOOLLY APHID *p.106;* BLACK BEAN APHID *p.107;* BLACKFLY *p.108;* CEDAR APHID

*p.115*; CHERRY BLACKFLY *p.116*; CONIFER APHIDS *p.119*; CURRANT BLISTER APHID *p.121*; CYPRESS APHID *p.121*; GREEN SPRUCE APHID *p.134*; HONEYSUCKLE APHID *p.137*; JUNIPER APHID *p.140*; LETTUCE ROOT APHID *p.145*; LUPIN APHID *p.146*; MEALY CABBAGE APHID *p.146*; MEALY PLUM APHID *p.146*; MELON COTTON APHID *p.147*; PEACH APHID *p.156*; PEAR BEDSTRAW APHID *p.157*; PLUM LEAF-CURLING APHID *p.161*; ROOT APHIDS *p.170*; ROSE APHIDS *p.170*; ROSY APPLE APHID *p.173*; WATER LILY APHID *p.188*; WILLOW BARK APHID *p.190*; WOOLLY APHID *p.192*.

## APPLE AND PEAR CANKER

**SYMPTOMS** Areas of bark sink inwards (*see* ***p.36***), usually starting near a wound or bud. The bark becomes discoloured and then shrinks and cracks, forming concentric rings of flaky bark. The branch may become swollen around the cankered area. As the canker enlarges, the shoot may be girdled and so foliage and growth above it starts to die back. Stems and trunks may be attacked and very occasionally semi-mature fruits are infected and show rotting. In summer raised, white fungal pustules develop on the canker, and in winter raised, red fruiting bodies develop. Apples and pears are frequently affected; poplars (*Populus*), beeches (*Fagus*), willows (*Salix*), hawthorn (*Crataegus*) and sorbus may also succumb.
**CAUSE** The fungus *Nectria galligena*, which is spread mainly during the spring as wind-borne spores enter through wounded areas such as leaf scars, pruning cuts, scab infection, frost crack, attack by woolly aphid or other wounds.

**CONTROL** Prune out whole spurs or branches where possible. On larger branch or trunk infections, carefully pare away all infected and damaged bark and wood, cutting back to completely clean tissue, and treat the wound with a proprietary wound paint, such as one containing cresylic acid. Dispose of prunings and parings carefully. Sprays used for the control of apple scab and apple powdery mildew (*see p.101*) should also help suppress canker. Improve general growing conditions, in particular soil drainage, since unsuitable conditions such as wet soil may aggravate the disease. Avoid growing apples known to be particularly susceptible, such as 'Cox's Orange Pippin', 'James Grieve', 'Worcester Pearmain' and 'Spartan'. The following cultivars seem to show a degree of resistance under most circumstances: 'Laxton's Superb', 'Newton Wonder', 'Bramley's Seedling' and 'Lane's Prince Albert'.

## APPLE BITTER PIT

**SYMPTOMS** The skin of the apple fruit may develop sunken brown spots, each less than 1mm in diameter (*see* ***p.48***). The flesh is spoiled by numerous pale brown spots or freckles. Affected apples may have an unpleasant, slightly bitter taste. Large fruits are especially susceptible, as are fruits produced by heavy-cropping trees. The symptoms may appear while the fruits are on the tree, but often only develop during storage.
**CAUSE** Calcium deficiency in the fruit. This results in the collapse of small groups of cells within the flesh, so that

brown flecks develop. Calcium levels in the soil may be perfectly adequate, but in dry conditions, the tree cannot take up the amount it needs. Large fruits and heavy-cropping trees are more prone to this disorder because of their greater demand for calcium. Too high a concentration of magnesium or calcium in fruits may also be a cause of bitter pit.
**CONTROL** Keep apple trees well watered and mulched to maintain soil moisture levels. Maintain even growth by feeding with a balanced fertilizer; avoid excessive use of high-nitrogen feeds. Spray developing fruits with calcium nitrate solution from early summer until early autumn. The cultivars 'Bramley's Seedling', 'Discovery' and 'Crispin' may be damaged by this chemical, so care should be taken when treating them.

## APPLE BLOSSOM WEEVIL

**SYMPTOMS** Apple flower buds start to develop but the petals fail to open and turn brown (*see* ***p.46***). If these "capped" blossoms are pulled apart, a creamy-white grub, up to 7mm long, or the pupal stage, will be seen. Damaged blossoms cannot produce fruits.
**CAUSE** A weevil, *Anthonomus pomorum*, which lays eggs in the unopened buds during mid- and late spring. The adult weevil is brown with whitish markings on its upper surface and is 4–6mm long.

ACTUAL LENGTH: 4–6MM

**ADULT APPLE BLOSSOM WEEVIL**

**CONTROL** Light infestations have little impact on the yield of fruits. Spraying with pirimiphos-methyl or fenitrothion when flowers are at the green cluster stage is only worthwhile if heavy attacks have occurred in previous years.

## APPLE CAPSID

**SYMPTOMS** Ripe apples have raised yellowish-brown corky bumps or scabby areas on the skin (*see* ***p.47***). Leaves at shoot tips may be distorted, with many small holes.
**CAUSE** Apple capsid, *Plesiocoris rugicollis*, which is a bright green sap-feeding insect up to 7mm long. It injects a toxic saliva into the plant as it feeds on the foliage and immature fruits during the early summer. This kills some plant cells and, as the developing leaves grow and expand, these dead areas tear into small holes. Fruits respond to the capsid's saliva by developing raised bumps or scabby areas.
**CONTROL** The fruit damage is only skin-deep and light attacks can be tolerated. If damage has been extensive in previous years, spray with dimethoate, heptenophos with permethrin, fenitrothion or pirimiphos-methyl at petal fall.

## APPLE FROST DAMAGE

**SYMPTOMS** Russeting of the skin, usually on the most exposed face of the fruit. The more protected faces are usually unharmed. Damaged areas may develop a corky, roughened appearance and the fruit may become deformed and uneven in growth. Flowers and young foliage subjected to frost wither and die off rapidly.

**CAUSE** Frost on young, unhardened plant tissues. This is exacerbated if the plant is growing in a position where, following overnight frosts, it is then subjected to rapid warming by early morning sun.

**CONTROL** Rarely a recurrent problem, unless the tree is in an unsuitable position. Seasonal use of horticultural fleece or similar material draped over trees will prevent injury.

## APPLE FRUIT SPLIT

**SYMPTOMS** Shallow cracks occur, usually only just penetrating into the flesh and most commonly running round the stalk end of the fruit (see **p.48**). These cracks may heal over and a corky layer forms, but if they occur when the fruit is more mature, the fruit usually succumbs to secondary infections.

**CAUSE** An erratic supply of water is the most common cause. The splitting occurs when the tree puts on a sudden spurt of growth, when moisture becomes available after a period of drought.

**CONTROL** Keep trees regularly watered and apply a mulch to moist soil to try to retain soil moisture.

## APPLE LEAF MINER

**SYMPTOMS** Silvery-white or brown lines appear on the foliage of apple and cherry trees and ornamental *Malus* and *Prunus* (see **p.19**), particularly in late summer. If the leaf miners have finished feeding and are pupating, hammock-like cocoons of white silk may be attached to the undersides of leaves.

**CAUSE** Small green caterpillars of a tiny moth,

*Lyonetia clerkella*, live within the leaves, and eat out meandering mines or tunnels, which show as white or brown lines.

**CONTROL** Although leaf miners are sometimes abundant, this pest is active too late in the growing season to have much effect on the tree's health or ability to produce fruit. If young trees are attacked earlier in the summer, they could be sprayed with pirimiphos-methyl or malathion when leaf-mining damage is seen.

## APPLE POWDERY MILDEW

**SYMPTOMS** Leaves are covered in a powdery white fungal growth (see **p.25**). Young leaves breaking from buds which were infected in the previous year often show the symptoms. If attacked early on, the foliage rarely reaches full size, but is distorted and dies prematurely.

**CAUSE** The fungus *Podosphaera leucotricha*, which overwinters in the apple buds or on the young stem growth (as silvery white patches on the bark). This fungus may also occasionally attack pears, quinces and medlars, but does not cause serious damage. Disease spread and development is most rapid in hot summers and if there are heavy dews at night.

**CONTROL** Keep apples well watered, as a dry soil seems to encourage powdery mildew development. Apply a mulch to the moist soil to preserve soil moisture. Prune out severely infected growth and mildewed stems. Prune established trees to improve air circulation within the crown. Spray with a suitable fungicide such as carbendazim or bupirimate

with triforine. Grow cultivars showing good resistance, such as 'Blenheim Orange', 'Discovery', 'Greensleeves' or 'Worcester Pearmain'.

## APPLE SAWFLY

**SYMPTOMS** Apple fruitlets fall off in early to mid-summer with a maggot hole in the side (see **p.48**). This hole is filled with excrement produced by a white caterpillar-like larva, or maggot, with a brown head. Mature fruits may be misshapen, with a broad ribbon-like scar on the skin, often running up from the eye end (opposite the stalk) to the middle of the fruit (see **p.47**).

**CAUSE** A sawfly, *Hoplocampa testudinea*, lays eggs in the fruitlets at flowering time. After hatching, the larva bores into the fruitlet and at first tunnels just beneath the skin before boring into the core. Fruitlets damaged in this way drop off, and when fully fed the larvae leave them (creating the exit hole) and go into the soil to pupate. However, if the larva dies before it can bore into the core, the fruit will reach maturity. The early feeding beneath the skin causes fruits to be misshapen with the scar mentioned above.

**CONTROL** Pick off and destroy fruitlets showing signs of damage before the larvae escape into the soil. Spray with pirimiphos-methyl, permethrin or fenitrothion if sawfly has affected a significant proportion of the crop in previous years.

## APPLE SCAB

**SYMPTOMS** Blackish-brown, scabby patches develop on the fruits, with similar but more greenish-grey spots on

the leaves (see **p.16**). In severe cases the fruit may be almost entirely covered with scabby patches, and be small and misshapen. Infected fruits may crack or split and so become infected with secondary organisms such as brown rot. Affected leaves fall early and may also show extensive infection on the petiole (leaf stalk). A very similar disease affects poplars.

**CAUSE** The fungus *Venturia inaequalis*, which overwinters on scabby patches on the young stems and also on fallen infected leaves. Scab can occur on ornamental and edible *Malus* species and cultivars. It is usually most prevalent in damp seasons and on trees with crowded branches.

**CONTROL** Rake up and dispose of affected leaves as they fall. Prune out cracked or scabby shoots to limit the overwintering places of the fungus. Spray with a fungicide containing carbendazim, mancozeb or bupirimate with triforine. Grow apple cultivars showing resistance to this infection, such as 'Discovery', 'Lane's Prince Albert', 'Gavin' or 'Sunset'.

## APPLE SCALD

**SYMPTOMS** Reddish or brown areas develop on the skin of the fruit (see **p.47**), on surfaces that are subjected to the full impact of the sun. In extreme cases the flesh beneath may be damaged, but the discoloration is usually limited to the skin.

**CAUSE** The scorching or scalding is caused by strong sunlight on the tender skin of young apples.

**CONTROL** In temperate zones, scald is generally only seen in unusually hot, bright summers, so controls are not really justified.

## APPLE SOOTY BLOTCH

**SYMPTOMS** Sooty, greenish-black spots develop on the fruit skin, usually on apples but occasionally on plums. They do not penetrate into the flesh, however, and can be scraped off. The fruit's development is not affected and it is not distorted.
**CAUSE** The fungus *Gloeodes pomigena*, which overwinters on young stems and which is most common in trees growing in shade, and in cool, damp conditions.
**CONTROL** Although common in some seasons, it is rarely widespread and usually only occasional. No specific control measures are recommended, but trees sprayed for apple scab (*see p.101*) will only rarely develop sooty blotch.

## APPLE SUCKER

**SYMPTOMS** Flattened, pale green insects, up to 2mm long, occur on apple blossom trusses in the spring (*see p.43*). Heavy infestations kill the flowers and prevent fruit setting.
**CAUSE** Sap-feeding, aphid-like insects, *Psylla mali*, that hatch from overwintering eggs as apples come into leaf. Most of the damage is caused by the immature nymphs described above, during the blossom period. The adult stage resembles a winged aphid and is present after petal fall.

ACTUAL SIZE ●

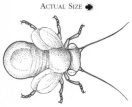

**APPLE SUCKER**

**CONTROL** Use a tar oil wash in winter against the egg stage, or spray with dimethoate, heptenophos with permethrin, pirimiphos-methyl or fenitrothion before flowering, at green bud stage.

## APPLE WATER CORE

**SYMPTOMS** This disorder is also known as "glassiness" because affected fruits develop a glassy, almost wet appearance to the flesh. This is usually most apparent close to the core, but may occur throughout. Symptoms may decrease when apples are stored, but more often affected fruits become brown and start to disintegrate.
**CAUSE** Can be encouraged by unsuitable cultural conditions, including excessive pruning and thinning, high temperature, high light intensity and extensive damage or removal of bark on the trunk or branches.
**CONTROL** Avoid excessive thinning or pruning. Other factors are usually impossible to control.

## AQUILEGIA SAWFLY

**SYMPTOMS** Pale green caterpillar-like larvae, up to 18mm long, feed on the leaf margins of aquilegias. They can quickly defoliate plants, leaving just the leaf stalks and flower stems. Several generations occur between mid-spring and early autumn.
**CAUSE** A sawfly, *Pristiphora aquilegiae*, which lays small batches of eggs on the undersides of leaves. The adults have black bodies 6mm long, two pairs of transparent wings and reddish-brown legs.
**CONTROL** Watch out for signs of leaf damage on aquilegias, especially in the spring and early summer, when defoliation will have a more severe effect on the plant's health. Either remove the larvae by hand or spray with derris, pyrethrum, permethrin, bifenthrin or pirimiphos-methyl.

## ARABIS MOSAIC VIRUS

**SYMPTOMS** Yellow mosaic-like flecking and discoloration on leaves, not only of arabis but of many plants. Infected plants are often stunted and distorted, and commonly fail to reach maturity.
**CAUSE** Arabis mosaic virus, which is one of the most common and widespread viruses infecting garden plants, both edible and ornamental.
**CONTROL** There is no control measure available. Affected plants should be removed and disposed of promptly.

## ARBUTUS LEAF SPOT

**SYMPTOMS** Numerous small spots develop on the leaves of strawberry trees (*Arbutus*). Each has a dry, grey-white central area bordered by a dark purple band (*see p.14*). The vigour of the tree is not affected.
**CAUSE** The fungus *Septoria unidonis*, which is common and widespread.
**CONTROL** No control is available, nor is one really required. Fallen leaves should be cleared up.

## ARUNCUS SAWFLY

**SYMPTOMS** Pale green caterpillar-like larvae, up to 20mm long, feed on the edges or central areas of the leaves of goatsbeard (*Aruncus dioicus*). Several generations occur between late spring and early autumn and plants can be stripped of foliage.
**CAUSE** A sawfly, *Nematus spiraeae*, sometimes known as the spiraea sawfly because *Aruncus* was at one time classified as a *Spiraea* species. This sawfly does not attack any of the plants currently known as *Spiraea*. The adults have black and yellow bodies 6mm long, two pairs of transparent wings and legs that are mainly yellow.
**CONTROL** Watch out for signs of leaf damage, especially in the spring and early summer, when defoliation will have a more severe effect on the plant's health. Either remove the larvae by hand or spray with derris, pyrethrum, permethrin, bifenthrin or pirimiphos-methyl.

## ASH CATKIN GALL MITE

**SYMPTOMS** Instead of producing the usual ash seeds or keys, the flowers become converted into lumpy, brown, woody galls. These persist on the tree and can be a distinctive feature on infested ash trees in the autumn and winter after leaf fall.
**CAUSE** Microscopic gall mites, *Eriophyes fraxinivorus*, which suck sap from flowers in the spring, and secrete chemicals that convert the flowers into galls. Individual trees growing close together can vary considerably in the extent to which they are affected, indicating that some trees are more susceptible than others.
**CONTROL** Apart from reducing the number of seeds produced by ash trees, this mite has no adverse effect on their growth. There is no effective chemical control available to amateur gardeners.

## ASH SUCKER

**SYMPTOMS** In early summer the margins of ash leaves become purplish-red, thickened and curled over. On the undersides of the curled leaves will be some flattened aphid-like insects which secrete a white waxy material from their bodies.
**CAUSE** The ash sucker, *Psyllopsis fraxini*, is an aphid-like, sap-feeding insect. The damage is caused by the wingless immature nymphs; the adults, which resemble winged aphids, are present during the summer but have no noticeable effect on the tree.
**CONTROL** Tall trees cannot be treated and spraying is not usually required, even on saplings. Although the distorted leaves may cause concern, ash suckers do not have a serious impact on the tree's health or vigour.

## ASPARAGUS BEETLE

**SYMPTOMS** Asparagus plants lose their foliage as adult beetles and their larvae eat the leaves and outer bark from the stems (*see **p.40***). The damage to the bark causes stems to dry up and turn yellowish-brown above the point where bark has been removed from all round the stem. The adult beetles are 6–8mm long, black with six yellow blotches on their wing cases, and a reddish thorax. They emerge from the soil in late spring and begin laying eggs on the stems and foliage. The larval stage is creamy-black in colour, up to 10mm long, with three pairs of legs towards the head end. There are two generations between late spring and early autumn.
**CAUSE** Both the adults and larvae of a leaf beetle, *Crioceris asparagi*.

ACTUAL LENGTH: 6–8MM

**ADULT ASPARAGUS BEETLE**

**CONTROL** Burn old stems at the end of the year to destroy some of the over-wintering beetles. Watch for signs of infestation from late spring onwards. Hand-removal of adults and larvae is feasible on small asparagus beds; otherwise spray with pyrethrum, permethrin or bifenthrin. Asparagus flowers attract bees, so when plants are in flower, spray at dusk if insecticides are needed.

## ASTER WILT

**SYMPTOMS** Rapid wilting of the plant, followed by death. The stem base is blackened and shrinks inwards. A faint, furry, pale pink or white fungal growth may develop on the affected areas.
**CAUSE** The fungus *Fusarium oxysporum* f. sp. *callistephi*, or fusarium wilt. It is most common in heavy or wet soils but can attack asters in any soil, and even in seed trays. The fungus is usually introduced on infected seedlings or plants, but may also be present in the soil. It may also be seed-borne.
**CONTROL** Promptly remove and dispose of infected plants and the soil in the vicinity of their roots. Use a fresh site for future plantings of asters and only plant perfectly healthy-looking plants. Discard any with dead roots or damaged stem bases. Avoid planting asters on heavy soils, or improve soil texture before planting. Grow china asters that show resistance, such as

'Duchess', 'Ostrich Plume', 'Milady Mixed', 'Starlight' or 'Super Sinensis'.

## AZALEA GALL

**SYMPTOMS** Pale green, waxy growths develop in place of foliage, or occasionally in place of petals. Sometimes they have a pinkish-red colour. The galls later turn white as a powdery spore layer develops on their surface, and at the end of the summer they wither and turn brown. The infected plant continues to grow and flower normally. Symptoms may occasionally develop on rhododendrons.
**CAUSE** The fungus *Exobasidium vaccinii*, which is common on outdoor and indoor Indian hybrid azaleas (*R. indicum*). Spores are carried on air currents and by insects, and months may elapse between infection and the appearance of symptoms.
**CONTROL** Remove galls promptly, before the spore layer develops.

## AZALEA WHITEFLY

**SYMPTOMS** White-winged insects, 2mm long, readily fly up from the undersides of the leaves of evergreen azaleas during the summer. The immature nymphs are flat, oval, whitish-green, scale-like creatures that live on the undersides of leaves. They are present for most of the year, as this pest overwinters in the nymphal stage. Both adults and nymphs excrete a sugary substance called honeydew that makes the upper leaf surfaces sticky and allows the growth of sooty moulds.
**CAUSE** A whitefly, *Pealius azaleae*, which feeds by sucking sap from the leaves of evergreen azaleas.

**CONTROL** The adult stage, which is reached in mid-summer, is more susceptible to insecticides than the eggs or nymphs. When adult whitefly are seen, spray the underside of the foliage thoroughly with pyrethrum or insecticidal soap, or with an insecticide containing permethrin, bifenthrin or pirimiphos-methyl.

## BACTERIA

**SYMPTOMS** The symptoms of bacterial infection vary with both the plant part and the type of plant infected, as well as the precise bacterium involved. Leaves usually become spotted, often with a yellow "halo" around the edge of the leaf spot. Fleshier plant parts such as stems, rhizomes, tubers, roots and bulbs may discolour and disintegrate rapidly, often becoming slimy in texture and giving rise to unpleasant odours. Specific bacterial infections such as bacterial canker and fireblight have individual entries in this section.
**CAUSE** Various species of bacteria may be involved in general infections, probably the most common being species of *Erwinia*. Many bacterial infections occur initially on injured tissues but then spread into other areas of the host plant. Bacteria may be spread by rain splash, by insects and animals, and on pruning tools; some, such as crown gall, may spread through soil water movement.
**CONTROL** The precise method of control varies, but prompt removal of all infected areas

and adjacent plant tissue is essential. Copper-based fungicides can be used to control bacterial canker of stone fruits.

- *See also* BACTERIAL CANKER *below;* BACTERIAL SOFT ROT *below;* BEAN HALO BLIGHT *p.105;* FIREBLIGHT *p.127;* LEAF SPOT (BACTERIAL) *p.143.*

## BACTERIAL CANKER

**SYMPTOMS** Clearly defined areas of bark flatten and sink inwards (*see **p.35***). Amber-coloured, resin-like ooze may appear, closely associated with injured bark. Only *Prunus* are commonly affected; symptoms are most obvious on cherry trees, far less so on plums. Buds at the tips of affected branches may fail to open, or leaves may appear but later wither and die back, due to the ringing effect produced by the expanding canker. Leaves on affected limbs are often yellowed and small in size. Foliage may develop shot-hole symptoms (*see p.175*). **CAUSE** The bacterium *Pseudomonas mors-prunorum* (on plum and sweet cherry) and *Pseudo-monas syringae* (on peach, apricot, cherry and plum). Most infections occur during wet, windy weather in autumn, or in damp conditions in spring, when young stems and leaves are readily infected. Spring infections usually start on the leaves and then spread to the wood. Most wood infections occur through injured bark, such as that caused by pruning, frost crack or leaf fall. Infection during summer is rare. **CONTROL** Prune out infected areas during the summer months. Spray infected trees with a copper-based fungicide, such as Bordeaux

mixture, or with copper oxychloride. Sprays should be applied once at the end of summer, once in early autumn and once in mid-autumn. Consider growing cultivars which show a degree of resistance, such as the cherries 'Merton Glory', 'Merton Premier', 'Merla' and 'Merpet', or plum cultivars 'Marjorie's Seedling' and 'Warwickshire Drooper'.

## BACTERIAL LEAF SPOT

*See* LEAF SPOT (BACTERIAL) *p.143.*

## BACTERIAL SOFT ROT

**SYMPTOMS** Roots, tubers, rhizomes and fruits are generally the most susceptible areas of host plants. Discoloured areas develop and, as the tissue beneath deteriorates, the injured area may sink inwards. Infection usually spreads rapidly and can prove fatal or seriously damaging if its progress is not halted. **CAUSE** Various bacteria may be involved, most commonly *Erwinia atroseptica* and its various pathovars or strains. The bacteria usually enter via wounds and are easily spread on pruning tools, by handling or by insects and other animals. **CONTROL** Control is rarely possible as the infection spreads so rapidly. It may, however, be possible to limit the damage caused by very prompt removal of infected areas together with the adjacent tissue. Disinfect pruning tools by using a garden disinfectant.

- *see also* BRASSICA BACTERIAL SOFT ROT *p.110.*

## BANDED PALM THRIPS

**SYMPTOMS** Indoor plants, especially those which have relatively tough leaves, such as ficus, schefflera, monstera and palms, develop a silvery-brown discoloration of the upper leaf surface (*see p.22*). Close examination will reveal narrow-bodied insects up to 2mm long, either creamy-yellow in colour or banded black and white. **CAUSE** Sap-feeding thrips, *Parthenothrips dracaenae*, which are creamy-yellow as immature nymphs. The adults are darker in colour, with dark bands on pale wings. The wings are folded over the thrips' bodies when not in use, which gives the adults their striped appearance. **CONTROL** Spray the foliage thoroughly with pyrethrum, permethrin or bifenthrin. Two or three applications at fortnightly intervals may be needed to break the pest's life cycle. Damaged leaves will not regain their normal colour but new growth will stay clean once thrips have been controlled.

## BARK BEETLES

**SYMPTOMS** Many woody plants can be affected by bark beetles, especially trees or shrubs that are already in poor health, that have dead limbs, or have recently died. The larvae feed between the bark and the wood and eat out a series of channels that can be seen if the bark is peeled away (*see p.33*). The tunnelling pattern is often distinctive for a particular species of bark beetle; it consists of a maternal tunnel made by the adult female, in which eggs are laid, and larval tunnels which radiate away from the original tunnel. Fully-fed grubs

pupate at the ends of the tunnels and adult beetles chew their way out through the bark. **CAUSE** Many different species of bark beetle can occur in gardens, on conifers and deciduous trees and shrubs. Most are harmless, as they are secondary pests on plants that are already dying from other causes. The most notorious bark beetle is *Scolytus scolytus*, the elm bark beetle. This breeds in dying elm trees, including those infected with Dutch elm disease (*see p.124*). When adult beetles emerge in early summer they have spores of the fungal disease on their bodies and can carry it to other elms in the area.

ACTUAL LENGTH: 2–5MM

**ADULT BARK BEETLE**

**CONTROL** Prune out and burn affected branches, which will be weak or dying, and check to see if another problem, such as honey fungus (*see p.136*), is present. Insecticides are not effective against bark beetles, since for much of the year they are protected under the bark.

## BASAL ROT

*See* FOOT AND ROOT ROTS *p.128.*

## BAY SUCKER

**SYMPTOMS** During the summer, leaves at the shoot tips of sweet bay (*Laurus nobilis*) develop thickened, pale yellow margins which curl over (*see p.26*). Often just one half of the leaf is affected, and later the damaged parts dry up and turn brown. Immature insects feed underneath the curled leaf margins but emerge when ready to moult to the adult stage. These fully-grown nymphs are grey, flattened insects which secrete a fluffy, white waxy material from their bodies. The adults are winged, aphid-like insects, 2mm in length, which may be seen on the shoot tips during summer.

**CAUSE** A sap-feeding insect, *Trioza alacris*, which has two or three generations between late spring and mid-autumn.

**CONTROL** Deal with light infestations by picking off affected leaves, or spray with dimethoate, pirimiphos-methyl or malathion when signs of recent damage are seen. If sprayed leaves are to be used in cooking, follow the insecticide manufacturer's instructions on harvest intervals.

## BEAN ANTHRACNOSE

**SYMPTOMS** Sunken, brown longitudinal marks develop on the stems of affected dwarf or runner beans. Leaf veins may develop a red coloration and may subsequently turn brown, wither and die off. A reddish-brown spotting develops on infected pods and, during wet weather, a pink-coloured slimy fungal growth may develop.

**CAUSE** The fungus *Colletotrichum lindemuthianum*, which is usually seed-borne. The fungal spores spread, by rain or irrigation splash, from young infected seedlings onto adjacent plants.

**CONTROL** Remove infected plants promptly, and do not save their seed. Chemical

control is rarely justified. Grow cultivars showing some resistance, such as 'Aramis' or 'Rido Kenyan'.

## BEAN CHOCOLATE SPOT

**SYMPTOMS** Chocolate-brown spots develop on the upper surfaces of broad bean leaves (*see* **p.12**). Chocolate-brown streaks may also appear on the stems, pods and flowers, and the seed coat may be similarly affected. Adjacent spots or streaks may coalesce to form large blotches. Severely infected plants may die off, and even a mild infection reduces yields.

**CAUSE** The fungus *Botrytis fabae*, which is encouraged by damp or humid air. The fungus may overwinter on the remains of infected plants and may also be seed-borne.

**CONTROL** Grow beans on a properly drained site, avoiding the use of high nitrogen fertilizers (which increase the production of soft, infection-prone growth). Apply sulphate of potash to harden growth slightly. Space plants wider apart than normal (more than 23cm) to improve air circulation, and ensure good weed control between the rows – this should help to minimize humidity and damp air around the plants. Spray affected plants with carbendazim.

## BEAN HALO BLIGHT

**SYMPTOMS** Small, angular spots appear on the leaves of dwarf French or runner beans, at first appearing water-soaked, and then darkening, each one surrounded by a bright yellow "halo". Leaves then develop yellowing between

the veins (interveinal chlorosis); the entire affected leaf may yellow and die. Growth is affected and yields are reduced. If pods or stems become infected, they develop grey, apparently water-soaked patches.

**CAUSE** The bacterium *Pseudomonas phaseolica*, which attacks dwarf French and runner beans. Water splash (from rain or overhead watering) spreads the bacteria. Initial infection usually starts from infected seed.

**CONTROL** Pick off affected leaves promptly, and avoid overhead watering. Remove and burn all affected plants at the end of the season, discarding their seed. Consider growing the dwarf French bean cultivars 'Forum' or 'Red Rum', which both show some resistance to this infection.

## BEAN RUST

**SYMPTOMS** Numerous dark brown pustules develop on the underside of the leaves of French and runner beans. Occasionally stems and pods are infected. Rarely, the white cluster cup form of the fungus appears beneath the leaves. Severely infected leaves wither and in extreme cases cropping is reduced.

**CAUSE** The fungus *Uromyces appendiculatus*, which is most common in warm, damp summers.

**CONTROL** Remove affected leaves promptly.

## BEAN SEED BEETLES

**SYMPTOMS** Holes (from which adult beetles have emerged) are seen in dry beans and peas kept for seed (*see* **p.51**). Several plants are vulnerable, but in gardens it

is mainly broad beans that are affected. While the beetle grub is present within the seed, a pale circular patch on the seed coat indicates where it has eaten a cavity in the seed's cotyledon.

**CAUSE** Several species of seed beetle, up to 3mm long, can attack pea and bean seeds. On broad beans it is usually *Bruchus rufimanus*. This lays eggs in the bean pods as the seeds are forming. The small grubs go unnoticed in bean seeds that are cooked for eating, but their presence becomes obvious in beans kept for next year's seed when the adult beetles emerge.

ACTUAL SIZE

**ADULT BEAN SEED BEETLE**

**CONTROL** There is no effective treatment available to amateur gardeners. Feeding damage by the larvae is to the seed's food reserve (known as the cotyledons) rather than the embryo, so infested seeds can still germinate.

## BEAN SEED FLY

**SYMPTOMS** Germinating seeds of runner and French beans are eaten by white, legless fly larvae. Seedlings are sometimes killed before the shoots emerge from the soil, but more usually ragged leaves and stems damaged by the soil-dwelling larvae appear above soil level (*see* **p.51**). Such plants will usually survive although they may make slow growth initially, especially if the growing points have been killed, causing sideshoots to develop.

**CAUSE** Maggots of a fly, *Delia platura*, which is similar in appearance to a house fly.
**CONTROL** Slow-germinating seeds are most at risk, so avoid sowing under adverse conditions such as when the soil is cold or very wet. The female flies are attracted to soil that contains fresh organic matter, so apply this in autumn rather than spring. To avoid bean seed fly problems, beans can be germinated in pots or seed trays, and transplanted after the first true leaves have expanded. For direct sowing, treat the seed hole with pirimiphos-methyl or lindane dust if this pest has caused losses in previous years.

## BEECH BARK DISEASE

**SYMPTOMS** Usually seen in association with beech bark scale (*see below*). Small areas of bark (a few centimetres in diameter) die and may appear as slight depressions (*see p.35*). They may then exude a black, tar-like substance formed from the tree's sap. As the infection spreads, larger areas of bark are killed and the leaves become yellowed. Leaves produced subsequent to the infection are often unusually small and yellow. Tiny red fungal bodies the size of pinheads develop on affected bark. Trees decline in vigour and may be killed, often snapping off part way up the trunk.
**CAUSE** The fungus *Nectria coccinea*, which infects beech trees via wounds in the bark created by beech bark scale. Young beeches are only rarely infected.
**CONTROL** Attempt to control beech scale to limit infection and spread of the fungus.

Fell severely infected trees, to minimise spread and because the weakened state of their wood may render them dangerous.

## BEECH BARK SCALE

**SYMPTOMS** A white powdery substance appears in crevices in the bark of beech trees, especially on the trunk and larger branches (*see p.35*). Small orange insects up to 1mm long live underneath this waxy exudation.
**CAUSE** A sap-feeding scale insect, *Cryptococcus fagisuga*, also called the felted beech coccus. It has little direct effect on the tree, but it is believed that heavy infestations make trees more susceptible to a form of canker known as beech bark disease (*see above*).
**CONTROL** Spray reachable parts of the tree in mid- to late winter with tar oil wash.

## BEECH GALL MIDGE

**SYMPTOMS** In late summer, yellowish-brown, cylindrical structures, or galls, which may be hairy or smooth, develop on the upper surfaces of beech leaves. These galls, each containing a single fly larva, readily break off and drop to the ground in the autumn.
**CAUSE** Small flies – gall midges – lay eggs in leaves in mid-summer. The hairy galls are caused by *Hartigiola annulipes* and in some years are abundant; the larger smooth-surfaced galls are caused by *Mikiola fagi*, which is uncommon in Britain but can be frequent elsewhere in Europe.
**CONTROL** Apart from creating galls, these midges have no effect on the health of beech trees, so their presence can be tolerated.

## BEECH LEAF-MINING WEEVIL

**SYMPTOMS** In spring, beech tree foliage develops many small, round holes eaten out by the adult weevils. These are black, 2mm long, and jump like flea beetles when disturbed. The larval stage feeds as a leaf miner, making a brown linear mine from the centre of the leaf, where the egg was laid, to the leaf tip. A large, dried-up brown area develops at the leaf tip where most feeding occurs (*see p.24*).
**CAUSE** A weevil, *Rhynchaenus fagi*, which in some years is so abundant that nearly every leaf is affected by larvae. This can be unsightly but the tree is able to withstand attack.

ACTUAL SIZE ●—

**BEECH LEAF-MINING WEEVIL**

**CONTROL** Nothing can be done on mature trees because of the problem of spraying large specimens. Clipping beech hedges will remove many of the damaged leaves and new growth produced after late spring will not be attacked. Chemical treatment is generally not worthwhile.

## BEECH WOOLLY APHID

**SYMPTOMS** Pale yellow insects covered with a fluffy, white waxy material (*see p.32*) form dense colonies on the shoot tips and undersides of beech leaves during late spring to mid-summer, after which their numbers decline. The foliage becomes sticky with the aphids' sugary excrement, called honeydew,

and this encourages the growth of sooty moulds. Winged aphids are present in mid-summer and fly up when beech hedges are clipped. Because they are white, they are sometimes mistaken for whiteflies.
**CAUSE** A sap-feeding aphid, *Phyllaphis fagi*, a specific pest of beech trees.
**CONTROL** Tall trees cannot be sprayed effectively and so the aphid must be tolerated. On hedges look for signs of aphid activity in late spring and spray with dimethoate, heptenophos with permethrin, pirimiphos-methyl or pirimicarb before heavy infestations develop.

## BEEFSTEAK FUNGUS

*See* FISTULINA HEPATICA *p.127*.

## BEES

*See* BURROWING BEES *p.112*; LEAF-CUTTING BEES *p.142*; NECTAR ROBBING *p.151*.

## BEET LEAF MINER

**SYMPTOMS** Leaves of beetroot and spinach beet develop large brown areas (*see p.17*) where internal tissues have been eaten by white maggots.
**CAUSE** Larvae of the leaf-mining fly *Pegomya hyoscyami*, which has two generations during summer.
**CONTROL** With light infestations, remove affected leaves or crush the larvae inside the leaf mines; otherwise spray with pirimiphos-methyl or malathion, particularly if young plants are attacked.

## BEETLES

In terms of named species, beetles are the most numerous insects in the world. They are characterized by their modified fore wings, hard opaque structures (the elytra) that form a covering for the abdomen. Some beetles, such as ladybirds (*see p.141*) and ground beetles (*see p.134*), are beneficial as predators of garden pests. There are also a number of beetles which attack various parts of plants.

**Root damage** – *see* CHAFER GRUBS *p.116*; TURNIP GALL WEEVIL *p.184*; VINE WEEVIL *p.186*; WIREWORMS *p.191*.

**Stem damage** – *see* BARK BEETLES *p.104*; SHOT-HOLE BORERS *p.175*.

**Foliage damage** – *see* ASPARAGUS BEETLE *p.103*; BEECH LEAF-MINING WEEVIL *p.106*; COLORADO BEETLE *p.118*; FIGWORT WEEVIL *p.126*; FLEA BEETLES *p.127*; LEAF WEEVILS *p.143*; LILY BEETLE *p.145*; MINT BEETLE *p.148*; PEA AND BEAN WEEVIL *p.155*; VIBURNUM BEETLE *p.185*; VINE WEEVIL *p.186*; WATER LILY BEETLE *p.188*; WILLOW LEAF BEETLES *p.190*.

**Flower damage** – *see* APPLE BLOSSOM WEEVIL *p.100*; POLLEN BEETLES *p.162*.

**Seed and fruit damage** – *see* BEAN SEED BEETLES *p.105*; NUT WEEVIL *p.151*; RASPBERRY BEETLE *p.167*; STRAWBERRY SEED BEETLE *p.179*.

## BEGONIA POWDERY MILDEW

**SYMPTOMS** Greyish-white, powdery fungal growth appears in distinct spots or patches on the upper leaf surface (*see p.23*). Infected leaves either become dry and brown or, more commonly with the fleshy-leaved types, yellowed and soggy. Stems and flower-heads may also be affected. The disease is more common on indoor begonias; plants grown in the open are rarely affected.

**CAUSE** The fungus *Microsphaera begoniae*, which is encouraged by dry conditions around the plant's roots, and by moist air around the foliage.

**CONTROL** Pick off affected leaves promptly. Keep plants adequately moist, and improve air circulation to reduce humidity. Spray with a suitable fungicide such as carbendazim, penconazole, bupirimate with triforine or sulphur. Unless the disease is noticed early, control is very difficult.

## BIG BUD MITES

**SYMPTOMS** Buds become abnormally enlarged and rounded, and fail to develop into shoots and leaves (*see p.41*). Each big bud will contain many hundreds of microscopic, elongate white mites. Dried-up big buds can be found at any time of year, but they become swollen over the winter months.

**CAUSE** Several species of gall mite cause this problem. The mites most frequently found on garden plants are *Cecidophyopsis ribis* on blackcurrant, *C. psilaspis* on yew and *Phytocoptella avellanae* on hazel.

**CONTROL** Big bud mites on yew and hazel will kill some buds, but generally not enough to affect the plants' growth, so they can be tolerated. Blackcurrant big bud mite is more serious, since apart from causing a loss of vigour, it can spread a virus-like disease called currant reversion (*see p.121*). The chemicals available to amateur gardeners will not control big bud mites; the fungicide carbendazim gives blackcurrants some protection against attack if the bushes are sprayed as the first flowers open, with

two further treatments at two-week intervals. Removing big buds from lightly infested bushes in late winter will limit infestations. Scrap badly affected bushes after fruiting and replace in the autumn.

## BIRCH WITCHES' BROOMS

**SYMPTOMS** Extensively branched twigs develop, clustered closely together to form "witches' brooms" scattered throughout the crown of the tree. The leaves on the brooms turn yellow and bear a layer of fungal spores on their lower surface. During the winter, the growths may, at first, be mistaken for birds' nests and can spoil the outline of the tree or hedge, but during the summer they appear as dense, leafy areas and so are less obtrusive. The growths are particularly susceptible to frost and so may be killed during the winter.

**CAUSE** The fungus *Taphrina betulina*, which is spread by water splash and on air currents. *Betula pendula* and *B. pubescens* are most frequently attacked. Similar witches' brooms may develop on birch as a result of an infestation of the mite *Eriophyes rudis*.

**CONTROL** Prune out the growths, if desired. Spread is very slow, but removal helps limit the problem. No chemical control is available.

## BIRD DAMAGE ON FRUITS

**SYMPTOMS** Apples, pears and plums show bird pecks on the ripening fruit (*see p.47*). Wasps often extend the damage. Smaller fruits such as strawberries, cherries and currants are eaten whole.

**CAUSE** Various birds, including blackbirds, bullfinches and starlings.

**CONTROL** Small trees and bushes can be netted or grown in a fruit cage. On large trees, bags made from muslin or old nylon tights drawn over the better fruit trusses will protect them from birds.

## BITTER PIT

*See* APPLE BITTER PIT *p.100*.

## BITTER ROT

*See* GLEOSPORIUM ROT *p.133*.

## BLACK BEAN APHID

**SYMPTOMS** Dense colonies of black insects, up to 2mm long, mass on the shoot tips and undersides of leaves (*see p.32*) of broad beans and other plants, including globe artichokes, nasturtiums, dahlias and poppies, during the summer. Affected plants are weakened and bean pods fail to develop. The same pest may occur on viburnum and mock orange (*Philadelphus*) during the spring, causing leaf curling at the shoot tips.

**CAUSE** A sap-feeding aphid, *Aphis fabae*.

ACTUAL SIZE

**BLACK BEAN APHID**

**CONTROL** Inspect susceptible plants at regular intervals and spray with heptenophos with permethrin, dimethoate, pirimicarb, permethrin, bifenthrin, pirimiphos-methyl or malathion. Pirimicarb is selective for aphids, leaving other pests and most beneficial insects unharmed.

The organic insecticides pyrethrum, derris and insecticidal soaps are also effective if applied thoroughly before heavy infestations have developed.

## BLACK LEG OF CUTTINGS

**SYMPTOMS** Before or shortly after roots start to form, the base of an affected cutting darkens and atrophies; the upper parts start to discolour and then die off.
**CAUSE** Various soil- and water-borne fungi, in particular those responsible for causing damping off (*see p.122*) in seedlings. They are usually introduced on dirty tools, pots or trays, in unsterilized compost or by the use of non-mains water.
**CONTROL** Ensure good hygiene when taking cuttings. Use only clean or sterilized materials and tools, and mains water. As a preventive, treat new cuttings with a rooting powder containing a fungicide.

## BLACKBERRY CANE SPOT

**SYMPTOMS** Silvery-grey, lens-shaped spots with purple margins appear on the canes. Pinprick-sized, black pycnidia, or fungal fruiting bodies, develop in the centre of each lesion. In severe cases canes may die back.
**CAUSE** The fungus *Elsinoe veneta*.
**CONTROL** Prune out severely infected canes.

## BLACKCURRANT GALL MIDGE

**SYMPTOMS** Leaves of blackcurrants remain small and crumpled (*see p.26*), due to whitish maggots up to 2mm long feeding on leaves

as they develop at shoot tips. Affected leaves dry up and die. Three generations occur during the summer, with symptoms first appearing in late spring at flowering time. Shoot tips may be killed, causing side-shoots to grow.
**CAUSE** The leaf distortion is induced by chemicals secreted by the larvae of a tiny fly, *Dasineura tetensi*. While the maggots pupate, the plant may produce a couple of normal leaves before the next generation affects more new leaves.
**CONTROL** The first generation in late spring is potentially the most harmful. Sprays of dimethoate, fenitrothion or pirimiphos-methyl applied shortly before the first flowers open will give some protection (and also control aphids and capsid bugs). Some blackcurrant cultivars, such as 'Ben Connan' and 'Ben Sarek', are resistant.

## BLACKFLY

Blackfly is a common name given to several species of black aphids.
• *See also* APHIDS *p.99;* BLACK BEAN APHID *p.107;* CHERRY BLACKFLY *p.116.*

## BLEEDING CANKER

**SYMPTOMS** Yellow or brown substances ooze from patches of dead or dying bark on the trunks or large limbs of trees. The ooze may dry on the bark, leaving crusty deposits. Affected bark often cracks, allowing other decay organisms to enter. Stems completely ringed by the lesions will die back. Horse chestnuts, limes, silver birch and apple trees are particularly prone to attack.
**CAUSE** Various species of the fungus *Phytophthora* (*see p.159*).

**CONTROL** There is no cure. An inspection should be made by an experienced arboriculturist to determine the tree's health and safety.

## BLINDNESS OF BULBS

**SYMPTOMS** Bulbs produce healthy, normal foliage, but fail to flower. Either no buds are formed at all, or buds form, but are dry and virtually empty of petals (*see p.41*).
**CAUSE** Occasionally due to large bulb fly infestation or basal rot (on *Narcissus*), but more often due to cultural problems. An inadequate supply of moisture and nutrients to the bulb causes it to decline in size and fail to produce buds, or a full complement of petals. Excessively crowded clumps (which are competing for moisture and nutrients), naturalized bulbs which have not been fed and bulbs growing in dry situations or suffering from drought are also likely to be affected. Narcissi are frequently affected, especially multi-headed or double forms.
**CONTROL** Divide and replant congested clumps, preferably into a fresh, well-fertilized site. Ensure an adequate supply of water during dry weather. Feed established clumps in the spring with a complete fertilizer. Treat with a foliar feed whilst in leaf. Do not tie up or cut off bulb foliage for at least six weeks after flowering is complete.

## BLIND SHOOTS

**SYMPTOMS** Occasionally, flowering shoots of roses and other flowering shrubs develop normally, but fail to develop a terminal flower bud (*see p.40*). The plant is

perfectly healthy, and other shoots flower normally.
**CAUSE** Unknown; possibly a cultural disorder or adverse weather conditions.
**CONTROL** Prune back the blind shoot to a strong bud or pair of buds. Normal flowering shoots should grow from these buds.

## BLOSSOM END ROT

**SYMPTOMS** A sunken patch develops at the blossom end (base) of developing fruits. The base of the fruit becomes tough and leathery and darkens to brown or black as cells collapse. Not all fruits on a truss will necessarily be affected, and some trusses on a plant may remain perfectly healthy. Tomatoes, especially those growing in containers, and occasionally sweet peppers are affected.
**CAUSE** Dry conditions around the plant's roots, which prevent the plant from taking up sufficient calcium. The low calcium content of the fruits causes cells to collapse and discolour. This disorder is also encouraged by a very acidic growing medium.
**CONTROL** Ensure an adequate, regular supply of moisture at all times, especially on fast-growing and heavy-cropping plants. If blossom end rot does develop, pick off affected fruits and improve the watering routine. Grow small-fruiting tomatoes, as these are rarely affected. Grow plants in open ground or in larger containers where the soil or compost is less prone to drying out.

## BLOSSOM WILT

**SYMPTOMS** Blossom withers and dies shortly after appearing. Dead trusses of blossom remain on the tree.

Infection may then spread to adjacent leaves which wilt, turn brown and die, remaining attached to the branch. Pinprick-sized, raised, buff-coloured fungal pustules develop on infected areas, and localized dieback may occur (see **p.41**). *Malus*, *Pyrus* and *Prunus*, including cherry, plum, apple, apricot and peach, may be attacked. Occasionally *Amelanchier* is affected.

**CAUSE** The fungus *Sclerotinia laxa* (on all except apple) and *S. laxa* f. *laxa* (on apple) which may overwinter as cankers on infected stems, or pustules on infected flowers or foliage. The spores are wind-borne and spread is most rapid during damp weather.

**CONTROL** Spray with copper oxychloride just before flowering. Prune out infected flower trusses, preferably before the infection spreads to the foliage or into the spur.

## BLUE MOULD ON BULBS

**SYMPTOMS** Bulbs in store are most likely to develop symptoms, but they may also occur while bulbs are in the ground. Reddish-brown lesions develop on the side of the affected bulb or corm, and pink fungal bodies may subsequently develop beneath the skin. A pale, bluish-green fungal growth appears on top of the lesion as it increases in size (see **p.54**). On bulbous irises this may be accompanied by rapid softening and rotting.

**CAUSE** Various species of the *Penicillium* fungi. These fungi are all encouraged by over-warm, moist storage conditions and poor air circulation. The fungus is believed to enter the bulbs whilst they are in the ground, but the symptoms will only develop if storage condions are unsuitable.

**CONTROL** Avoid purchasing infected bulbs or those which are damaged, as these too are very susceptible. Store bulbs for as short a time as possible, and in a cool but frost-free, well-ventilated spot.

## BLUEBELL RUST

**SYMPTOMS** Dark brown fungal pustules develop in elliptical, often concentric, formations on the leaves. Flower stems may also be attacked. Leaves yellow and wither prematurely and if infections are recurrent and severe, flowering may be somewhat reduced. Scillas and grape hyacinths (*Muscari*) may also be affected.

**CAUSE** The fungus *Uromyces muscari*. Infection is encouraged by damp growing conditions.

**CONTROL** Rarely necessary, but remove infected foliage to limit the spread of the infection.

## BLUSHING BRACKET

See DAEDALEOPSIS CONFRAGOSA *p.122*.

## BOLTING

**SYMPTOMS** Affected vegetable plants produce flowers and seeds prematurely, before the crop has developed fully or been harvested (see **p.44**).

**CAUSE** Various factors may be involved; these may include exposure to low temperatures at a certain growth stage (often induced by a cold, late onset of spring) which then induces flowering, or by excessively dry soil conditions.

**CONTROL** Avoid growing early varieties of susceptible crops, as these are usually more prone to the problem. Check seed catalogues for cultivars showing resistance to bolting.

## BORON DEFICIENCY

**SYMPTOMS** Symptoms vary with the crop affected, but root vegetables such as beetroot, turnip, carrot, radish and swede are most susceptible. Cauliflower, cabbage, celery, pears and strawberries occasionally develop symptoms.

**Beetroot** – Roughened, cankered areas develop on the sides of the roots. In severe cases, the subsequent rotting spreads to cause a heart rot, which shows as brown ring development in the inner root tissues and the crown. Foliage is unusually small and may die.

**Cabbage** – Leaf distortion and hollow areas in the stems.

**Carrot** – Splitting to expose the greyish and sometimes split central core, as well as yellow and pink discoloration of the older foliage and splitting of the leaf stems.

**Cauliflower** – This shows curd browning, poor curd development and roughening of the main stem, leaf stalks and midribs.

**Celery** – Transverse cracks in the outer surface of the leaf stalk, after which the inner tissues turn reddish brown. Leaves may occasionally show distortion.

**Pears** – Distorted fruits with brown, hardened flecks in the flesh, often combined with slight shoot dieback.

**Radishes** – Splitting and dull skin, often with woody flesh.

**Strawberries** – Small leaves and general stunting, with puckering and yellowing at the leaf tips. The berries are small, pale in colour and with a characteristic "waist" close to the calyx.

**Swede and turnip** – These show "brown heart", which can be seen when the roots are cut across to reveal grey or brown discoloured areas, often in concentric rings, in the lower parts of the root.

**CAUSE** This disorder often occurs if the soil has been limed excessively or if it is allowed to become very dry. Boron is readily leached out of the soil, so heavy rainfall may encourage this deficiency on light soils.

**CONTROL** For vegetables, apply about 35g per 20m² of borax before sowing or planting (mix borax with horticultural sand to make even application possible). For pear trees, spray at petal fall with 70g borax in 22 litres of water (2½oz in 5 gallons), together with a wetting agent.

## BOTRYTIS

See GREY MOULD *p.134*.

## BOX SUCKER

**SYMPTOMS** The shoots of box are stunted, with closely-spaced cupped leaves that resemble cabbages (see **p.26**). Droplets of liquid with a white, waxy coating can be tipped out of the shoot tips during late spring. This is excrement produced by the sucker's nymphal stages.

**CAUSE** The immature nymphs of a sap-feeding insect, *Psylla buxi*, which attack the new growth in spring. The nymphs are flattened, pale green insects up to 2mm long. Winged adults are present in mid-summer, but they do not affect the plant's growth.

**CONTROL** If the box plant has grown to the desired height and is being clipped to keep it in shape, something that restricts shoot extension is not really a problem. Protect young plants that need to grow larger by spraying with dimethoate, heptenophos with permethrin or pirimiphos-methyl in mid-spring.

## BRACKET FUNGUS

**SYMPTOMS** Fungal outgrowths develop on large limbs or the main trunk of affected trees (see **pp.37, 38**). They may appear throughout the tree, but are most common on the trunk, particularly towards the base. Some may be found growing from the buttress roots, or even attached to roots at some distance from the trunk. Some are annual, but most are perennial and quite woody in texture. Most have a tough leathery or woody upper surface, with spores being produced from pores on the undersurface. Most bracket fungi cause damage to trees – usually, extensive deterioration of the structural woody tissues. Infected trees often become dangerous because they are weakened structurally, and may suffer wind snap or wind throw.
**CAUSE** Various fungi, the most common being *Fistulina hepatica*, species of *Ganoderma, Innonotus hispidus, Laetiporus sulphureus, Meripilus giganteus, Phaeolus schweinitzi, Phellinus pini, Piptoporus betulinus* and *Sparrisis crispa*.

**BRACKET FUNGUS**

**CONTROL** Precise identification is essential, together with examination of the tree and consideration of its environment. As infected trees may be hazardous, employ an arboriculturist to examine the tree. Removal of the fungal brackets may limit spread, but will not cure the problem as the deterioration continues within.

## BRASSICA BACTERIAL SOFT ROT

**SYMPTOMS** Patches of foliage or areas within the heart of the plant, often towards the stem base, discolour. Rapid disintegration of the tissues follows, usually accompanied by a foul smell.
**CAUSE** Various bacteria may be involved, but *Erwinia carotovora* is the most common. The bacteria usually gain entry through injured areas of the plant, when it has been attacked by pests such as slugs, snails and caterpillars.
**CONTROL** Avoid initial injury, where possible. It is rarely possible to save an infected plant because the soft rot is generally only noticed when in a fairly advanced state, but prompt removal of affected parts when the first symptoms occur may prevent its development.

## BRASSICA DARK LEAF SPOT

**SYMPTOMS** Numerous round, brown to grey spots develop on leaves (see **p.13**), each with concentric rings of pinprick-sized fungal bodies. Infected areas of the leaf are killed and may drop away, leaving holes and a rather ragged-looking leaf.
**CAUSE** The fungus *Alternaria brassicola*, which is encouraged by wet weather

and poor air circulation, and thrives on plants which are grown rather soft.
**CONTROL** Remove infected leaves promptly. Burn debris at the end of the season. Improve air circulation by spacing plants well and weeding between the rows. Avoid overhead watering and excessive use of high-nitrogen fertilizers.

## BRASSICA DOWNY MILDEW

**SYMPTOMS** Yellow blotches appear on upper leaf surfaces, each corresponding with a patch of greyish-white fungal growth on the lower leaf surface. Seedlings and young plants are more prone than mature plants, but all stages can be attacked. Severe infection may cause extensive leaf yellowing and the death of seedlings.
**CAUSE** The fungus *Peronospora parasitica*, which is encouraged by mild, damp weather and humid conditions. It has a wide host range, attacking many cruciferous plants, including some ornamentals and weeds. The spores can be spread by wind to nearby plants, and infection of young plants may also occur through the roots from infection within the soil.
**CONTROL** Improve air circulation by wide plant spacing and regular weeding. Remove affected leaves as seen. Spray with a fungicide based on mancozeb.

## BRASSICA LIGHT LEAF SPOT

**SYMPTOMS** Discoloured circles of leaf tissue develop, becoming covered in fluffy, off-white fungal growth. In extreme cases, spots merge, causing extensive damage.

**CAUSE** The fungus *Cylindrosporium concentricum*.
**CONTROL** Remove affected leaves promptly, and dispose of all debris at the end of the season. Rotate crops, or at least allow a two-year break between brassica crops.

## BRASSICA RING SPOT

**SYMPTOMS** Purple-brown spots with yellow edges develop on leaves, the older, outer leaves being most susceptible. Well-defined concentric ringing of the tissue is usually visible, and the spot may be covered in numerous, black, raised, pin-prick-sized, fungal fruiting bodies. This disease may occur on all brassicas but is most common and damaging on Brussels sprouts, where it may also attack the buttons.
**CAUSE** The fungus *Mycosphaerella brassicicola*, which is encouraged by damp weather conditions.
**CONTROL** Remove severely affected leaves and clear up all plant debris at the end of the season.

## BRASSICA WIRE STEM

**SYMPTOMS** Leaves on brassica seedlings, especially those of cauliflower and broccoli, become yellowed and narrowed, or may die back.
**CAUSE** A deficiency of molybdenum, which is most common in acid soils.
**CONTROL** Lime the soil to reduce acidity, and water affected plants or seed rows with sodium molybdate.

## BROOM GALL MITE

**SYMPTOMS** During early summer, buds on broom (*Cytisus*) fail to develop into normal shoots and leaves. They proliferate into green cauliflower-like growths which are covered in silvery hairs (see **p.39**). Later in the summer, these galled buds dry up and become greyish-brown. They remain on stems over winter, and will usually persist on the plant for several years. Infestations increase in successive years until the shrub's flowering is significantly reduced.

**CAUSE** A microscopic gall mite, *Eriophyes genistae*, which lives and feeds within the affected buds.

**CONTROL** None of the insecticides available to amateur gardeners is effective against this mite. Remove badly infested plants and replant with clean stock in the autumn. Limit light infestations by picking off galled buds.

## BROWN CORE OF PRIMULAS

**SYMPTOMS** Primula and polyanthus plants show signs of poor growth, and leaves may be yellowed. Plants wilt readily, especially during warm weather, but even when there is plenty of soil moisture available. Plants wither and die, usually within a single season. The root system deteriorates, leaving a reduced number of roots and little fine root growth. If roots are cut lengthwise, a distinct reddish-brown core of discoloration is seen running down them.

**CAUSE** The fungus *Phytophthora primulae*, which is soil-borne and builds up in the soil when susceptible plants are grown on the same site for a number of years in succession.

**CONTROL** Remove infected plants promptly. There is no soil treatment available, but growing fresh plants on a new site is usually successful. Take care not to spread soil from around infected plants to other parts of the garden.

## BROWN ROT

**SYMPTOMS** A soft, brown area develops on the fruit skin of tree fruits (see **p.47**) and also in the flesh beneath when it starts to deteriorate. The whole fruit is ultimately affected. Raised creamy-white pustules appear on the rotted areas, in roughly concentric rings (see **p.49**). Infected fruits either fall off the tree, or may become "mummified" and dry and remain attached to the branches. Most tree fruits can be affected, especially apple, pear, plum, peach and nectarine.

**CAUSE** The fungi *Sclerotinia fructigena* (on apples) and *S. laxa* (other host plants), which gain entry through injured skin. Bird pecks, codling moth exit holes, frost crack, cracking due to irregular growth and fungal scab infections are all common causes of the initial injury. The spores are readily spread by insects or birds, by rain splash and by fruits coming into contact with a source of infection.

**CONTROL** Avoid, or minimize, possible causes of injury to fruits by taking appropriate control or preventive measures. Remove infected fruits promptly, as well as fallen ones, and dispose of them well away from the garden, or burn them. Prune out all mummified fruits, together with a short section of the spur to which they are attached.

## BROWN SCALE

**SYMPTOMS** Brown, shell-like, convex objects, up to 5mm in length, are attached to the stems of many ornamental shrubs and fruit bushes (see **p.36**), especially those growing against walls or in other sheltered places. The honeydew excreted by the insects makes the foliage sticky and allows the growth of sooty moulds.

**CAUSE** A sap-feeding insect, *Parthenolecanium corni*, which usually has one generation a year. Its eggs are laid under the scale.

**CONTROL** The newly-hatched scale nymphs are the most vulnerable stage in the insect's life cycle, feeding on the stems and undersides of leaves in early to mid-summer. Spray in mid-summer with malathion. In winter, fruit trees and bushes can be sprayed with a tar oil wash against overwintering nymphs.

## BROWN-TAIL MOTH

**SYMPTOMS** Greyish-black, hairy caterpillars, up to 40mm in length, feed together in colonies under silk webbing on hawthorn, cherry, apple, blackthorn, blackberry, rose and other trees and shrubs. The caterpillars have red and white markings on their bodies and a pair of orange warts on the upper surface at the rear end. They overwinter as small larvae inside dense, white silk webs spun at shoot tips.

**CAUSE** Caterpillars of a moth, *Euproctis chrysorrhoea*, which during late spring and early summer cause severe defoliation. In Britain, the moth is mainly found on the south and east coasts of England, but sometimes occurs further inland.

**CONTROL** Prune out the overwintering silk nests after leaf fall. Wear gloves to avoid contact with the hairs, as some people develop an intensely irritating rash when the caterpillar hairs come into contact with their skin. From mid-spring to early summer spray infested plants with the bacterium *Bacillus thuringiensis*, or use an insecticide such as permethrin, bifenthrin or pirimiphos-methyl. Young caterpillars tend to be more susceptible than the older larvae.

## BRYOBIA MITES

**SYMPTOMS** Black-bodied mites, up to 1mm in length, with four pairs of pinkish legs, occur on the upper surface of leaves on ivies, primulas, polyanthus, gooseberry and other plants during late winter to early summer. Leaves develop a fine, pale mottling on the upper surface.

**CAUSE** One of several species of sap-feeding mites, *Bryobia* species.

**CONTROL** Spray thoroughly with bifenthrin, malathion or pirimiphos-methyl when bryobia mites are seen. Heavy infestations may need three applications at two-week intervals.

## BUD DROP

**SYMPTOMS** Buds form, usually to full size, but drop while still apparently healthy. The plant is otherwise healthy.

**CAUSE** A period of dry weather while the flower buds are forming. On very susceptible plants, such as camellia and rhododendron, just a few consecutive days of drought at the end of summer when the buds are

forming are enough to do damage. Plants growing in containers are particularly vulnerable because they are more prone to dry conditions around the roots. Feeding too late in the growing season (after early summer) may also induce bud drop the following spring.
**CONTROL** Water thoroughly and regularly and never let the soil or compost dry out completely. Apply a deep mulch around the root area. Feed no later than early to mid-summer and avoid excessive feeding.

## BUD, DRY

**SYMPTOMS** Buds form, but may be undersized and are usually dry and brown; the plant is otherwise healthy.
**CAUSE** As with bud drop (*see above*), a period of dry weather while the flower buds are forming. On very susceptible plants, such as rhododendron, just a few consecutive days of drought at the end of summer are enough. Plants growing in containers are particularly vulnerable because they are more prone to dry conditions around the roots.
**CONTROL** Water thoroughly and regularly and do not let the soil dry out completely. Apply a deep mulch around the root area.

## BUFF-TIP MOTH

**SYMPTOMS** Colonies of black and yellow chequered caterpillars, up to 50mm in length, feed together on leaves of deciduous trees and shrubs, including lime, oak, hornbeam, birch, hazel, rose and cherry, from mid- to late summer. Unlike other gregarious moths, buff-tip larvae produce no silk webbing.

**CAUSE** Caterpillars of a moth, *Phalera bucephala*, which defoliate branches on their host plants.
**CONTROL** Buff-tip moth is only likely to harm young trees or shrubs. If necessary, prune out infested shoots or spray with the bacterium *Bacillus thuringiensis* or an insecticide such as permethrin, bifenthrin or pirimiphos-methyl.

## BULB SCALE MITE

**SYMPTOMS** Daffodil and amaryllis bulbs grown indoors for early flowering are much more likely to show symptoms of this pest than similar bulbs planted in gardens. The whitish-brown mites are less than 1mm long and barely visible to the naked eye. Their feeding causes stunted and distorted growth. Flower stems and leaves have saw-tooth scarring along the edges and show a sickle-shaped curvature.
**CAUSE** Sap-feeding mites, *Steneotarsonemus laticeps*, which live in the neck area of dormant bulbs. These should not be confused with *Rhizoglyphus* species mites, which are larger, broader, pearly-white mites often found feeding on rotting bulbs. These are not pests, as they feed on damage initiated by some other cause.
**CONTROL** None of the chemicals available to amateur gardeners will control bulb scale mite. Infested bulbs should be thrown away.

## BULLFINCHES

Bullfinches (*Pyrrhula pyrrhula*) feed on wildflower seeds for most of the year, but turn to flower buds when the seed supply runs

out during the winter. During late autumn to spring they will peck out and eat the flower buds of plum (*see p.46*), pear, cherry, gooseberry, almond, apple and forsythia. The brown outer bud scales are discarded and are easily seen when the ground below the plant is covered by snow or frost. Buds at the shoot tips often survive and so, in the spring, shoots which are mostly devoid of flowers will have clusters of blossom at the tips.
**CONTROL** Only netting will give complete protection; bird-repellent sprays are unreliable and may need frequent applications.

## BURROWING BEES

**SYMPTOMS** Small, conical heaps of fine soil appear in lawns or on bare soil during late spring and summer (*see p.58*). They are distinguished from ant nests by the hole at the top of the heap where the bee enters its nest. Each nest consists of a vertical shaft in the soil; this shaft has several side-chambers which the bee provisions with nectar and pollen as food for its grubs.
**CAUSE** Many different species of solitary bees nest in this way in gardens. A common species is a ginger-brown bee, *Andrena fulva*, which is active in late spring. Although all burrowing bees are solitary, with each female constructing her own nest, the nests may be clustered together and the same piece of ground may be used by the bees in successive years.
**CONTROL** Control measures are not necessary. Burrowing bees cause no real damage to lawns, and they are of benefit as pollinating insects

for fruit trees and bushes. The females have stings, but lack aggressive instincts and do not attack humans.

## CABBAGE MOTH

**SYMPTOMS** All types of brassica are damaged during the summer by yellowish-brown or green caterpillars, up to 42mm in length. They eat holes in the foliage (*see p.30*), especially the heart leaves of cabbages. Cabbage moth caterpillars also enter onion leaves, making holes and causing them to flop over.
**CAUSE** Caterpillars of a moth, *Mamestra brassicae*, which has one generation a year.
**CONTROL** Check brassicas regularly in summer and take control measures as soon as caterpillars are seen; they are difficult to deal with once they grow larger and go into the heart leaves. Use an insecticide containing pyrethrum, permethrin, bifenthrin, fenitrothion or pirimiphos-methyl, or the biological control *Bacillus thuringiensis*.
• *see also* CABBAGE WHITE BUTTERFLY (*facing page*); LEEK MOTH *p.144*.

## CABBAGE ROOT FLY

**SYMPTOMS** All brassicas and allied vegetables such as turnip, swede and radish are attacked. Some related ornamentals, including stocks, alyssum and wallflowers, are sometimes damaged. Affected plants grow slowly and tend to wilt on sunny days. Brassicas are particularly vulnerable as seedlings, especially shortly after transplanting, when they can be killed. White, legless maggots, up to 9mm in length, eat the finer roots until just a rotting stump is left. The maggots also bore into the swollen roots of radish, swede and turnip.

**CAUSE** Maggots of a fly, *Delia radicum*, which has three generations from mid-spring to early autumn, hatching from eggs laid in the soil.

**CONTROL** Established plants usually tolerate cabbage fly attacks: it is seedlings and transplants that need protection. Dust the seed rows or the base of transplants with lindane or pirimiphos-methyl. Infestations can be reduced without using insecticides by placing squares or discs made of carpet underlay, cardboard or roofing felt, about 10cm in diameter, on the soil around the stems of transplants. The female flies deposit their eggs on the discs, instead of in the soil; they then dry up before hatching.

## CABBAGE WHITE BUTTERFLY

**SYMPTOMS** Holes are eaten in the leaves of brassicas and some ornamental plants such as nasturtium. Caterpillars of the large cabbage white butterfly are yellow with black markings. They have hairy bodies, up to 40mm long, and feed mainly on the outer leaves of brassicas. Small cabbage white butterfly larvae are up to 25mm long and feed mainly in the hearts of cabbages and other brassicas. They are pale green with a velvety appearance, unlike the smooth and largely hairless caterpillars of cabbage moth (*see facing page*).

ACTUAL BODY LENGTH: 20MM

**LARGE CABBAGE WHITE**

**CAUSE** Large cabbage white, *Pieris brassicae*, and small cabbage white, *P. rapae*, which each have two generations a year, with caterpillars occurring from spring to early autumn.

**CONTROL** Check brassicas regularly from early summer to early autumn and take control measures as soon as caterpillars are seen. They are difficult to deal with once they have grown larger and have gone into the heart leaves. Use the bacterium *Bacillus thuringiensis*, or an insecticide such as pyrethrum, permethrin, bifenthrin, fenitrothion or pirimiphos-methyl.

## CABBAGE WHITEFLY

**SYMPTOMS** Brassicas, such as Brussels sprout, kale, cabbage, cauliflower and broccoli, have 2mm long, white-winged insects on the undersides of the leaves (*see p.32*). The adult whiteflies are active throughout the year and readily fly up when the plants are disturbed. The immature nymphs are flat, oval, scale-like insects that remain immobile on the lower leaf surface. Heavy infestations make the foliage sticky with the whiteflies' sugary excrement, and sooty moulds will grow on this substance.

**CAUSE** The adults and nymphs of a sap-feeding insect, *Aleyrodes proletella*. This insect has several generations during the summer, and overwinters as an adult whitefly.

**CONTROL** Light infestations can be tolerated, but if large numbers of whitefly are present and the foliage is becoming soiled with sooty moulds, spray with permethrin, bifenthrin, insecticidal soap or pirimiphos-methyl. Several treatments at weekly intervals may be necessary.

## CACTUS CORKY SCAB

**SYMPTOMS** Buff to tan corky-brown markings develop on the skin and may become sunken (*see p.22*). Central areas often become grey or dark brown. Adjacent areas of the plant usually retain their normal green colour. Cacti and many succulents can be affected, especially *Opuntia* and *Epiphyllum*.

**CAUSE** Unsuitable growing conditions, in particular excessively high humidity or light levels.

**CONTROL** Improve growing conditions by moving the plant to a slightly cooler position and improving ventilation to reduce humidity. Take care not to alter conditions too dramatically or rapidly or the plant may suffer even more.

## CALCIUM DEFICIENCY

**SYMPTOMS** These vary according to the plant and plant part affected. The most common symptoms are blossom end rot and bitter pit. Other symptoms may include: browning of the internal tissues of Brussels sprouts' buttons and the heads of cabbage; cavity spot of carrots; stunting and blackening of central leaves on celery; tip burn of lettuce, and the production of rolled leaves and spindly

shoots on potato. On other plants the most common symptom is poor growth of young shoots and leaves, often combined with some curling and shoot or leaf deterioration.

**CAUSE** A deficiency of calcium, either in the soil or compost, or because of an inadequate uptake of calcium due to a low moisture content of the soil. This disorder is most common on acidic soils or growing media.

**CONTROL** Ensure an adequate supply of moisture and encourage soil moisture retention by mulching, where appropriate. Where possible on acid soils lime the soil to raise the pH level.

• see also APPLE BITTER PIT *p.100*; BLOSSOM END ROT *p.108*; CARROT CAVITY SPOT *p.114*.

## CAMELLIA GALL

**SYMPTOMS** Large, pale green swellings, often forked or hand-like, develop in place of the leaves (*see p.26*). Each gall may measure several centimetres in width and up to about 20cm in length. The surface soon becomes covered in a layer of spores, giving the gall a creamy-white appearance. Small shrivelled pieces of leaf may protrude from the tips of the galls. There are rarely more than one or two galls per plant. The overall vigour and flowering of the camellia is not affected.

**CAUSE** The fungus *Exobasidium camelliae*, which seems to be encouraged by damp weather. The spores are probably spread by insects and by rain splash.

**CONTROL** Remove galls, preferably before the spore layer develops.

## CAMELLIA YELLOW MOTTLE

**SYMPTOMS** Irregular blotching and mottling of camellia leaves, with bright yellow or near-white markings on the dark green surface (*see p.22*). Occasionally whole branches may bear completely white or strongly chlorotic leaves. The overall vigour of the plant seems not to be affected.

**CAUSE** An organism known as camellia yellow mottle, which is virus-like in its structure and behaviour. This organism is systemic within the plant, so propagation should not be carried out using material from any part of an affected plant.

**CONTROL** Prune out all affected stems.

## CANKER

**SYMPTOMS** The precise symptoms of the various cankers (see references below) vary with both the host plant and the causal organism. The bark is usually raised and often roughened and may split to reveal the tissues beneath. In some cases ooze may develop around the cankered areas. As cankers enlarge, they often ring the infected stem and so growth above the canker deteriorates and may die.

**CAUSE** Various fungal and bacterial organisms, most of which have a fairly restricted and specific host range.

**CONTROL** Remove affected areas as soon as possible and, where possible, consider treating with an appropriate chemical. Improve growing conditions.
• *see also* APPLE AND PEAR CANKER *p.100*; BACTERIAL CANKER *p.104*; MULBERRY CANKER *p.149*; NECTRIA CANKER *p.151*; SEIRIDIUM CANKER *p.174*; PHOMOPSIS CANKER *p.159*; POPLAR CANKER *p.162*.

## CAPSID BUGS

**SYMPTOMS** Leaves at the shoot tips of many plants, such as fuchsias, caryopteris, roses, hydrangeas, currants, forsythias, dahlias and chrysanthemums, are misshapen, with many small holes (*see p.29*). Flowers of dahlia and chrysanthemum open unevenly and those of fuchsia (*see p.46*) abort at an early stage.

**CAUSE** Pale green insects, such as the common green capsid, *Lygocoris pabulinus*, and tarnished plant bug, *Lygus rugulipennis*, both of which are about 6mm long when adult. They suck sap from the shoot tips, and secrete a toxic saliva that kills some of the plant cells. Later, as leaves develop from the damaged shoots, these dead areas tear into many small holes.

ACTUAL LENGTH: 5–6MM

**COMMON GREEN CAPSID**

**CONTROL** Damage can occur at any time between late spring and late summer. Check vulnerable plants during the summer and spray with dimethoate, fenitrothion or pirimiphos-methyl as soon as capsid bugs or signs of their feeding are seen.
• *see also* APPLE CAPSID *p.100*.

## CARNATION TORTRIX MOTH

**SYMPTOMS** Pale green caterpillars up to 18mm long, with brown heads, feed between two leaves bound together with silk webbing, or under a curled leaf margin similarly held in place by

silk. The caterpillars graze away the inner surfaces of their hiding places, causing the remaining damaged tissues to dry up and turn brown (*see p.20*). This pest occurs on a wide range of shrubs and herbaceous plants, both in gardens and in greenhouses.

**CAUSE** Caterpillars of a small, brown-winged moth, *Cacoecimorpha pronubana*. This moth has several generations a year, and its caterpillars can be found at any season on plants growing indoors or in sheltered places.

**CONTROL** The fact that the caterpillars are concealed within their hiding places makes them difficult to reach with insecticides. Some control can be achieved by looking for bound-up leaves and squeezing them to crush the caterpillars and pupae. If plants are too heavily infested for this to be feasible, spray with pirimiphos-methyl, permethrin or bifenthrin.

## CARROT CAVITY SPOT

**SYMPTOMS** Lens-shaped spots of discoloration develop on carrot roots. Cracks and associated craters develop beneath. Secondary organisms may then invade and cause rapid deterioration of the root.

**CAUSE** The precise cause is uncertain, but a deficiency of calcium is believed to be the main cause. This deficiency causes groups of cells to collapse, so causing discoloration and crater formation.

**CONTROL** Ensure an adequate supply of moisture. Where feasible on acid soils, lime the soil to raise the pH level to about 6.5.

## CARROT FLY

**SYMPTOMS** The roots of carrots and related plants, such as parsley, parsnip and celery, are tunnelled by slender, creamy-yellow maggots up to 9mm long. These cause rusty brown lines on the outside of the tap roots where tunnels close to the surface have collapsed (*see p.51*). Damaged roots are susceptible to secondary rots, which can affect their storage during the winter.

**CAUSE** Larvae of carrot fly, *Psila rosae*, which has two or three generations between late spring and early autumn. Carrot is the favourite host.

**CONTROL** Seedlings of carrots sown after late spring will miss the first generation of larvae; carrots harvested before late summer will miss the second generation. Growing susceptible plants on a fresh site under horticultural fleece prevents female flies getting close enough to lay eggs. Lindane or pirimiphos-methyl dust can be applied to seed drills to protect seedlings; mature plants can be watered with the liquid formulation of pirimiphos-methyl at spray strength in mid-summer against the second generation of maggots. Some carrots, such as 'Sytan' and 'Flyaway', are less susceptible to damage than most other cultivars.

## CARROT MOTLEY DWARF VIRUS

**SYMPTOMS** Foliage of carrots and parsley develops distinct yellow and pink hues (*see p.25*), and is stunted and sometimes twisted. Leaf stalks may show pronounced twisting. Affected plants are stunted, and carrots form little if any edible root. Root tips and root hairs die back.

Similar foliage discoloration may be seen when these plants are suffering from a deficiency of magnesium, but the other symptoms described are not seen.

**CAUSE** Carrot motley dwarf virus is usually found together with carrot red leaf virus, and it is these two viruses, in combination, which are responsible for the symptoms. The viruses are both spread by aphids.

**CONTROL** Dig up and dispose of infected carrots. Attempt thorough aphid control to minimize the risk of infection development and spread.

## CATERPILLARS

Caterpillars are the larval stage of butterflies and moths. There is much variety in their colour, size and degree of hairiness, but they generally have an elongate tubular body shape with a distinct head capsule. They have three pairs of jointed legs at the head end and two to five pairs of clasping legs (prolegs) on their abdomen. Larvae of sawflies look similar but have more than five (usually seven) pairs of prolegs. Most pest species feed on leaves, sometimes gregariously under silk webbing. Others live in soil and feed on roots, or bore into stems, or feed inside leaves as leaf miners, or feed on fruits, berries and seeds.

**Root damage** – see CUTWORMS *p.121*; SWIFT MOTHS *p.180*.
**Stem borers** – see CURRANT CLEARWING MOTH *p.121*; LEOPARD MOTH *p.144*.
**Leaf miners** – see APPLE LEAF MINER *p.101*; LABURNUM LEAF MINERS *p.141*; LEEK MOTH *p.144*; LILAC LEAF MINER *p.145*; PYRACANTHA LEAF MINER *p.166*.
**Leaf damage** – see CABBAGE WHITE BUTTERFLY *p.113*;

CABBAGE MOTH *p.112*; CARNATION TORTRIX MOTH *p.114*; CHINA MARK MOTH *p.116*; DELPHINIUM MOTH *p.123*; ELEPHANT HAWK MOTH *p.125*; MULLEIN MOTH *p.149*; VAPOURER MOTH *p.185*; WINTER MOTH *p.191*; YELLOW-TAIL MOTH *p.192*.
**Gregarious leaf feeders, often with silk webbing** – see BROWN-TAIL MOTH *p.111*; BUFF-TIP MOTH *p.112*; COTONEASTER WEBBER MOTH *p.119*; GYPSY MOTH *p.134*; JUNIPER WEBBER MOTH *p.140*; LACKEY MOTH *p.141*; SMALL ERMINE MOTHS *p.176*.
**Fruit, seed and berry damage** – see CODLING MOTH *p.118*; PEA MOTH *p.155*; PLUM MOTH *p.161*; TOMATO MOTH *p.182*. Other species can sometimes be removed by hand, especially on small plants. Many caterpillars feed after dark and are easier to find by torchlight. Other treatments are the use of the biological control *Bacillus thuringiensis*, or spraying with pyrethrum, permethrin or bifenthrin when signs of caterpillar feeding are seen.

## CATS

Despite their value as pets, cats can be a nuisance in gardens. They soil flower borders and vegetable plots and may damage plants as they attempt to bury their droppings. Their urine can scorch foliage, and the bark of trees and shrubs is damaged when cats use them as scratching posts.

**CONTROL** In localities where cats are numerous, it is difficult to stop them being a problem. Proprietary animal repellent sprays and powders often fail to give reliable long-term protection. Devices which emit ultrasonic sounds that are virtually inaudible to humans but sound loud to

cats can encourage them to go elsewhere. However, as cats become used to the noise some may turn a deaf ear to it. Cats looking for toilet areas are attracted to loose dry soil or mulch-covered flower beds. Close planting helps to cover up the soil surface with vegetation, making it less accessible.

## CAVITY DECAY

**SYMPTOMS** Once exposed to the elements and open to infection by fungal spores, damaged areas on trees may soon deteriorate and form a cavity. The decay may then spread further into the trunk or limb, which may result in sufficient structural damage to render the tree dangerous.
**CAUSE** Initial injury may result from storm damage, poor pruning or other physical injury.
**CONTROL** Prevent cavity formation wherever possible. When cavities develop, the rotting wood should not be removed and the healthy wood exposed: this was once believed to be the best method of dealing with wood rotting, but it has since been determined that this action simply promotes the breaching of the barriers which trees form in response to injury, and more extensive rotting subsequently develops. Trees with cavities should be examined regularly to determine their condition and safety.

## CEDAR APHID

**SYMPTOMS** The foliage and stems of cedars become thickly coated with a black, sooty mould that grows on the sugary honeydew excreted by the aphids (see **p.32**). Heavy infestations cause the foliage to turn

yellow in colour and drop off in mid-summer.
**CAUSE** A greyish-brown, sap-feeding aphid, *Cedrobium lapportei*, up to 2mm in length, which can form dense colonies at the leaf bases in late spring to early summer.
**CONTROL** Spray small trees with pirimicarb, malathion, pirimiphos-methyl or heptenophos with permethrin when aphids or sooty moulds are detected. Damage to large trees has to be tolerated but, fortunately, infestations resulting in defoliation are infrequent occurrences.

## CELERY LEAF MINER

**SYMPTOMS** The leaves of celery and related plants such as parsnip, parsley and lovage develop brown dried-up patches (see **p.19**). This is where the interior of the leaf has been eaten by one or more white maggots up to 7mm long. Heavy attacks give the plants a scorched appearance, with celery plants producing thin, bitter-tasting stems.
**CAUSE** Larvae of a leaf-mining fly, *Euleia heraclei*; it has two generations a year, with damage occurring from late spring to late summer.
**CONTROL** With light infestations, pick off the damaged portions of the leaves, otherwise spray with malathion.

## CELERY LEAF SPOT

**SYMPTOMS** Pinprick-sized spots develop on the leaf surface (see **p.12**), then even smaller black fungal fruiting bodies develop on the spots. The infection may spread to the leaf stalks, and severely affected leaves die off.

**CAUSE** The fungus *Septoria apiicola*, which is spread by handling, by insects and by water splash. In dull, damp weather, or on damp, misty nights, the spores are spread very readily. The infection may also be seed-borne and the fungus may overwinter on infected crop debris.

**CONTROL** Grow plants with care to ensure that they are vigorous and so better able to compensate for any losses. Always use disease-free seed which has been treated against leaf spot, from a reliable company. Spray young plants with a copper-based fungicide. Practise crop rotation (*see pp.92–96*) to minimize the risk of disease carry-over from year to year. Destroy all infected plant debris at the end of the season.

## CENTIPEDES

Centipedes are yellowish-brown animals with many body segments, each of which has one pair of legs. This distinguishes them from the superficially similar millipedes which have two pairs of legs per segment. Centipedes also have longer antennae on their heads, and are more active animals, which run away if disturbed. They are predators of small soil-dwelling animals and may help control soil pests.

ACTUAL LENGTH: 25–60MM

CENTIPEDE

## CHAFER GRUBS

**SYMPTOMS** Plants may suddenly wilt. Those species of chafers that eat grass roots cause yellow patches in lawns (*see p.58*). Heavy infestations attract foxes, badgers, rooks, magpies and crows, which scratch out the loosened turf to feed on the grubs during late summer to spring.

**CAUSE** Plump, white-bodied grubs with brown heads and three pairs of legs that live in the soil and feed on roots (*see p.53*). Their bodies are curved like the letter C. They can kill small plants and gnaw cavities in root vegetables. Several species of chafers occur in gardens. The largest species is the cock chafer or maybug, *Melolontha melolontha*, which has larvae up to 40mm long. It mainly occurs in cultivated soil such as flower-beds and vegetable gardens. Lawns are attacked by the garden chafer, *Phyllopertha horticola* and Welsh chafer, *Hoplia philanthus*. The latter are particularly associated with sandy soils and are smaller, with grubs up to 20mm long. When fully fed, the grubs pupate in the soil and emerge as adult beetles in late spring to early summer.

ACTUAL LENGTH: UP TO 40MM

COCK CHAFER GRUB

**CONTROL** Chafer grubs are difficult to kill with insecticides once they have grown large enough to cause damage. Infestations in flower borders and vegetable gardens are generally light, and it is feasible to search for the grubs in the soil near plants which have suddenly wilted. Problems in lawns are sporadic, but can be devastating when they do occur. A pathogenic nematode, *Steinernema carpocapsae*, can be watered into the soil in late summer while it is moist and in the temperature range 14–20°C (57–68°F). The nematodes enter the bodies of the grubs and kill them by releasing bacteria that cause a fatal infection. Insecticides such as lindane or pirimiphos-methyl can be used against young larvae in mid-summer.

## CHERRY BLACKFLY

**SYMPTOMS** Leaves at the tips of shoots become curled in late spring, and dense colonies of black aphids are visible on the leaf undersides (*see p.24*). By mid-summer some of the affected leaves will turn brown and dry up.

**CAUSE** A sap-feeding aphid, *Myzus cerasi*, which overwinters as eggs on fruiting cherries and some ornamental species such as *Prunus avium* and *P. padus*. The Japanese flowering cherries are generally not attacked.

**CONTROL** On fruit trees, spray with a tar oil wash in winter to control the eggs. Look for aphids on new growth from mid-spring onwards and spray with pirimicarb, malathion, pirimiphos-methyl, permethrin or bifenthrin before leaf curling has become extensive. Systemic insecticides may damage flowering cherry foliage.

## CHICKEN OF THE WOODS

*See* LAETIPORUS SULPHUREUS *p.141.*

## CHINA ASTER FUSARIUM WILT

*See* ASTER WILT *p.103.*

## CHINA MARK MOTH

**SYMPTOMS** Water lily pads and other floating leaves of pond plants have oval holes up to 15mm in length (*see p.30*). The removed leaf pieces are spun together with silk by a small caterpillar to form a protective case around it. The caterpillars become fully fed in summer.

**CAUSE** Caterpillars of the china mark moth, *Elophila nymphaeata*. The adult moths have brown wings with white blotches and they rest on pond-side vegetation.

**CONTROL** Insecticides cannot be used in ponds as they will kill fish and other pond life. Search for the caterpillars' leaf cases, which are often on the undersides of floating leaves, and remove them by hand.

## CHIRONOMID MIDGES

**SYMPTOMS** The leaves of water lilies and other plants with floating foliage, such as water hawthorn, frog-bit and *Potamogetum* species, have a nibbled appearance around the edges. The foliage goes yellow and rots. Slender, transparent larvae, up to 10mm long, can be found feeding in damaged tissues.

**CAUSE** Maggots of certain chironomid flies (*Cricotopus* species). These seem to be a particular problem in new ponds or recently cleaned established pools. Usually, once the pond has settled down and regained its complement of predatory insects, the midge population is kept at a non-damaging level.

**CONTROL** Insecticides cannot be used because of the danger to fish and other pond life. Remove damaged leaves and introduce fish if there are none in the pond.

## CHLOROSIS

**SYMPTOMS** Yellowing of plant tissue, most commonly in the foliage. The loss or deterioration of the green pigment (chlorophyll) may also allow other pigments, which are usually masked by the green, to be seen. In addition to yellowing, red or orange discoloration may also develop.

**CAUSE** Many quite unrelated factors may be involved. Most common are deficiencies of iron and manganese (so-called lime-induced chlorosis), nitrogen and magnesium. Waterlogging, low temperatures, virus infection or weedkiller contamination are also fairly commonly responsible. The precise patterning of the chlorotic symptoms and their distribution within the plant may help to determine which factor is responsible.

**CONTROL** Take appropriate action to remedy any mineral deficiencies, sources of contamination or cultural problems.

## CHOCOLATE SPOT

*See* BEAN CHOCOLATE SPOT *p.105.*

## CHRYSANTHEMUM EELWORM

**SYMPTOMS** Foliage on chrysanthemums, Japanese anemones, penstemons and many other herbaceous plants progressively turns brown and dries up from the base of the plants upwards. Spread of the microscopic eelworms causing the damage within the leaf is restricted by the larger leaf veins, therefore recently infested leaves have brown wedges or islands separated from green healthy parts by

the leaf veins. Eventually the entire leaf will turn brown or black.

**CAUSES** Microscopic nematodes or eelworms, *Aphelenchoides ritzemabosi*, which feed within the foliage. Although present on suitable host plants all year, infestations develop most rapidly during damp weather in late summer and autumn as the foliage starts to age.

**CONTROL** None of the insecticides available to amateur gardeners is effective. Destroy badly affected plants. Dormant chrysanthemum stools dipped in hot water will produce shoots for cuttings that are free of eelworm. The old stems are cut down and soil washed from the stool before immersing it for five minutes in water held at 46°C (115°F). The amount of heat is critical, as too much will damage the plant, while too little will allow eelworms to survive.

## CHRYSANTHEMUM LEAF MINER

**SYMPTOMS** Sinuous white or brown lines develop in the leaves of chrysanthemums (*see **p.25***), gerberas, cinerarias, marguerites, pyrethrum and other related composite flowers. If an infested leaf is held up to the light, a grub or its pupa may be seen at the end of its tunnel. Small, pale, slightly raised round scars appear on the upper surface of leaves at shoot tips where the adult flies have been feeding.

**CAUSE** Maggots of a leaf-mining fly, *Phytomyza sygenesiae*, which has two generations during the summer on garden plants, but can breed continuously throughout the year in greenhouses.

**CONTROL** Deal with light infestations by picking off mined leaves, or crush the grubs at the ends of their tunnels. Heavy infestations may need spraying two or three times at two-week intervals with pirimiphos-methyl or malathion.

## CHRYSANTHEMUM PETAL BLIGHT

**SYMPTOMS** Water-soaked lesions or smaller, brown spots develop on the petals of chrysanthemums (*see **p.45***) and occasionally other members of the Compositae family, and on anemones. Infected flowers deteriorate rapidly.

**CAUSE** The fungus *Itersonilia perplexans*, which is encouraged by wet weather and by humid conditions. It is less common than ray blight (*see below*).

**CONTROL** Remove infected plant parts or whole plants immediately. Improve air circulation and attempt to lower humidity. Spray with mancozeb.

## CHRYSANTHEMUM RAY BLIGHT

**SYMPTOMS** Red-brown spots develop on the petals; as the spots increase in number and size, the petals are killed off and the whole flower head starts to decay. Rotting of shoot tips and brown spots on the foliage may also be seen. Occasionally infections occur at the stem base, causing the whole plant to be stunted and discoloured, and eventually to die. Similar symptoms may be caused by chrysanthemum petal blight (*see above*) or grey mould (*see p.134*).

**CAUSE** The fungus *Didymella chrysanthemi*, which can

overwinter on infected plants and plant debris. Conditions of high humidity, especially common in glasshouses, encourage disease development and spread.

**CONTROL** Remove and dispose of infected plants immediately and spray plants with mancozeb. Decrease humidity by opening vents and windows to improve air circulation.

## CHRYSANTHEMUM WHITE RUST

**SYMPTOMS** Pale buff or cream-coloured raised, warty pustules develop on the lower leaf surface (*see **p.14***). The upper surface is marked with yellow pits, each pit corresponding with a pustule. As pustules age they become pale to tan-brown, and leaves may become distorted and die.

**CAUSE** The fungus *Puccinia horiana*, which is encouraged by moist, humid atmospheres. It may persist from year to year on over-wintering plants and stools, and can infect indoor and outdoor chrysanthemums.

**CONTROL** Remove infected leaves promptly. Spray all chrysanthemums in the vicinity with a suitable fungicide such as penconazole or bupirimate with triforine, giving at least three applications. Destroy severely infected plants and do not propagate from them. Improve air circulation to decrease humidity.

## CLEMATIS GREEN PETAL

**SYMPTOMS** The flowers of clematis open to reveal a distinct green coloration. This is especially clear on pale-flowered or white-petalled clematis. It is usually only the first flush of flowers that is affected.

**CAUSE** The most common causes are adverse growing or cultural conditions at a critical stage in the flower's development. Green flowers frequently develop a few weeks after there has been an unusually cold spell of weather. If cold damage is responsible, subsequent flushes of flowers are perfectly normal. However, if the problem lasts throughout the season, and reappears the following year, then the development of green pigments in the petals could be due to the presence of a mycoplasma infection.

**CONTROL** Protect plants against excessively cold conditions, using horticultural fleece or similar material. If the plant has a mycoplasma infection it should be destroyed, as there is no cure available.

## CLEMATIS WILT

**SYMPTOMS** Leaves and shoots wilt and die back. Wilting usually starts at the shoot tips, leaf stalks blacken where they join the leaf blade and young foliage wilts. New, healthy shoots may appear from below soil level or from nodes beneath the wilted area. A small spot of discoloration may develop under the lowest pair of wilting leaves. Waterlogging, graft failure, insect or slug damage may cause wilting too, but in these cases affected clematis do not show the other symptoms.

**CAUSE** The fungus *Ascochyta clematidina*.

**CONTROL** Prune out affected stems, cutting right back into healthy tissue, even cutting below soil level if necessary. Mildly infected plants should reshoot.

## CLICK BEETLES

Click beetles are elongated brown beetles, 8–20mm long. If a click beetle is placed on its back it will right itself by jumping into the air with an audible clicking noise, hence its common name. Click beetles cause no damage as adult insects, but several species are the adult stage of soil pests known as wireworms (*see p.191*).

ACTUAL LENGTH: 15–40MM

**CLICK BEETLE LARVA (WIREWORM)**

## CLUBROOT

**SYMPTOMS** Roots become swollen, and the whole root system may become distorted (*see p.55*). The foliage is pale, chlorotic or pinkish in colour and wilts readily, especially during hot weather, but may even wilt when the soil is moist. Plants fail completely or produce a poor crop. Members of the Cruciferae may be attacked, in particular brassicas. Certain ornamentals, including wallflowers, stocks and candytuft, and also some weeds, including shepherd's purse, charlock and wild radish, are susceptible.

**CAUSE** The slime mould *Plasmodiophora brassicae*, which is soil-borne and can remain viable in the soil for in excess of 20 years, even in the absence of susceptible host plants. Infection occurs via the root hairs. When swollen roots disintegrate, they release many spores. The pathogen is easily introduced into a garden, on infected plants, on soil adhering to boots, tools and wheelbarrows and in soil, manure or garden compost. Acidic and waterlogged soils are most prone to clubroot.

**CONTROL** Improve soil drainage and lime the soil to deter the pathogen. Burn all infected plants promptly, certainly before the roots start to disintegrate. Avoid spreading the spores around the garden on tools or boots, or by moving potentially infective soil. Raise your own plants, rather than buying them in; otherwise only buy plants from a reputable source and check them carefully before planting. Practise good weed control. If crops of brassicas need to be grown on infested soil, dip the roots of young plants in a suitable formulation of carbendazim or thiophanate-methyl before planting. Raise plants in proprietary compost in individual pots of at least 5cm diameter. Plant out when plenty of roots have formed. Choose resistant vegetable cultivars such as calabrese 'Trixie', Chinese cabbage 'Harmony', kale 'Tall Green Curled' and swede 'Marian'.

## CODLING MOTH

**SYMPTOMS** In mid- to late summer, fruits of apple and pear are tunnelled by small white caterpillars with brown heads (*see p.47*). By the time the fruit is ripe, the caterpillars have usually finished feeding and have sought overwintering sites under loose flakes of bark. The caterpillar feeds in the core and makes an exit tunnel leading either to the eye end or elsewhere on the fruit surface.

**CAUSE** Caterpillars of a small moth, *Cydia pomonella*, which is generally more widespread and troublesome than the other apple fruit pest, apple sawfly (*see p.101*). The latter attacks

apples at the fruitlet stage while codling moth damages the ripening fruits in late summer.

**CONTROL** Suitable insecticides are permethrin, bifenthrin, fenitrothion or pirimiphos-methyl, which are applied in early summer with a second treatment three weeks later. Pheromone traps (*see p.86*) are available for detecting when adult codling moths are active and likely to be laying eggs. Accurate timing of insecticides is important, since it is necessary to kill the young caterpillars before they have bored into fruits. Pheromone traps can catch enough male moths to reduce the mating success of females on isolated apple trees and so may reduce the number of maggoty apples.

## COLLAR ROT

*See* PHYTOPHTHORA *p.159*.

## COLLETOTRICHUM

**SYMPTOMS** Leaf and occasionally stem spotting, developing into lesions that may become sunken. Stems may become ringed, causing the whole plant, or parts of it, to die back.

**CAUSE** Various species of the fungus *Colletotrichum*, some of which are believed to be seed-borne, and which can all be spread by water splash.

**CONTROL** Remove affected plant parts promptly, and entire plant if stem infections are present. Do not save seed from infected plants.

## COLORADO BEETLES

**SYMPTOMS** The foliage of potato, tomato, aubergine and pepper plants is eaten by pale yellowish-orange beetles, up to 10mm in length, with five black stripes on each wing case. Rotund, orange-red grubs with black heads and two rows of black spots along the sides of the bodies also attack leaves and can cause extensive defoliation.

**CAUSE** A leaf beetle, *Leptinotarsa decemlineata*, which originates from North America but is now found throughout Europe, except in Britain and Ireland.

ACTUAL LENGTH: 10MM

**COLORADO BEETLE**

**CONTROL** Strains of the beetle which have gained resistance to some insecticides can occur. If Colorado beetles are found in Britain, they must be reported to the Ministry of Agriculture, Fisheries and Food, which will organize the necessary measures to eradicate them. Elsewhere, infested plants can be sprayed with permethrin, pirimiphos-methyl or fenitrothion when the adults or larvae are seen.

## CONIFER APHIDS

**SYMPTOMS** Colonies of greyish-black or brown insects develop on the shoots, branches and sometimes trunks of conifers. Spruce (*Picea*), pine (*Pinus*), juniper (*Juniperus*), cypress (*Cupressus*) and Leyland cypress (x *Cupressocyparis leylandii*) are liable to be attacked by the aphids which, at 3–6mm long, are significantly larger than most aphids on other plants. The foliage and stems become sticky with honeydew excreted by the aphids from late spring to mid-summer.

A black, sooty mould grows on the honeydew, and heavy infestations can cause foliage to turn yellow and dry up.
**CAUSE** Several species of aphids, mainly *Cinara* species.
**CONTROL** Only small trees and hedges can be treated effectively. Spray with pirimicarb, pirimiphos-methyl, malathion, permethrin or bifenthrin in mid- to late spring before heavy infestations build up.
• *see also* CEDAR APHID *p.115*; CYPRESS APHID *p.121*; GREEN SPRUCE APHID *p.134*; JUNIPER APHIDS *p.140*.

## CONIFER RED SPIDER MITE

**SYMPTOMS** Foliage becomes finely mottled and yellowish-brown in mid-summer, and may fall prematurely.
**CAUSE** Careful examination with a hand lens will show tiny yellowish-green mites with dark markings crawling amongst the foliage on fine silk webbing. Orange-red spherical eggs will also be found on the leaves and stems. Spruce is the most susceptible conifer, especially *Picea albertiana*, but this mite also attacks juniper, thuya, cedar and cypress.
**CONTROL** Watch for signs of infestation from late spring onwards. If necessary, spray thoroughly with malathion, pirimiphos-methyl or bifenthrin on two or three occasions at weekly intervals.

## CONIFER WOOLLY APHIDS

*See* ADELGIDS *p.98*.

## CONIFERS, NEEDLE RUSTS OF

*See* NEEDLE RUSTS OF CONIFERS *p.151*.

## COPPER DEFICIENCY

**SYMPTOMS** Blue-green or slightly yellow discoloration of the foliage, occasionally followed by dieback. Symptoms are not easy to recognize or differentiate from other problems, and are fairly rare, being most likely on peaty soils.
**CAUSE** A deficiency or poor availability of copper in soil.
**CONTROL** Treatment with a complete fertilizer may help to rectify the problem, but other possible causes for the symptoms shown should be investigated, too.

## CORAL SPOT

**SYMPTOMS** Bright, coral to orange raised pustules appear on dead, woody stems and branches (*see* **p.41**). The pustules may only appear after the stem has been dead for several weeks. Many different plants may be affected, but currant bushes (*Ribes*), elaeagnus, magnolias and maples (*Acer*) are very susceptible. Dieback occurs, and if the infection spreads right down into the crown, the whole plant may be killed.
**CAUSE** The fungus *Nectria cinnabarina*, which may also live on dead logs, twigs, woody debris and old pea sticks. Spores are produced all year, and water, either as rain or irrigation splash, is the main method of dispersal. The fungus enters a plant via a wound or colonizes a dead snag left by physical injury or poor pruning.
**CONTROL** Prune out all dead or dying stems promptly, cutting well back into sound, healthy wood. Remove and burn wood bearing the pustules. Dispose of garden debris that may be harbouring the disease.

## COTONEASTER WEBBER MOTHS

**SYMPTOMS** In early summer, the stems of cotoneaster are covered with silk webbing spun by numerous dark brown caterpillars up to 12mm long (*see* **p.37**). The leaves dry up and turn brown where they have been grazed by the caterpillars. Less extensive webbing and feeding damage also occurs in late summer, caused by the newly-hatched caterpillars before they hibernate for the winter and resume their activities in spring.
**CAUSE** Caterpillars of two moths, *Scythropia crataegella*, which also attacks hawthorn, and *Numonia suavella*. Both have one generation a year and their caterpillars pupate amongst the silk webbing.
**CONTROL** Prune out light infestations, otherwise spray with permethrin, bifenthrin or pirimiphos-methyl, or use the biological control *Bacillus thuringiensis*, when webbing is seen.

## CRANE FLIES

Crane flies, or daddy-long-legs, have soil-dwelling larvae – leatherjackets – that feed on roots and the stem bases of small plants. The adults do not feed.
• *see also* LEATHERJACKETS *p.143*.

ACTUAL LENGTH: 20–30MM

**CRANE FLY**

## CROWN GALL

**SYMPTOMS** Rounded swellings or galls – which are off-white when young and

harden and turn brown with age – develop on woody or herbaceous plants. Galls are very variable in size, ranging from less than one centimetre to more than 30cm in diameter. They usually develop in small groups. Crown gall may appear on the roots or stem base (the crown of the plant). Less frequently, symptoms of an aerial form may show at some distance up the stem or trunk (see **p.38**). Galls on woody plants are persistent, those on herbaceous plants are softer and more short-lived. They soon disintegrate, often allowing entry of secondary organisms in the process. The plant's overall vigour is rarely affected. However, if galls ring the stem or trunk, or cause extensive disruption of water and nutrient flow, the whole plant may decline. If stem or root splitting occurs due to the erupting out of the galls, secondary organisms may gain entry and cause dieback.
**CAUSE** The bacterium *Agrobacterium tumifaciens*, which is soil-borne, and capable of living saprophytically (without a living host) in the soil. The bacteria gain entry via wounds and then stimulate rapid proliferation of cells, so forming the galls. Gall growth is greatest when the plant is growing rapidly, so infections that occur at the very end of the growing season may only produce symptoms the following year.
**CONTROL** Improve soil drainage. Check all plants carefully before planting. Keep root and stem injury to a minimum. Remove infected plants and grow root crops (except beet) on the site for

at least a year to minimize the risk of subsequent infections, as these crops are not susceptible to crown gall.

## CROWN ROT

**SYMPTOMS** Deterioration and rotting of the tissues at the crown (the junction of the stems and roots) of the plant. Herbaceous plants are very prone, but other woodier hosts, such as wisterias, may also succumb. It may then cause deterioration of the stems and foliage and the whole plant can die.
**CAUSE** Various soil- and water-borne fungi and bacteria, often several in combination.
**CONTROL** If dealt with very promptly it may be possible to save an infected plant. Remove affected areas, cutting well back into healthy tissue.

## CUCKOO SPIT

**SYMPTOMS** Blobs of white frothy liquid appear on the stems of many plants in early summer (see **pp.39, 43**). A yellowish-green insect lives in the froth where it feeds by sucking sap.
**CAUSE** The immature nymph stage of froghoppers such as *Philaenus spumarius*. Once they have reached the adult stage in early summer, they stop producing froth and become less noticeable.
**CONTROL** Spraying is often unnecessary, since cuckoo spit is only a temporary nuisance on plants, and the nymphs cause little harm unless they are feeding at the shoot tips, when they may cause some distorted growth. Picking off the pests by hand is usually sufficient to control most cases of infestation.

## CUCUMBER FOOT AND ROOT ROT

See FOOT AND ROOT ROTS *p.128*.

## CUCUMBER GUMMOSIS

**SYMPTOMS** Discrete sunken grey-black spots, often with a dense, furry, grey fungal growth, develop on the fruit. A viscous liquid may ooze from these lesions, and severely infected fruits may split. Stem and leaf lesions may also develop but are less common.
**CAUSE** The fungus *Cladosporium cucumerinum*, which is encouraged by cool, damp conditions.
**CONTROL** Remove infected areas promptly. Improve ventilation and, if feasible, raise temperatures. Grow resistant cultivars such as 'Birgit', 'Femspot', 'Jazzer' or 'Petita'.

## CUCUMBER MOSAIC VIRUS

**SYMPTOMS** Plants are stunted and show distinct yellow mosaic patterning (see **p.21**), usually on stunted, deformed foliage (see **p.25**). Flowering is reduced or non-existent, and plants may die completely. If fruits are produced on cucurbits (marrows, courgettes or cucumbers), they are small, pitted and strangely dark green with bright yellow blotches. Their texture is hard and the fruits inedible. On ornamentals, flower breaking (changes in colour on clearly defined areas on the petals) may occur.
**CAUSE** The cucumber mosaic virus, which has an extremely wide host range, including many common garden plants, food crops and weeds. Virus particles

are readily spread by handling and by sap-feeding pests, in particular aphids.
**CONTROL** There are no cures available and infected plants should be destroyed. Practise aphid control and minimize the risk of infection spread. Remove weeds. Avoid unnecessary handling of plants and always handle healthy plants before touching infected or suspect ones. After handling infected plants, wash hands thoroughly. Grow resistant varieities where available, such as aubergine 'Bonica'; courgette 'Defender', 'Supremo' or 'Tarmino'; cucumber 'Bush Champion', 'Crispy Salad', 'Jazzer', 'Petita', or 'Pioneer'; marrow 'Badger Cross' or 'Tiger Cross'.

## CURRANT AND GOOSEBERRY LEAF SPOT

**SYMPTOMS** In early summer, very small dark brown or black spots develop on the leaves of currants, goose-berries and, occasionally, ornamental currants (*Ribes*). The spots may enlarge and join together, and the infected leaf turn brown. Affected leaves fall prematurely and the bush may be extensively defoliated by mid-summer. In extreme cases, fruits shrivel.
**CAUSE** The fungus *Drepanopezizsa ribis*. It produces spores from the leaf spots that are spread by wind and rain. The fungus overwinters on the leaf spots on fallen leaves.
**CONTROL** Regularly remove and dispose of fallen leaves to prevent the fungus overwintering. Spray with a suitable fungicide such as carbendazim, copper oxychloride, mancozeb or bupirimate with triforine, ensuring that the undersides

of the leaves are treated thoroughly. Feed and maintain plants well, as this appears to make them less prone to attack.

## CURRANT BLISTER APHID

**SYMPTOMS** Leaves at the shoot tips become crumpled, with raised areas which have small pale yellow insects on the undersides. The affected leaves are yellowish-green or have a reddish coloration (*see* **p.27**), especially on redcurrants.
**CAUSE** Sap-feeding aphids, *Cryptomyzus ribis*, which are active on the foliage between bud burst and mid-summer. After that they develop wings and migrate to the wildflower known as hedge woundwort (*Stachys sylvatica*). There is a return migration in autumn when overwintering eggs are laid on currant bushes.
**CONTROL** A tar oil wash applied in early to mid-winter will kill the egg stage. Alternatively, spray with dimethoate, heptenophos with permethrin, pirimicarb or fenitrothion between bud burst and flowering.

## CURRANT CLEARWING MOTH

**SYMPTOMS** Red- and black-currant stems are tunnelled by white caterpillars, up to 15mm long, with brown heads (*see* **p.33**). Their feeding activities sometimes cause shoots to die but more usually the presence of a caterpillar is detected when a weakened stem snaps. This is likely to happen during windy weather, especially when stems are heavy with berries. A blackened tunnel can be seen in the centre of attacked stems.

**CAUSE** Caterpillars of a clear-wing moth, *Synanthedon tipuliformis*. The name clearwing is derived from the fact that the moth's wings are largely devoid of coloured scales and are therefore transparent, except at the wing tips. This feature, combined with a black and yellow-striped abdomen, gives this day-flying moth a wasp-like appearance.
**CONTROL** There is no effective chemical treatment, as it is not possible to reach caterpillars inside stems. Fortunately, this pest does not usually damage all of the stems on a bush. During the winter, tunnelled stems can sometimes be detected by gently bending them, causing them to snap at the weak point. These stems should be cut back to sound wood and the prunings burnt to destroy the caterpillars.

## CURRANT REVERSION

**SYMPTOMS** Plants produce leaves with slight yellowing, and unusually small main veins. The leaves are smaller and have fewer lobes than healthy ones. Flowering and ultimately cropping of the plant is also reduced.
**CAUSE** The blackcurrant reversion virus, which is spread largely by big bud mite (*see* p.107) and may also be transmitted by grafting. Symptoms of mite infestation, such as swollen buds, may also be present.
**CONTROL** Remove infected plants, as the problem will not decline, and infected plants may act as a source of infection for other, healthy, bushes. Always buy black-currants from a reputable source and ensure that they are certified free from pests and diseases on purchase.

## CUSHION SCALE

**SYMPTOMS** Leaves of camellias, hollies, rhodo-dendrons, evergreen azaleas and euonymus develop sooty mould on the upper surfaces, especially during the winter months. Flat, oval, yellow-brown objects are attached to the undersides of the leaves, and in early summer the females deposit eggs under a broad, white, waxy band (*see* **p.15**).
**CAUSE** Sap-feeding scale insects, *Pulvinaria floccifera*.
**CONTROL** Spray the undersides of the leaves thoroughly with malathion in mid-summer to control the more vulnerable newly-hatched scale nymphs.

## CUTWORMS

**SYMPTOMS** Small plants in flower beds and vegetable gardens wilt and die. Examination of the plants shows that the roots just below soil level have been severed, or the outer bark has been gnawed away. Root vegetables, such as potatoes and carrots, may have cavities eaten in them. Caterpillars, often creamy-brown or sometimes greenish-brown in colour and up to 40mm long, are often found in the soil near recently damaged plants (*see* **p.53**). At night these caterpillars will feed above soil level on the foliage of low-growing plants.
**CAUSE** The soil-dwelling caterpillars of several moths, including large yellow under-wing (*Noctua pronuba*), lesser yellow underwing (*N. comes*), turnip moth (*Agrotis segetum*), heart and dart moth (*A. exclamationis*) and garden dart (*Euxoa nigricans*).
**CONTROL** Cutworms tend to work their way along row

crops such as lettuce, so it is worthwhile sifting the soil near a damaged plant to find the culprit. These caterpillars are more troublesome in weedy plots, so good cultivation does help. The older larvae are tolerant of insecticides, but young larvae can be killed with lindane or pirimiphos-methyl. One of the pathogenic nematodes, *Steinernema carpocapsae*, can be used in moist, well-drained soil during the summer months.

## CYPRESS APHID

**SYMPTOMS** Conifers, such as *Cupressus* species, *Chamaecyparis lawsoniana*, x *C. leylandii* and *Thuja*, develop sooty mould on the foliage and stems. This grows on the sticky honeydew excreted by the aphids, which are brownish-black and up to 4mm long. Heavy attacks during the early summer cause shoots to dry up and turn yellow-brown in mid-summer. On hedges, the damage is generally worse on the lower parts, with some green growth left at the top.
**CAUSE** A sap-feeding aphid, *Cinara cupressi*, which is active during summer.
**CONTROL** Look for aphids on the stems from late spring onwards and if seen, spray with pirimicarb, malathion, pirimiphos-methyl, permethrin or bifenthrin. Once foliage has started to discolour, the process cannot be reversed, but in time plants usually recover. Damaged trees and hedges should be fed with a general fertilizer in spring and kept watered during dry spells to help them regain vigour.

## DAEDALEOPSIS CONFRAGOSA

**SYMPTOMS** Otherwise known as the blushing bracket, this is a common bracket fungus. It attacks deciduous trees, in particular alders, birches and willow. The bracket-shaped, semi-circular fruiting bodies measure between 5 and 12cm in diameter (see *p.37*). The upper surface is smooth or slightly ridged, and reddish-brown in colour. The pores are circular but become elongate and red. The bracket has a corky, almost rubbery texture. It causes a white rot of the woody tissues and is generally associated with wounded or weakened trees.
**CAUSE** The fungus *Daedaleopsis confragosa*.
**CONTROL** Remove brackets to prevent spores spreading to other parts of the tree, or to other susceptible trees. Affected trees should be inspected by a professional arboriculturist to determine the extent of decay, and the overall health of the tree.

## DAHLIA SMUT

**SYMPTOMS** Pale brown, elongate spots with yellow margins develop on leaves in mid-summer. The lower, older leaves are affected first, but younger leaves may then be affected too. The spots darken and increase in size and may join together, often killing the leaf.
**CAUSE** The fungus *Entyloma calendulae* f. *dahliae*, which may live in the soil around infected plants and on plant debris. It can persist in the soil in the absence of living dahlias for about five years.

**CONTROL** Remove and dispose of severely infected dahlias, and plant replacements on a fresh site. The fungus does not seem to be present in the tubers, so provided any affected leaves are removed and the tubers appear clear of infected debris, they can be stored and should produce healthy plants the following year.

## DAMPING OFF

**SYMPTOMS** Seedlings flop over, often showing discoloration and deterioration around the stem base (see *p.34*). Initially the basal lesions may appear water-soaked. The infection spreads rapidly, causing patches of seedlings to die off. An entire tray of seedlings may be killed within a few days. Fluffy white fungal growth may be seen on the surface of the dead and dying seedlings. If infection occurs before germination is complete, the seedlings only emerge patchily.
**CAUSE** Several soil- and water-borne fungi, in particular species of *Pythium*, *Phytophthora* and *Rhizoctonia*. Infection is encouraged by over-wet compost and prolonged raised temperatures. Seedlings which are not given adequate light, and which are from seed sown too densely, are particularly prone to infection. Damping off is most common when hygiene is poor – if pots, trays and implements are not properly scrubbed and cleaned, if unsterilized compost is used, or if non-mains water is used.
**CONTROL** Observe strict hygiene and only use clean trays, pots and implements and proprietary or sterilized

compost. Do not use anything other than mains water, especially not water collected in a butt. Sow seedlings thinly and ensure they have adequate light and are not kept too warm for any longer than is needed to encourage germination. Drench compost with a copper-based fungicide at sowing and treat seedlings with the same fungicide at intervals throughout their development.

## DEER

Several species of deer cause problems in gardens. The most widespread is roe deer (*Capreolus capreolus*), but muntjac (*Muntiacus muntjak*), fallow deer (*Dama dama*) and red deer (*Cervus elephus*) also cause problems in some areas. A wide range of herbaceous and woody plants are eaten, with damage mainly occurring between dusk and dawn. Sometimes it is blooms rather than foliage that is eaten, as with tulips, but more usually whole shoots are lost. On woody plants, such as roses, the stems are partly bitten through cleanly but there remains a ragged edge to the stem (see *p.41*). This is because deer have no front teeth in their upper jaw and they cut off woody shoots by biting and tugging. The trunks of trees and side-branches can be damaged by male deer rubbing their antlers against the bark, a habit known as fraying.
**CONTROL** It can be difficult to exclude deer from gardens. They are agile animals good at jumping over obstacles, and a fence about 2m high is necessary to keep roe deer out. Proprietary animal repellents often give disappointing results; the

effects tend to be short-lived and they may simply encourage deer to feed on something else in the garden. Deer are inquisitive feeders, attracted to new plantings, so it is worthwhile giving new plants some protection. Although deer eat almost anything, it is a good idea to look at other gardens in the locality to see which plants tend to be left alone.

## DELPHINIUM BLACK BLOTCH

**SYMPTOMS** Black blotches develop on leaves (see *p.12*), rapidly enlarging and spreading to the stems, petioles and flower buds. Entire leaves may turn black and die off.
**CAUSE** The bacterium *Pseudomonas delphinii*, spread from the soil or by rain or watering splash.
**CONTROL** Remove and dispose of infected plants promptly and avoid over-head watering. Plant new delphiniums on a fresh site.

## DELPHINIUM LEAF MINER

**SYMPTOMS** The leaves of delphinium and monkshood develop large, brown, dried-up areas extending from the tips towards the leaf stalks (see *p.18*). White maggots up to 3mm in length may be found feeding within the mined areas.
**CAUSE** Larvae of a leaf-mining fly, *Phytomyza aconiti*, which has two generations during the summer months.
**CONTROL** For light infestations, the damaged parts of leaves can be picked off; otherwise spray with pirimiphos-methyl or malathion when damage is first noticed, in late spring to early summer.

# DELPHINIUM MOTH

**SYMPTOMS** During the spring, leaves at the shoot tips of delphinium and monkshood are spun together with silk threads by small green caterpillars heavily marked by black spots. Later the caterpillars feed openly on the expanded foliage. They are greenish-white by the time they pupate, in a pale yellow cocoon spun on the underside of leaves during late spring to early summer. The mature caterpillars are up to 40mm long, but often escape notice as they remain inactive on the underside of the foliage during the day.
**CAUSE** Caterpillars of a moth, *Polychrysia moneta*, which has one generation a year and overwinters as young larvae in the old plant stems.
**CONTROL** It is the young larvae feeding at the tips of new shoots in the spring that cause most harm. Look for bound-up leaves and remove the caterpillars by hand. For heavy infestations, spray with permethrin, bifenthrin, pirimiphos-methyl or fenitrothion.

# DELPHINIUM VIRUSES

**SYMPTOMS** Yellowing of the foliage, usually as mosaic patterns, mottling or vein-clearing (yellowing immediately adjacent to the leaf veins). Infected plants may be stunted and fail to flower properly, often dying within a year or so of the first sign of the symptoms.
**CAUSE** Various viruses may be involved, either singly or in combination. The most common is cucumber mosaic virus (*see p.120*). The viruses involved are not all spread in the same manner, but common means of spread include handling, and vectors include sap-feeding pests, such as aphids, and soil-inhabiting nematodes.
**CONTROL** Remove infected plants promptly and avoid handling other plants after handling one which shows symptoms. Do not propagate from infected plants.

# DIANTHUS ANTHER SMUT

**SYMPTOMS** Anthers of pinks and carnations (*Dianthus*) become swollen and darken as they fill with masses of black-purple spores. These are released in great quantity as the anthers rupture. Stems become stunted.
**CAUSE** The fungus *Ustilago violacea*, which appears to be present throughout the plant, although only producing symptoms on the anthers and flower stems.
**CONTROL** Dispose of infected plants promptly. Do not propagate from any part of an infected plant. Remove all plant debris and scrub down the greenhouse when plants are grown under glass. Do not grow *Dianthus* on the site for at least five years.

# DIANTHUS FUSARIUM WILT

**SYMPTOMS** Carnations (*Dianthus*) start to fade, and the leaves wilt and develop a greyish or straw-coloured discoloration. The leaves may also develop a purple-red discoloration. The whole plant dies, but the symptoms rarely occur over the whole plant simultaneously.
**CAUSE** The fungus *Fusarium oxysporum* f. sp. *dianthi*, which has thick-walled resting spores which enable it to remain dormant in the soil for many years, even in the absence of carnations in the area.

**CONTROL** Remove infected plants promptly and grow new plants on a fresh site. Do not propagate from infected plants.

# DIDYMELLA STEM ROT

**SYMPTOMS** Black-brown, sunken blotches develop on the stems of tomatoes (*see **p.33***), and occasionally aubergines, at soil or ground level, often combined with the development of adventitious roots (those above soil level). The older leaves turn yellow. A blackish-brown rot, usually at the calyx end, develops on the fruit. Infected areas later develop numerous tiny, black, pinprick-sized fungal fruiting bodies on the surface.
**CAUSE** The fungus *Didymella lycopersici*, which can be seed-borne, but is usually spread from plant to plant by rain or watering splash, or by handling. The fungus can overwinter on debris from infected plants or in the soil.
**CONTROL** Remove and dispose of infected plants promptly; do not compost them. Clear up all crop debris at the end of the season and grow tomatoes on a fresh site. As aubergines may also occasionally be affected, choose a fresh site for these, too.

# DIEBACK

**SYMPTOMS** The stems of plants start to die back. Dieback may start either at the tip, the base, or part-way up the stem; it rarely affects all stems simultaneously. Dark blotches or sunken patches may develop at the point where the deterioration starts. Leaves wilt, yellow and die off. The symptoms may spread down into the base of the plant, or its crown, so causing the whole plant to die.
**CAUSE** Various fungi, some of which can actively invade healthy stems, and others which only enter the plant through wounds or other points of damage. Dieback, especially from stem tips downwards, may also be caused by cultural problems, including poor establishment of young plants, drought and waterlogging.
**CONTROL** Prune out affected stems to a point well below any signs of damage. Improve cultural conditions to ensure increased plant vigour and plenty of replacement growth.

# DOGS

Although not pests, dogs can cause problems in gardens. Their urine can scorch foliage, with bitches causing circular brown patches in lawns, and dogs damaging conifers and other plants that they cock their legs against.
**CONTROL** Spraying the foliage with water to dilute the urine will prevent damage, but this needs to be done soon after the dog has urinated. Proprietary animal repellent products are available from garden shops but these do not usually give reliable long-term protection.

# DOGWOOD ANTHRACNOSE

**SYMPTOMS** Spots on the leaves and bracts of *Cornus* in late spring or early summer. These symptoms usually appear first towards the base of the tree, and then progress upwards. Cankers may develop on branches. As the disease develops, extensive dieback may occur and the whole tree may even be killed. The precise range of dogwoods affected is still under investigation, but *Cornus florida* and *C. nuttallii* are known to be particularly susceptible. *Cornus sanguinea* and *C. alba* are also affected, but on these, dieback is restricted to weak shoots.

**CAUSE** The fungus *Discula destructiva*, which is encouraged by cool, damp weather and which is most prevalent in areas of high rainfall.

**CONTROL** Prune out infected stems immediately to try to save the affected plant.

## DOUGLAS FIR ADELGID

**SYMPTOMS** Small, fluffy white balls of waxy threads appear on the undersides of leaves on Douglas fir (*Pseudotsuga menziesii*) during the summer. Heavy infestations may cause some yellowing of the foliage, which can also be discoloured by the growth of sooty moulds on the pests' excrement.

**CAUSE** *Adelges cooleyi*, a blackish-brown sap-feeding insect, up to 2mm long, with an aphid-like appearance. It can be found on Douglas firs throughout the year, and may also occur on its alternative host plant, Sitka spruce (*Picea sitchensis*), where it causes an elongate swollen gall at the shoot tips. The spruce generation is not normally a problem in gardens.

**DOUGLAS FIR ADELGIDS ON STEM AND NEEDLES**

**CONTROL** Large trees cannot be sprayed adequately and the presence of adelgids has to be tolerated. On small trees, some protection can be given against over-wintering adelgid nymphs by spraying with malathion or pirimiphos-methyl during mild, dry weather in late February. At that time they are not protected by a waxy coating and are therefore more vulnerable to sprays.

## DOWNY MILDEW

**SYMPTOMS** Yellow or otherwise discoloured areas develop on upper leaf surfaces. Each corresponds to a slightly fuzzy, greyish-white or, occasionally, purplish fungal growth beneath. The infection may spread, causing large areas or even the entire leaf to discolour and die. A wide range of plants may be affected, but downy mildew is especially common on young plants and those growing in moist environments. The symptoms may be confused with those of powdery mildew infection.

**CAUSE** Several different fungi, in particular species of *Peronospora*, *Bremia* and *Plasmopara*. These include *Peronospora parasitica* (downy mildew of cruciferae), *Peronospora destructor* (on onion), *Peronospora farinosa* f. sp. *spinaceae* (on spinach), *Peronospora viciae* (on pea), *Peronospora sparsa* (on rose), *Bremia lactucae* (on lettuce) and *Plasmopara viticola* (on grape).

**CONTROL** Remove infected leaves promptly and improve air circulation around plants by increased spacing and good weed control. Increase ventilation in a greenhouse.

Avoid overhead watering. Spray infected plants with a suitable fungicide such as mancozeb.

• *see also* BRASSICA DOWNY MILDEW *p.110*; GERANIUM DOWNY MILDEW *p.131*; HEBE DOWNY MILDEW *p.135*; LETTUCE DOWNY MILDEW *p.144*; ONION DOWNY MILDEW *p.153*; PANSY DOWNY MILDEW *p.154*.

## DROUGHT

**SYMPTOMS** Symptoms vary depending on the plant affected and whether the drought has been recurrent or occasional. Common symptoms include poor growth and stunting, wilting of the foliage (*see p.28*) and, in extreme cases, other plant parts such as flower stems, stems and flowers (*see p.44*). Prolonged drought may also cause poor flowering, bud drop and the formation of small fruits. Drought followed by sudden watering or rainfall may cause splitting or cracking of fruits and stems.

**CAUSE** Most commonly, inadequate rainfall or watering; occasionally, an inability of the plant to take up adequate moisture because of a damaged root system. Very free-draining soils, such as thin soils over chalk or soils with a high sand content, are particularly prone to drying out. Plants with restricted root runs, such as those growing in containers, are also particularly prone.

**CONTROL** Take any measures possible to ensure that the soil or compost never dries out completely. Where feasible, grow plants in open ground, or protect containers from the effects of direct sunlight. Improve soil moisture retention by incorporating bulky organic

materials such as compost into light soils. Apply a mulch to improve soil moisture retention and reduce loss by evaporation.

## DRY ROT OF BULBS AND CORMS

**SYMPTOMS** Foliage turns brown and dies, often with a clear dark discoloration just above soil level. Crocus, freesias, gladioli and montbretias (*Crocosmia*) are particularly affected. Whole plants may die back. The corms develop numerous small sunken lesions, each of which bears many pinprick-sized black dots or fungal fruiting bodies.

**CAUSE** The fungus *Stromatinia gladioli*, which can overwinter on stored corms and which may persist in the soil as small, hard, black resting bodies (sclerotia). These can remain viable in the soil for many years even in the absence of host plants.

**CONTROL** Always check bulbs are healthy and free from the small black sclerotia before planting. Remove any infected plants immediately. Plant replacement plants, and any others which are susceptible, on a fresh site.

## DUTCH ELM DISEASE

**SYMPTOMS** Areas of the tree's crown show wilting and leaf yellowing. Affected leaves then turn brown and die. Dieback of twigs and branches usually follows and young shoots may curl over to form "shepherds' crooks". The whole tree may be killed by the end of the first summer. If the bark on affected stems and branches is peeled away, longitudinal, discontinuous brown staining is apparent. If stems or branches are cut across, they may show rings of brown staining in the current year's growth. Feeding damage caused by the elm bark beetle may be seen in the crotches of young stems. The "galleries" of tunnels caused by the beetle which spreads the infection show beneath the bark on larger stems and trunks. Symptoms are much more obvious and the disease more damaging during hot, dry weather, as these conditions exacerbate the problems caused by the fungus blocking the tree's vascular system.

**CAUSE** The fungus *Ceratocystis ulmi*, which may cause this disease on elms and on *Zelkova*.

**CONTROL** There is currently no reliable chemical control method. Research is, however, being carried out into the breeding of resistant elms. Remove infected trees and dispose of by burning.

## EARTHWORMS

Earthworms (*see* **p.61**) are generally beneficial animals that improve soil condition. Their tunnels provide aeration and drainage pathways through the soil, and worms help incorporate organic matter by pulling dead leaves into the soil. They can, however, be a nuisance when they excrete small piles of soil onto the surface of lawns (*see* **p.58**). These worm casts can spoil the appearance of a fine lawn, as well as making the surface muddy and slippery. When lawns are mowed, worm casts become smeared,

and this creates patches of bare soil that lawn weeds can colonize. Not all earthworms produce worm casts. The culprits are usually the *Allolobophora* species of earthworms. They are mainly a problem during spring and autumn when warm, moist soil conditions encourage worm activity in the upper layers of the soil; drought or cold sends worms deeper and reduces surface casting.

**CONTROL** There are currently no chemicals approved for earthworm control available to amateur gardeners. Worm casts can be dispersed by brushing the lawn with a besom broom or using an up-turned wire rake. This is best done when worm casts are dry and crumbly. During wet periods, walking on the lawn should be kept to a minimum to avoid smearing the worm casts.

## EARWIGS

Earwigs, *Forficula auricularia*, are up to 14mm long, tawny brown in colour and easily recognised by the pair of pincers on the rear end of the body. They eat ragged holes in the young foliage and flowers of many plants (*see* **p.46**), especially clematis, dahlias and chrysanthemums. During the day they hide in dark crevices or amongst the foliage and petals of host plants. They emerge and feed after sunset, so a torchlight inspection is often needed to find the culprits.

ACTUAL LENGTH: UP TO 16MM

**EARWIG**

**CONTROL** Vulnerable plants can be given some protection by placing earwig traps made of flower pots inverted on top of bamboo canes and placed level with the plant tops. The pots are loosely stuffed with hay or straw and will be used as daytime hiding places by earwigs. Inspect the pots daily and destroy any earwigs found inside. If earwigs are numerous and still cause damage, spray the plants with pirimiphos-methyl, fenitrothion, permethrin or bifenthrin at dusk on a mild evening.

## EELWORMS

Eelworms or nematodes are mostly microscopic worm-like animals, some of which are important plant pests. Chrysanthemum eelworms (*see p.117*), narcissus eelworms (*see p.150*), onion eelworms (*see p.153*), phlox eelworms (*see p.159*), potato cyst eelworms (*see p.163*) and root knot eelworms (*see p.170*) all feed within their host plants, but there are others, such as *Pratylenchus, Longidorus, Trichodorus* and *Xiphinema* species, that live in the soil and attack the root hairs of plants. The last three are able to transmit some viral diseases of ornamentals and soft fruits.

**CONTROL** There is no effective chemical control for soil-dwelling eelworms available to gardeners. Parts of gardens from which virus-infected plants have been removed should not be replanted with the same types of plant, otherwise they may quickly be infected by the eelworms. Not all eelworms are pests however; the vast majority feed on bacteria, fungi and other micro-organisms, while some

are predators. Slugs (*see p.176*) and vine weevils (*see p.186*) can be controlled by beneficial eelworms.

## ELEPHANT HAWK MOTH

**SYMPTOMS** Various garden plants, but particularly fuchsias, are defoliated (*see* **p.30**) by green or, more usually, brownish-black caterpillars. When fully grown these are up to 80mm long and as thick as a man's finger. They have two pairs of eye spot markings on the front end of the abdomen. When the caterpillar is disturbed it retracts its head into its body, causing the front of the abdomen to swell into a false head and making the eye spots more prominent. In this state the caterpillar can be mistaken for a small snake.

**CAUSE** Caterpillars of a hawk moth, *Deilephila elpenor*, which are active during mid- to late summer. Feeding is generally done at night, and the caterpillars rest on the underside of leaves during the day. Despite their size, the caterpillars often go unnoticed until they have completed their feeding and are wandering about looking for somewhere to pupate.

**CONTROL** Damage to garden plants is infrequent. If the caterpillars are found they can be removed by hand and perhaps transferred to one of their wildflower hosts such as rose bay willow herb or Himalayan balsam.

## ELM GALL MITE

**SYMPTOMS** Elm leaves develop hard yellowish-green swellings on the upper surface (*see* **p.16**) during mid- to late spring that persist throughout summer.

**CAUSE** A microscopic gall mite, *Aculus ulmicola*, which begins feeding on the foliage in the spring. Chemicals injected into the leaves by the mites induce the galls to develop. They are hollow and the mites spend the summer months feeding and breeding within these structures before emerging in late summer and over-wintering under bud scales. **CONTROL** Amateur garden pesticides will not control gall mites. Fortunately, apart from creating galls, this mite does little harm to elm trees and so can be tolerated.

## ENCHYTRAEID WORMS

These small animals are related to earthworms and are also sometimes known as "white" or "pot" worms. They have white bodies which are up to 20mm long and 1mm thick. They can be mistaken for eelworms or nematodes; but where these have completely smooth, unsegmented bodies, enchytraeid worms have segmented bodies, visible with the aid of a hand lens. Enchytraeids cause no harm as, like earthworms, they feed on decaying organic matter in the soil. They are sometimes abundant in soil that has been heavily dressed with farmyard manure or similar composts. They may also be found amongst the roots of a dead plant but would not be responsible for the plant's condition.

## EUCALYPTUS SUCKER

**SYMPTOMS** Grey or yellowish-orange insects, up to 2mm in length, cluster at the shoot tips of eucalyptus. The adults resemble winged aphids and are the overwintering stage.

They lay yellow eggs at the shoot tips in early spring, and all stages in the life cycle can be found together during the summer. The nymphal stages are wingless, flattened insects that secrete wispy white wax filaments from their bodies. Heavy attacks cause stunted growth, and sooty moulds grow on the suckers' excrement. **CAUSE** A sap-feeding insect, *Ctenarytaina eucalypti*, which originates from Australia but is now wide-spread in European gardens. **CONTROL** Infestations on mature trees have to be tolerated because of the difficulty of spraying thoroughly. Small trees can be treated with malathion, pirimiphos-methyl, permethrin or bifenthrin.

## EUONYMUS SCALE

**SYMPTOMS** This is a specific pest of evergreen euonymus, particularly *Euonymus japonica*. The foliage becomes heavily coated with elongate, whitish scales about 2mm long (*see p.21*), which are the males. The females mainly cluster on the stems and are brownish-black, pear-shaped objects, up to 3mm in length, which are well camouflaged against the bark. Heavy infestations cause yellowish discoloration of the foliage, plants lack vigour and die back. **CAUSE** A sap-feeding scale insect, *Unaspis euonymi*. In Britain it is mainly confined to the south coast of England, but can occur further inland. **CONTROL** The newly-hatched scale nymphs are the most vulnerable stage and are present in early summer and early autumn. Spray at those times with malathion if the scale is present.

## FAIRY RINGS

See TURF FAIRY RINGS *p.183*.

## FASCIATION

**SYMPTOMS** Shoots, or less commonly flower stems, become enlarged and flattened, developing a ribbon-like, often slightly curled appearance. Leaves, buds and flowers develop normally on the distorted stem (*see p.43*). Sometimes many stems are affected simultaneously, but more commonly the majority of the plant remains unaffected. Where weather is involved, stems on the more exposed areas of the plant are more likely to be affected. **CAUSE** Early injury to the growing point, such as that caused by frost or insect attack, slug damage or some form of mechanical injury. It has been suggested that viral or other infections may also be involved. **CONTROL** Although rather peculiar in appearance, fasciation is harmless and does not affect the overall growth and development of the plant. Prune out affected stems, if desired.

## FELT GALL MITES

**SYMPTOMS** Various trees and shrubs, particularly alder, beech, birch, crab apple, lime, maple and sycamore, develop a dense coating of hairs on the underside of the leaves. These hairs are usually creamy-white, but on purple-leaved trees such as copper beech, some maples and crab apples, the hairy patches will be reddish-pink (*see p.19*). Later in the

summer the hairs dry up and become brownish. On some plants, such as walnut and grape vine, there is an up-ward bulging of leaves where felt galls have developed on the undersides. **CAUSE** Microscopic mites of various species which are usually specific to a particular plant species. The mites feed by sucking sap from the foliage and while doing so they secrete chemicals which induce the growth of hairs that form felt galls. **CONTROL** Light infestations on small trees can be checked by picking off affected leaves. No pesticides are available for controlling gall mites, so heavier infestations have to be tolerated. Fortunately, apart from creating the galls, these mites have little impact on the host plant's growth.

## FELTED BEECH COCCUS

See BEECH BARK SCALE *p.106*.

## FIGWORT WEEVILS

**SYMPTOMS** Small beetles, which are mainly black and mottled with brownish-white markings, infest the shoot tips of plants such as figworts (*Scrophularia*), phygelius, buddleja and verbascum. The beetles have a snout-like projection on their heads and there is usually a black circular mark where the wing cases meet. Both the adults, which are about 4mm long, and their brownish-yellow slimy larvae feed on foliage and flowers. **CAUSE** Several species of weevil, *Cionus* species, which have two generations between late spring and late summer. When fully fed, the larvae pupate inside

brownish cocoons on the shoot tips and these can be mistaken for seed pods.
**CONTROL** Inspect plants for signs of infestation and if necessary spray with bifenthrin, pirimiphos-methyl or permethrin.

## FIREBLIGHT

**SYMPTOMS** Flowers wilt and wither, then die back, followed by adjacent leaves and stems shortly afterwards (see **p.35**). The plant may show extensive dieback and be killed within a few seasons. Infections usually occur scattered throughout the crown of the tree or shrub, healthy stems often growing side by side with infected ones. Infection commonly occurs at flowering time. The bark sinks inwards on young stems and, if pared back, a foxy-red discoloration is visible on the wood. Bacterial ooze may be produced from affected areas. The symptoms may be confused with those of blossom wilt (see p.108), but the infection only occurs in pome-fruit (apple-like) bearing members of the plant family Rosaceae.
**CAUSE** The bacterium *Erwinia amylovora*, which is encouraged by warm, wet weather. The bacteria usually enter the plant via the open flowers, but infection may also take place through stem injuries. Bacteria may be carried by rain splash and spread on pruning tools. In certain areas of the United Kingdom (Northern Ireland, Isle of Man and the Channel Islands) fireblight is still a notifiable disease, so cases in gardens should be reported to the local Ministry of Agriculture, Fisheries and Food.

**CONTROL** Prune out affected stems to a point at least 60cm into healthy wood. Dispose of all prunings by burning, and sterilize all tools after use on an infected stem. If the infection is widespread, or the plant small, it may be better to remove it entirely.

## FISTULINA HEPATICA (BEEFSTEAK FUNGUS)

**SYMPTOMS** Soft bracket fungi, usually growing singly and measuring 5–30cm in diameter, develop on branches, stems and tree stumps between late summer and late autumn. They are tongue-shaped or become semi-circular as they enlarge. Cream-coloured when young, they become reddish-brown and very juicy as they mature, the flesh often being marbled and resembling a slice of raw steak, giving rise to their common name. If cut or otherwise injured, blood-red juice is produced. The fruiting bodies produce spores, mainly from pores on the lower surface of the fruiting body. The wood on affected trees develops a streaky brown discoloration, called "brown oak" and later cubical cracking occurs. The fruiting bodies disintegrate with the onset of cold weather and frosts.
**CAUSE** The fungus *Fistulina hepatica*, which may enter trees via wounds and may cause the wood to become so brittle that it is weakened.
**CONTROL** Remove fruiting bodies in order to limit spore spread. The fungus itself will continue to remain active within the tree. An inspection of the tree to determine its health and safety is advisable.

## FLATWORMS

There are many flatworms native to Britain; most live in water but some occur in soil. They are mostly small animals which cause no damage to plants. In recent years two species from Australasia have become established in Britain and are causing concern because they prey on earthworms and can reduce them to very low numbers. The loss of earthworms leads to impaired drainage and aeration of the soil, with consequent effects on plant growth. The New Zealand flatworm (*Artioposthia triangulata*) is up to 17cm long when fully extended, and is purplish-brown on its upper surface, although the colour fades towards the sides of the body. It is widely distributed in Ulster and Scotland and is spreading southwards, particularly down the western side of Britain. The Australian flatworm (*Geoplana sanguinea*) mainly occurs in south-west England and is up to 10cm long. It is orange-brown in colour. Both species are mainly nocturnal and can be found during the day curled up under stones, logs or other objects which are in close contact with the soil. These animals are flat to the point of being ribbon-like and have entirely smooth, unsegmented bodies. They can be confused with soil-dwelling leeches, which are flattish but have fine lines running across their bodies to indicate their segmented body structure. Leeches also have a sucker underneath the rear end of their bodies, lacking in flatworms.
**CONTROL** There is currently no control for flatworms and it is likely that they will

become more widespread. They do not move far by their own efforts but can easily be transported amongst the roots of container-grown plants. When receiving or purchasing plants, inspect the rootballs carefully. Immersing the pots in water for a couple of hours will encourage flatworms to emerge and be seen. Infested plants should be rejected or repotted in clean compost after all the old soil has been washed from the roots to remove flatworms and their shiny black egg capsules.

## FLEA BEETLES

**SYMPTOMS** Small beetles, mostly 2mm long but sometimes 3–4mm, readily jump off foliage when disturbed. They are shiny black, sometimes with a yellow stripe running down each wing case, or can be metallic blue or yellowish-brown in colour. They eat small, rounded holes in the upper leaf surface (see **p.29**) which often do not go all the way through the leaf. The remaining damaged tissues dry up and turn brownish-white. Heavy attacks can kill seedlings and check the growth of older plants.
**CAUSE** There are many different species of flea beetle. On brassicas, including radish, swede and turnip, it is mainly *Phyllotreta* species and these also attack wallflowers, alyssum, stocks, aubrieta and nasturtiums. Potatoes and ornamental *Solanum* species are sometimes damaged by *Psylliodes affinis*. The damage is caused by the adult beetles, with the larvae living in the soil and feeding on roots. On fuchsias, both adults and larvae of a

metallic blue species, *Altica lythri*, feed on leaves at the shoot tips. Flea beetles are active between mid-spring and late summer.

ACTUAL SIZE

**FLEA BEETLE**

**CONTROL** Protect seedlings by sowing when weather and soil conditions will allow rapid germination and growth through the vulnerable seedling stage. Keep young plants watered during dry spells. If necessary, control with derris, bifenthrin, lindane or pirimiphos-methyl.

## FLIES

Flies are characterized by having only one pair of wings; the hind wings are reduced to small club-shaped structures called halteres. There are a number of fly pests of garden plants and in all cases damage is caused by the larval stage rather than the adults.

**Root and seedling pests** – *see* BEAN SEED FLY *p.105*; CABBAGE ROOT FLY *p.112*; CARROT FLY *p.114*; FUNGUS GNATS *p.129*; LEATHERJACKETS *p.143*; ONION FLY *p.153*.

**Bulb pests** – *see* NARCISSUS BULB FLY *p.150*.

**Foliage pests** – *see* BEECH GALL MIDGES *p.106*; BEET LEAF MINER *p.106*; BLACK-CURRANT GALL MIDGE *p.108*; CELERY LEAF MINER *p.115*; CHIRONOMID MIDGES *p.116*; CHRYSANTHEMUM LEAF MINER *p.117*; DELPHINIUM LEAF MINER *p.122*; GLEDITSIA GALL MIDGE *p.133*; HOLLY LEAF MINER *p.136*; LABURNUM LEAF MINERS *p.141*; VIOLET GALL MIDGE *p.186*.

**Flower pests** – *see* HEMERO-CALLIS GALL MIDGE *p.135*.

**Fruit pests** – *see* PEAR MIDGE *p.157*.

## FOMES POMACEOUS

**SYMPTOMS** Hoof-shaped or rounded fungal fruiting bodies develop on the bark of dead or living stems or trunks of various deciduous trees (*see p.39*). They have a woody texture and measure about 5cm in diameter. The upper surface is pale grey-brown and may show concentric furrows. The margin of the developing fruiting body is velvety grey, later becoming cinnamon-coloured. The wood within the stem may decay, turning the centre of the stem or trunk crumbly and white with a dark brownish-purple zone around the edge.

**CAUSE** The fungus *Fomes pomaceous* (syn. *Phellinus pomaceous*) which produces spores from minute pores on its undersurface. The spores infect via wounds caused by pruning or other mechanical injury.

**CONTROL** Remove fruiting bodies to minimize spread of spores. However, the decay will continue within the branch or trunk. Have the tree inspected to determine its overall health and safety.

## FOMES ROOT AND BUTT ROT

**SYMPTOMS** A dark reddish-brown, woody-textured bracket fungus (*see p.37*), usually measuring 5–10cm across but occasionally up to about 30cm in diameter, develops on the trunk or butt of trees, usually at ground level. Infection is usually seen on conifers but occasionally deciduous species such as birch, beech,

alder and oak are attacked. The lower surface of the bracket is creamy-white and it is from the minute pores on this undersurface that the spores are released. The fungus causes extensive deterioration of the woody tissues within the tree, causing rotting and deterioration of the wood. Trees are rendered unsafe, and may occasionally be killed. The fruiting bodies are fairly frost-resistant and usually perennial.

**CAUSE** The fungus *Heterobasidion annosum* (syn. *Fomes annosus*), which may infect stumps through cut surfaces, and then infects nearby trees by root contact.

**CONTROL** Trees are rendered unsafe, and because of the ease with which the infection spreads, should be felled promptly. Remove stumps, together with the soil in the immediate vicinity, to minimize the risk of spread.

## FOOT AND ROOT ROTS

**SYMPTOMS** Deterioration of the tissues around the stem base causes the upper parts of the plant to wilt, discolour and die back. Soft-stemmed plants, in particular seedlings and cuttings, are commonly affected, but most annual, biennial and herbaceous perennial plants may show symptoms. Tomatoes, cucumbers and melons, especially those growing in greenhouses, are frequently affected. The tissues at the stem base may darken, atrophy, become discoloured (usually darkening) and, occasionally soften. Root rotting may occur simultaneously (*see p.55*).

**CAUSE** Various fungi, often similar to or the same as those responsible for

damping-off disease of seedlings (*see p.122*). Other fungi, including various species of *Fusarium* and *Verticillium*, may cause both wilt and foot rot symptoms.

**CONTROL** Once foot rot symptoms have developed, there is no cure available for infected plants. It may, however, be possible to prevent spread of the infection by very prompt removal of any infected plants, together with the compost or soil around their roots. Unhygienic cultivation techniques, including the use of unsterilized compost, dirty trays, pots and tools and the use of non-mains water, are often responsible for introducing or spreading this type of infection.

## FORSYTHIA GALL

**SYMPTOMS** Numerous galls develop along the length of stems (*see p.40*). The whole or just certain areas of the plant may show the symptoms. Individual galls are usually up to 10–15mm in diameter, rough-surfaced and commonly near-spherical; occasionally several may be joined or fused together. They are the same colour as the forsythia stems. Overall growth and development of the plant is rarely affected but the plant's shape and outline can be spoiled.

**CAUSE** Unknown.

**CONTROL** If the forsythia becomes disfigured, prune out affected stems to a point several inches beneath the lowest gall.

## FOXES

Foxes are becoming increasingly at home in urban gardens and often make their dens under garden sheds. They cause damage by digging up or trampling plants. Dog foxes use their faeces and urine to mark out their territories and this can create an unpleasant nuisance. Fox urine can scorch the foliage of plants, especially conifers, and has a distinctive pungent smell.

**CONTROL** It is difficult to prevent foxes from coming into a garden. They are agile animals that are as good as cats when it comes to scaling walls and fences, but have the additional ability to dig underneath obstacles. Proprietary animal repellents have little lasting effect. In gardens where foxes are a problem, the use of bonemeal or dried blood should be avoided. Foxes dig up plants given these fertilizers because the smell makes them think there is food beneath.

## FROGHOPPERS

*See* CUCKOO SPIT *p.120.*

## FROST DAMAGE, APPLE

*See* APPLE FROST DAMAGE *p.100.*

## FROST DAMAGE (BUDS AND FLOWERS)

**SYMPTOMS** Buds, whether tightly closed or partially opened, discolour, usually turning brown and sometimes becoming soft and squashy. Petals of open blooms discolour, usually turning brown, and either retain their shape or wither (*see **p.44***). The buds or flowers on the more exposed areas of the plant are usually the worst affected. Frost may only damage buds or flowers at certain developmental stages, killing some,

damaging others and leaving some untouched. On fruiting plants, such as plums, there may be subsequent partial or total loss of crop.

**CAUSE** Frost, which is most likely to cause damage on slightly tender plants or those which have put on precocious growth. Late frosts, occurring once flower buds have opened, can also be very damaging. Plants which are subjected to the warming of early morning sun following overnight frosts are especially prone to injury.

**CONTROL** Choose planting positions carefully, especially for early flowering plants, or those known to be readily frost-damaged. Provide protection using horticultural fleece or similar material.

## FROST DAMAGE (FOLIAGE AND STEMS)

**SYMPTOMS** Foliage, usually towards the shoot tips and particularly on the more exposed parts of the plant, turns brown or black or appears scorched (*see **p.28***) and may wilt, wither and die back. Young stem growth adjacent to the frosted leaves may also die back. Isolated patches on otherwise sound stems may develop, but this is less common.

**CAUSE** Frost, which is most likely to cause damage on slightly tender plants or those which have put on precocious growth. Late frosts, occurring once flower buds have opened, can also be very damaging. Plants subjected to the warming of early morning sun following overnight frosts are especially prone to injury.

**CONTROL** Choose planting positions carefully, especially for plants known to be

readily frost-damaged. Provide protection using horticultural fleece or similar material. An autumn dressing of sulphate of potash helps to harden stem growth and so may help to minimize frost damage.

## FRUIT TREE RED SPIDER MITE

**SYMPTOMS** The foliage of apples and plums loses its glossy green colour and becomes increasingly dull and yellowish-green (*see **p.20***). There is a fine pale mottling of the upper surface and large numbers of tiny mites and spherical eggs (*see **p.40***) occur on the lower leaf surfaces. Heavy infestations can develop in hot, dry summers and result in early leaf fall.

**CAUSE** Sap-feeding mites, *Panonychus ulmi*, which are blackish-red in colour and less than 1mm long. This mite is generally not a problem on unsprayed trees where predatory mites and insects keep it at a low level. Spraying often kills beneficial mites and insects, allowing spider mites to breed unchecked. Overwintering eggs are laid in crevices on the underside of branches and can be so numerous that the bark has a reddish hue.

**CONTROL** Avoid unnecessary spraying against other pests as this may lead to spider mite problems. If spider mite needs controlling, spray thoroughly with bifenthrin, pirimiphos-methyl or fenitrothion. Three applications at weekly intervals may be needed.

## FUCHSIA RUST

**SYMPTOMS** Numerous patches of pale orange or deep yellow fungal pustules

develop on the lower leaf surface, older leaves towards the base of the plant usually showing symptoms first. The upper leaf surface shows corresponding areas of yellow or tan-brown discoloration (*see **p.18***), followed by necrosis. Infected leaves fall prematurely, leaving leggy plants which may be killed.

**CAUSE** The fungus *Pucciniastrum epilobii*, which is encouraged to spread and infect by moist or overcrowded growing conditions and high humidity. Ornamental and weed species of *Epilobium* (willowherbs) may also be infected and so may harbour the fungus. In addition, the fungus may develop on the foliage of firs (*Abies*) but infection is much more common on fuchsia than on either of the other genera.

**CONTROL** Remove infected leaves promptly. Spray with a fungicide containing mancozeb. Other rust-controlling fungicides should be avoided as they may cause damage to fuchsias foliage. Check willowherbs growing in the vicinity and remove them or control infections as appropriate.

## FUNGUS GNATS

**SYMPTOMS** Greyish-brown flies with slender bodies, 3–4mm long, run over the soil surface of pot plants or fly slowly and alight on leaves. The larvae are whitish maggots, up to 5mm long, with black heads (*see **p.53***). They live in the soil and feed mainly on decaying organic matter, such as dead roots, but may damage seedlings or the base of soft cuttings.

**CAUSE** Several species of fungus gnats occur in greenhouses and on house plants. Most are harmless but some *Bradysia* species damage fine roots and tunnel in soft stems.

ACTUAL LENGTH: UP TO 5MM

**FUNGUS GNAT LARVAE**

**CONTROL** The adult flies can be trapped by hanging sticky yellow strips close to soil level. A predatory mite, *Hypoaspis miles*, which feeds on the maggots can be purchased by mail order from biological control firms. On established plants, fungus gnats are a nuisance rather than a pest. If seedlings or cuttings are being damaged, treat the compost with spray-strength permethrin or lindane dust. Maintain good hygiene in the greenhouse and remove dead leaves and flowers from the soil surface.

## FUSARIUM BULB ROT

*See* GLADIOLUS CORM ROT *p.131*; NARCISSUS BASAL ROT *p.149*.

## FUSARIUM WILT

**SYMPTOMS** Soft-stemmed plants wilt, either in part or in their entirety. Woodier-stemmed plants may retain their overall shape, but foliage on affected stems wilts and withers. Internal staining of the vascular tissue (conducting elements in the stems and roots) may occur and is usually brown or black. This may be apparent just beneath the bark on the stems or within the central area of the roots. Death almost always follows, either rapidly or over several

seasons, depending on the plant affected. Cool, damp conditions may encourage very pale pink or white, slightly fluffy fungal growth from the infected tissues.
**CAUSE** Various types of the fungus *Fusarium*, most commonly *F. oxysporum*. The fungi involved are usually fairly host-specific, attacking only one genus or a few very closely related genera of plants. The fungi are responsible for the whole or partial blocking (usually by gum-like formations) of the vascular tissues, so the symptoms are very similar to those caused by drought. Dry soil conditions may encourage rapid wilt symptoms, but unlike plants suffering from drought, those affected by fungal wilts do not recover permanently when watered. The fungus may persist in plant debris and may also be capable of remaining viable in the soil for several years, even in the absence of any suitable host plants. In *Fusarium oxysporum*, the specialized spores responsible are known as chlamydospores.
**CONTROL** There is no chemical control available. Remove infected plants promptly to minimize spread. Where feasible, remove soil or compost from the immediate vicinity of the roots. Do not grow the same or closely related plants on the site again, or at least for a minimum of five years.

## GALL MIDGES

Gall midges are tiny flies, usually no more than 2–3mm long, and most of them have

larvae that feed on plants. The maggots are orange-white and secrete chemicals that affect the development of plant parts and cause them to produce galls that enclose the larvae. These midges lay their eggs on a particular part of their host plant, such as developing leaves, shoot tips, flower buds or fruits, and the galls that subsequently develop have a distinctive and characteristic shape that enables the pest to be recognised. Not all members of the gall midge family are pests; one species, *Aphidoletes aphidomyza*, has larvae that feed on aphids and is used to give biological control in greenhouses.
• *See also* BEECH GALL MIDGES *p.106*; BLACKCURRANT GALL MIDGE *p.108*; GLEDITSIA GALL MIDGE *p.133*; HEMEROCALLIS GALL MIDGE *p.135*; PEAR MIDGE *p.157*; VIOLET GALL MIDGE *p.186*.

## GALL MITES

These are microscopic animals that can be found on many plants, particularly trees and shrubs. The mites have tubular bodies with just two pairs of legs. Although too small to be seen without magnification, the effect that gall mites have on their host plants is often very characteristic. They feed mainly on foliage or in buds and they secrete chemicals that induce the plant to make abnormal growths. On foliage, gall mites cause raised structures, thickened and curled leaf margins, blisters, excessive growth of hairs or discoloured patches. Bud-feeding gall mites cause swollen buds which later dry up or proliferate into cauliflower-like structures, or shoots develop as witches' brooms. With a few

exceptions, most gall mites have little harmful effect on their host plants and the presence of galls can be tolerated.
• *See also* ACER GALL MITES *p.98*; BIG BUD MITES *p.107*; BROOM GALL MITE *p.110*; ELM GALL MITE *p.125*; FELT GALL MITES *p.126*; LIME NAIL GALL MITE *p.146*; MOUNTAIN ASH BLISTER MITE *p.149*; PEAR LEAF BLISTER MITE *p.157*; PLUM LEAF GALL MITE *p.161*; RASPBERRY LEAF AND BUD MITE *p.167*; RED BERRY MITE *p.168*; VINE ERINOSE MITE *p.186*; WALNUT BLISTER MITE *p.187*.

## GALL WASPS

Gall wasps are black or brown insects up to 5mm long, which are mainly associated with oaks and roses. The gall structures are produced by the plant in response to chemicals secreted by the larval stages, which are white, legless grubs. Although galls are sometimes numerous they have no harmful effect and so control measures are unnecessary.
• *See also* ACORN GALL WASP *p.98*; OAK GALL WASPS *p.152*; ROBIN'S PIN CUSHION *p.170*.

## GANODERMA

**SYMPTOMS** Large, brown or tan-brown, tough fungal brackets develop, most frequently towards the base of the trunk. The lower surface of the bracket is usually a cream or pale brown colour and perforated by numerous closely packed, tiny pores, through which the spores are released. Initially, brackets may be rounded in outline, but they become increasingly bracket-shaped and flattened as they enlarge. They are perennial and woody in texture. Large quantities of dark brown spores are produced and may cover the vegetation beneath. Affected trees may initially show little sign of deterioration, but the woody tissues within the trunk may be seriously weakened, rendering the tree potentially dangerous. The tree may become thin in the crown, with a sparse covering of leaves, and show signs of dieback.

**CAUSE** Various species of the fungus *Ganoderma*.

**CONTROL** Remove the bracket to prevent further dissemination of spores. The damage will, however, continue within the tree, so it must be inspected by a reputable tree surgeon or arborist to determine its overall safety. In most instances the tree has to be felled.

## GERANIUM DOWNY MILDEW

**SYMPTOMS** Pale green or light brown, often angular areas of discoloration develop on the leaves (*see* **p.25**) of *Geranium* species and hybrids. Severely affected leaves wither and die. Off-white, slightly fuzzy fungal patches develop on the lower leaf surface, corresponding with the discoloured areas. Symptoms may be confused with the much whiter fungal growth which occurs on upper leaf surfaces and stems and is due to powdery mildew (*see p.165*).

**CAUSE** A species of the fungus *Peronospora*, which is encouraged to spread by moist, humid conditions and poor air circulation.

**CONTROL** Remove infected leaves promptly. Spray with a fungicide containing mancozeb, taking care to spray lower leaf surfaces thoroughly.

## GERANIUM SAWFLY

**SYMPTOMS** This pest attacks *Geranium* species and hybrids, not pelargoniums. Rounded holes appear in the foliage and greyish-green caterpillar-like larvae, up to 12mm long, may be found on the underside of leaves, although the larvae often drop to the ground when disturbed.

**CAUSE** Larvae of a sawfly, *Protoemphytus carpini*, which has two generations between late spring and late summer.

**CONTROL** By late summer the foliage may be extensively holed but this has little effect on the plant's growth, so control may be unnecessary. If required, spray the undersides of the leaves with derris, pyrethrum, permethrin, bifenthrin or pirimiphos-methyl.

## GEUM SAWFLIES

**SYMPTOMS** Three species of sawfly occur on geums. One has leaf-mining larvae up to 4mm long that feed within leaves, causing the damaged areas to dry up and form large brown blotch mines. The other two have pale green larvae up to 13mm long with short, forked hairs on their upper surface. They feed on the young leaves in early summer and can cause severe defoliation.

**CAUSE** The leaf-mining sawfly is *Metallus lanceolatus* and it has two generations between late spring and late summer. The other two sawflies are *Claremontia waldheimii* and *Monophadnoides rubi*, both of which have a single generation, with larvae active in late spring to early summer.

**CONTROL** Light infestations can be picked off, or spray infested plants with derris, pyrethrum, permethrin, bifenthrin or pirimiphos-methyl. The last-mentioned would be best against geum leaf-mining sawfly larvae.

## GHOST SPOT

*See* GREY MOULD *p.134*.

## GLADIOLUS CORE ROT

**SYMPTOMS** A moist rot and blackening affects the central core of the corm. The base of the corm looks normal but is soft if squeezed. The rot spreads outwards until the whole corm becomes soft and brownish-black. Fluffy, white fungal growth develops beneath the outer scales. Black sclerotia (resting bodies) of the fungus develop on the surface in amongst the white fungal growth. Symptoms are usually noticed in storage. If a slightly diseased corm is planted, it may grow but produces only a few yellowed and stunted leaves, often with brown spots and red margins.

**CAUSE** The fungus *Sclerotinia draytonii*, which spreads rapidly during damp weather. The spores are spread by rain and water splash. The fungus may survive in the soil for many years, as the sclerotia are very resilient.

**CONTROL** Remove and dispose of plants showing foliage symptoms. Check all corms thoroughly before storing, and again at the end of storage before replanting. Allow plants to dry off naturally but quickly after lifting. After lifting, dip corms in a solution of carbendazim, or treat them with sulphur dust. Store corms in a cool, dry place. Plant corms on a fresh site each year.

## GLADIOLUS CORM ROT

**SYMPTOMS** Concentrically ridged brown markings develop on the corm, usually at its base and whilst in storage (*see* **p.55**). Crocuses and bulbous irises may also be affected. The corm itself dries out, becoming mummified but not developing fungal growth. If plants grow from infected corms, the foliage shows yellow flecking and later striping towards the tips of the outer leaves (*see* **p.55**). The discoloration spreads downwards and the leaves turn brown and die. Roots of affected corms blacken and may die back.

**CAUSE** The fungus *Fusarium oxysporum* f. sp. *gladioli*, which persists in the soil and usually enters the plant through the basal plate or roots of the corm.

**CONTROL** Dispose of plants showing foliage symptoms. Check all corms thoroughly before storing, and again at the end of storage before replanting. Dry plants off after lifting and dip corms in a solution of carbendazim, or treat them with sulphur dust. Store corms in a cool, dry place. Plant on a fresh site each year.

## GLADIOLUS DRY ROT

**SYMPTOMS** Corms develop numerous slightly sunken lesions which may join together. Rotting occurs and causes dark brownish-black areas of discoloration, often accompanied by black fungal resting bodies (sclerotia). If infected corms are planted, they may grow, but the foliage dies off and a dark brown rot occurs at the base of the foliage, later becoming covered in tiny sclerotia.

**CAUSE** The fungus *Stromatinia gladioli*, which can persist as sclerotia both on the corms and in the soil.
**CONTROL** Remove and dispose of plants showing foliage symptoms. Check all corms thoroughly before storing, and again at the end of storage before replanting. Dry plants off naturally after lifting and dip corms in a solution of carbendazim, or treat them with sulphur dust. Store corms in a cool, dry place. Plant on a fresh site each year.

## GLADIOLUS HARD ROT

**SYMPTOMS** Small, purple-brown spots form on leaves and develop grey centres bearing numerous pycnidia. Dark, reddish-brown to black areas appear towards the base of the corms. Affected areas of the corm often sink in and enlarge. Gladioli, crocus, acidantheras and freesias may be attacked.
**CAUSE** The fungus *Septoria gladioli*, which is encouraged by wet weather and is most severe on gladioli growing in poor soils.
**CONTROL** Dispose of affected corms; plant only those which seem perfectly healthy. Dip corms in a solution of carbendazim following the manufacturer's instructions.

## GLADIOLUS SCAB AND NECK ROT

**SYMPTOMS** Small, red-brown spots develop on leaves and these darken as they enlarge. Leaf tips become shrivelled. Pale yellow to golden, sunken lesions develop on the corms and produce a resin-like, bacterial ooze. In addition to gladioli, freesias and crocuses are commonly affected by this disease.

**CAUSE** The bacterium *Pseudomonas gladioli*, which is encouraged by wet weather.
**CONTROL** Do not plant affected corms. Dispose of any showing symptoms and once the disease has developed, do not grow susceptible plants on that site for at least five years.

## GLADIOLUS THRIPS

**SYMPTOMS** The foliage of gladioli develops a fine, pale mottling with tiny black excrement spots during mid- to late summer. Flowers also have pale flecking (*see p.45*) and in severe attacks, flower buds fail to open.
**CAUSE** Sap-feeding insects, *Thrips simplex*, which tend to be more troublesome in hot, dry summers. Elongate, narrow-bodied insects, up to 2mm in length, they live in the flower buds and at the bases of the leaves. Adult thrips are brownish-black while the immature nymphs are pale yellow.
**CONTROL** Watch for signs of infestation and spray with pirimiphos-methyl, malathion or dimethoate. Open flowers can be damaged by sprays, particularly if treated in sunny weather. At the end of the growing season burn infested foliage and seed heads. Some thrips over-winter on corms; store them in a cool, frost-free place, otherwise thrips continue feeding during the winter and weaken the corms.

## GLASSHOUSE LEAFHOPPER

**SYMPTOMS** A coarse, pale mottling occurs on the upper leaf surface (*see p.20*) of many indoor plants. This insect also attacks garden plants including primulas,

polyanthus, foxgloves (*Digitalis*) and nicotiana. The adults are pale yellow with greyish markings, 3mm long and broadest at the head end, tapering to a point behind. Adults readily jump off plants when disturbed and so it may be easier to find the less active, creamy-white, wingless immature nymphs. Both adults and nymphs live on the underside of leaves.
**CAUSE** Sap-feeding insects, *Hauptidia maroccana*, which breed continuously throughout the year in greenhouses and in sheltered places out of doors.
**CONTROL** Spray with permethrin, bifenthrin, heptenophos with permethrin, dimethoate or pyrethrum when leaf spotting is seen. Damaged leaves remain discoloured, but new growth remains green once the pest is controlled.

## GLASSHOUSE RED SPIDER MITE

**SYMPTOMS** Leaves develop a fine pale mottling (*see p.22*) and the foliage becomes firstly dull green, but later increasingly yellowish-white. These symptoms can be confused with a mineral deficiency but examination of the lower leaf surface, preferably with a x10 hand lens, will reveal tiny spider mites and their spherical eggs. The mites are less than 1mm long and have four pairs of legs; they are yellowish-green with two large dark markings towards the head end of the body. They only become orange-red in the autumn and winter when the adult females are hibernating in sheltered places. When plants are heavily infested, mites will crawl over a fine

silk webbing spun between the leaves and stems (*see p.43*). Leaves dry up (*see p.23*) and fall prematurely, so that only young leaves at the shoot tips remain.
**CAUSE** Sap-feeding mites, *Tetranychus urticae*, which attack a wide range of indoor and greenhouse plants. They also cause problems out of doors in warm dry summers, especially to beans, raspberries, strawberries, currants, roses and many other ornamental plants.

ACTUAL SIZE ▪

**RED SPIDER MITE**

**CONTROL** Red spider mites breed very rapidly under warm conditions, and there are strains which have gained resistance to some pesticides. Biological control with the predatory mite *Phytoseiulus persimilis* can give good results if it is introduced before a heavy infestation has developed. It also requires warm daytime temperatures and an absence of pesticides if it is to become established. *Phytoseiulus* can also be used on outdoor plants during the summer. Alternatively, plants can be sprayed with bifenthrin, pirimiphos-methyl, malathion or an insecticidal soap. Several applications will be necessary to break the red spider mite's life cycle.

## GLASSHOUSE WHITEFLY

**SYMPTOMS** White-winged insects, 2mm long, readily fly up when plants are disturbed. The adults and the immobile, whitish-green, oval, scale-like nymphs live on the undersides of leaves of many houseplants and greenhouse vegetables and ornamentals. The upper leaf surfaces become sticky with honeydew excretion and a black, sooty mould often develops on this substance. Glasshouse whitefly can spread to garden plants during the summer, but will not survive a cold winter.

**CAUSE** A sap-feeding insect, *Trialeurodes vaporariorum*, which breeds continuously throughout the year on indoor plants.

**CONTROL** Pesticide resistance is a widespread problem with this pest. Biological control with a tiny parasitic wasp, *Encarsia formosa*, is often the best remedy on indoor plants between mid-spring and mid-autumn when temperatures are high enough to suit the parasite. *Encarsia* must be introduced before a heavy infestation has developed, as it needs time to breed before it can start reducing the whitefly population. Sticky yellow traps hung just above plants will catch adult whiteflies and indicate when they are starting to appear in a greenhouse. Insecticidal soaps will reduce whitefly numbers and have no lasting impact on *Encarsia*. Other insecticides, such as permethrin, bifenthrin, pirimiphos-methyl and pyrethrum, are all dangerous to the parasite and may give poor control of whitefly if a resistant strain is present. Several applications at about five-day intervals will be required even if the whitefly is susceptible. Sooty mould can be removed easily from tomato and cucumber fruits by wiping them with a damp cloth.

## GLEDITSIA GALL MIDGE

**SYMPTOMS** Leaves at the shoot tips of honey locust (*Gleditsia triacanthos*) fail to expand into the normal pinnate form. The leaflets are folded over and make pod-like galls (see **p.27**) which contain up to seven whitish-orange maggots. These galls first appear in early summer; by late summer, infestations may be so severe that normal leaf production comes to a complete halt. The cultivar 'Sunburst' seems to be particularly susceptible and produces pale yellow galls, sometimes with a pink flush.

**CAUSE** A tiny fly, *Dasineura gleditchiae*, that originates from North America but arrived in Europe during the 1970s. Grey, and 2mm long, it has three generations, with female midges laying eggs in developing leaves at the shoot tips in late spring and mid- to late summer.

**CONTROL** Small trees can be sprayed with malathion, pirimiphos-methyl or fenitrothion when the midges are seen on the shoot tips, particularly against the first generation in late spring and early summer. Tall trees cannot be sprayed properly and this pest often has to be tolerated. Fortunately, the tree is able to make enough normal leaves to maintain growth before heavy infestations develop in mid- to late summer.

## GLEOSPORIUM ROT

**SYMPTOMS** Also known as bitter rot, this is seen as circular brown, somewhat sunken lesions on the surface of apples and occasionally quinces and pears while in store or, occasionally, while still on the tree. On close inspection

it may be possible to see many tiny fungal pustules arranged in concentric circles on the surface of the lesions. Occasionally a canker of old stems may also develop at the base of the spurs.

**CAUSE** The fungus *Glomerella cingulata*, which usually enters fruits through their lenticels (breathing pores in the skin).

**CONTROL** Remove infected apples promptly. Check apples in store regularly and remove any showing symptoms.

## GOOSEBERRY DIEBACK

**SYMPTOMS** Stems of gooseberries or sometimes blackcurrants die back. Occasionally the whole bush is affected, but usually only certain areas deteriorate while most of the bush appears perfectly healthy. Infection will spread, and if the main stem or base of the plant is affected, the whole plant may die back. During damp weather, fuzzy, grey fungal growth may develop around the damaged areas. Foliage on infected stems yellows, turns brown and dies back, falling prematurely. Any developing fruits shrivel and die.

**CAUSE** The fungus *Botrytis cinerea*. The spores are spread on air currents and by rain or water splash.

**CONTROL** Prune out infected areas promptly, removing any wood showing staining or discoloration and cutting back to a healthy-looking bud. The infection may also occur on the fruits, and these too should be removed promptly and the plant sprayed with a fungicide containing carbendazim to minimize the risk of spread of the infection.

## GOOSEBERRY MILDEW

See AMERICAN GOOSEBERRY MILDEW *p.98*.

## GOOSEBERRY RUST

**SYMPTOMS** In spring, reddish-orange pustules develop on the older, lower leaves and occasionally on the fruits and stems. In summer, the cluster cup stage of the fungus develops on the pustules, causing extensive distortion and curling of the leaves.

**CAUSE** The fungus *Puccinia caricina*, which may also attack sedges (its alternate host) and which is most damaging to gooseberries following dry weather in early spring.

**CONTROL** Remove infected areas promptly. Remove sedges growing in the vicinity, as these may harbour the infection. Spray with a suitable fungicide such as mancozeb.

## GOOSEBERRY SAWFLY

**SYMPTOMS** Gooseberry bushes become rapidly and severely defoliated (see **p.31**), often reduced to bare stems by the time the berries are ready for picking. Three species of sawfly can attack gooseberry. They have pale caterpillar-like larvae up to 20mm long, heavily marked with black spots in two of the species. Damage starts in mid- to late spring, but as there can be two or three generations, there may be further damage later in the season. Red and white currants are also attacked in a similar manner.

**CAUSE** Larvae of the common gooseberry sawfly (*Nematus ribesii*), the lesser gooseberry sawfly (*N. leucotrochus*) and the pale gooseberry sawfly (*Pristiphora pallipes*).

**CONTROL** Inspect gooseberry and currant bushes carefully for larvae from mid-spring. Eggs are often laid on the underside of leaves in the centre of bushes, so larvae may go unnoticed until they have eaten upwards and outwards and there are few leaves left. Spray affected plants with derris, pyrethrum, permethrin, bifenthrin, fenitrothion or pirimiphos-methyl. Use one of the first two if the fruit is ready for picking.

## GREENFLY

See APHIDS *p.99*.

## GREEN SPRUCE APHID

**SYMPTOMS** The foliage on spruce (*Picea*) develops a whitish-yellow mottling, turns brown and drops off during the late winter and spring. The new growth in the spring is not affected. Dark green aphids with red eyes can be found amongst the foliage, which is often blackened by sooty mould growing on the aphids' sugary excrement.
**CAUSE** A sap-feeding aphid, *Elatobium abietinum*. Unlike most greenfly, this species is active between autumn and spring and can be particularly damaging in mild winters.
**CONTROL** Spray small trees with pirimicarb, heptenophos with permethrin, malathion or pirimiphos-methyl in the autumn, and again if symptoms are noticed in the winter or spring. Badly defoliated plants will take several years to regain their appearance; encourage better growth by feeding with a general fertilizer in the spring and keep trees watered in dry summers.

## GREY MOULD

**SYMPTOMS** Grey or occasionally off-white or grey-brown fuzzy fungal growth develops on infected areas (*see p.34*). Grey mould is a very common fungus and, since it can live on most living and dead plant material, its spores are almost always present in the air. Grey mould may attack all above-ground parts of most plants, usually gaining entry via wounds or points of damage. Fruit infections may originate from infection via the open flowers, the fungus remaining dormant until the fruits start to ripen (*see p.50*). Prior to the development of the fungal growth, the infected plant tissue usually discolours, often turning brown and soft. Growth above points of infection may deteriorate, with leaves yellowing and wilting and flowers or fruits dying off (*see p.45*). Grey mould infection on petals or fruits may initially cause a hypersensitive reaction which results in colour changes but no rotting. On flowers, bleached white or pale brown spots (ghost spots) may form (*see p.42*). The fungus may also produce numerous small black sclerotia (resting bodies) which are capable of withstanding a wide range of growing conditions and then infecting plants when conditions are most favourable for its growth.
**CAUSE** The fungus *Botrytis cinerea*. Its spores are readily spread by rain or water splash and on air currents. The fungus may also carry over from year to year as sclerotia in the soil or on infected plant debris.
**CONTROL** Control is difficult, as the fungus is so widespread. Remove all dead and injured plant parts before they can become infected. Remove infected areas promptly, cutting right back into healthy growth. Do not leave plant debris lying around. Spray with a suitable fungicide such as carbendazim.

## GRIFOLA FRONDOSA

**SYMPTOMS** Overlapping, greyish-tan to honey-coloured fungal brackets develop around the base of a tree, on its roots or on tree stumps. The margin of each "frond" is white. The fruiting body may measure up to about 30cm across, the individual "fronds" measuring 20–60mm across. If bruised, the fungus has a distinct mouse-like smell. The fungus causes a butt rot of susceptible trees, and infected wood becomes soft and pulpy. White mycelium (a fungal sheet) develops in the decaying areas. The fruiting body usually appears in late summer or autumn and disintegrates with the first frosts.
**CAUSE** The fungus *Grifola frondosa*, easily confused with the much larger *Meripilus giganteus* (*p.147*).
**CONTROL** Affected trees may be rendered dangerous and so should be inspected by a reputable tree surgeon and felled if necessary.

## GROUND BEETLES

Ground beetles or carabids are mostly predatory insects that feed on a variety of insects and other invertebrate animals. They vary in size from 2–35mm and are mostly black, brown or bronze-green. They have long, slender legs that enable them to run rapidly when disturbed. Adult ground beetles live and feed mainly in or on the soil surface but some species will climb up plants at night to prey on aphids, small caterpillars and other insects. The larvae live in the soil and are also predatory. Some adult ground beetles feed on plant seeds and while most are harmless, damage can occur on strawberry fruits.
• *see also* STRAWBERRY SEED BEETLES *p.179*.

ACTUAL LENGTH: 2–30MM

**GROUND BEETLE**

## GUMMOSIS

See CUCUMBER GUMMOSIS, *p.120*.

## GYPSY MOTH

**SYMPTOMS** A wide range of trees and shrubs, including conifers, can be severely defoliated by caterpillars in early summer. The hairy, blackish-yellow caterpillars are gregarious and variable in colour but distinctive, due to a pair of blue warts on each of the front five segments of their bodies. The other segments have reddish warts. The caterpillars can be up to 70mm long.
**CAUSE** Caterpillars of a moth, *Lymantria dispar*. An Asiatic form of this moth has moved into Europe in recent years and is proving to be a more formidable pest than its European counterpart. It has a wider range of hosts and the females fly more readily, thus dispersing the pest more widely. Neither form of gypsy moth is established in Britain and suspected

sightings should be reported to the Forestry Commission or the local office of the Ministry of Agriculture, Fisheries and Food.

**CONTROL** Look for caterpillars, which hatch from overwintered eggs laid in batches on the bark, from mid-spring onwards. Small trees and shrubs can be treated with the biological control *Bacillus thuringiensis*, or sprayed with permethrin, bifenthrin, fenitrothion or pirimiphosmethyl. Small caterpillars are more susceptible than older larvae.

## HALO BLIGHT

*See* BEAN HALO BLIGHT *p.105.*

## HAWTHORN GALL MIDGE

**SYMPTOMS** In early summer, hawthorn shoot tips fail to extend and instead develop a rosette of small leaves. Several orange-white, legless larvae, up to 2mm in length, live within each rosette, which dries up and turns brown later in the summer.

**CAUSE** A small fly, *Dasineura crataegi*, that has several generations during the summer. Hawthorns grown as clipped hedges are always more heavily infested than uncut hawthorns.

**CONTROL** Little real damage is done by this pest, although the old, dried-up galls can be unsightly. There is no point in using insecticides; as only clipped hawthorns are particularly susceptible, the damaged shoots will be removed the next time the hedge or bush is trimmed.

## HEBE DOWNY MILDEW

**SYMPTOMS** Yellow or buff-coloured discoloration develops on the upper leaf surface, corresponding with off-white, fluffy fungal growth beneath. Infected leaves die and drop early, and may leave the plant with sparse foliage. Lower leaves and those in congested parts of the plant are most likely to be attacked.

**CAUSE** The fungus *Peronospora grisea*, which is encouraged by moist, humid conditions and poor air circulation.

**CONTROL** Remove infected leaves promptly and spray with a fungicide containing mancozeb. Destroy severely infected plants.

## HELLEBORE LEAF BLOTCH

**SYMPTOMS** Slate-grey or greyish-brown lesions develop on the leaves (*see* **p.20**). They may join together, leaving large leaf areas discoloured and dead. Flower stems and occasionally the flowers themselves may also be attacked, causing flower stems to flop over.

**CAUSE** The fungus *Coniothyrium hellebori*, which spreads especially rapidly in damp weather.

**CONTROL** Remove and dispose of affected leaves and other plant parts promptly. Even if the plant is left with only very sparse foliage, replacement leaves should soon be produced. In certain circumstances, *Helleborus orientalis* appears to show a good degree of resistance to this infection, and so in troublesome sites it may be preferable to concentrate on growing this rather than other species.

## HEMEROCALLIS GALL MIDGE

**SYMPTOMS** The flower buds of daylilies (*Hemerocallis*) become abnormally swollen but fail to open before going brown and drying up (*see* **p.45**). Petals inside a galled bud are thickened, with a crinkled appearance, and there are many white maggots, 2–3mm long, at the base of the petals.

**CAUSE** A tiny fly, *Contarinia quinquenotata*, that deposits its eggs in the developing flower buds during late spring and early summer. There is only one generation a year, so buds developing after the egg-laying period and flowering from mid-summer onwards escape damage.

**CONTROL** Pick off and destroy galled buds before the maggots escape into the soil to pupate. Chemical controls are likely to be ineffective since it is difficult to prevent the females from laying eggs, and the grubs are well protected inside the buds. By selecting hemerocallis cultivars with main flowering periods from mid-summer onwards, the problem can be largely avoided.

## HEMISPHERICAL SCALE

**SYMPTOMS** A wide range of indoor ornamental plants are susceptible to this pest. When mature, the scales (*see* **p.16**) have a dark brown, convex, hemispherical shell over their bodies. The shell is up to 4mm across and has an H-shaped ridge on it. The immature scales are flat, oval and yellowish-brown in colour. They occur on the underside of leaves and on stems. Their sugary excrement, called honeydew, falls down and makes upper leaf surfaces sticky. Sooty moulds often grow on this substance.

**CAUSE** A sap-feeding insect, *Saissetia coffeae*, which breeds continuously throughout the year on houseplants and in heated greenhouses.

**CONTROL** Heavily infested plants are best disposed of, especially ferns, which generally do not tolerate insecticides. Some scales and honeydew can be removed by wiping the foliage and stems with a damp cloth. Spray with insecticidal soap or malathion against the more vulnerable young scale nymphs. Several applications at seven-day intervals may be needed to break the insect's life cycle.

## HETEROBASIDION ANNOSUM

*See* FOMES ROOT AND BUTT ROT, *p.128.*

## HIGH TEMPERATURE INJURY

**SYMPTOMS** Injury may be temporary or result in tissue death and permanent damage. Scorching or scalding of leaves, fruits and flowers is common, often being seen as dry, papery, buff or brown areas. Wilting may occur if the plant is losing more water than it can take up, due either to excessive transpiration or dry soil conditions. Trees or shrubs which have thin bark may show bark death down the exposed side of their stem or trunk.

**CAUSE** Excessively high temperatures, or the rapid onset of high temperatures immediately following cool weather.

**CONTROL** There is rarely anything that can be done once injury has occurred, other than to ensure that further injury is prevented by improving ventilation, provision of shading, damping-down of the greenhouse floor or other means appropriate to the situation in which the plants are growing.

## HOLLY LEAF MINER

**SYMPTOMS** Leaves develop yellowish-purple blotches (*see p.17*) where the internal tissues have been eaten out by a fly maggot. Leaf mines can be seen throughout the year and may be abundant on hollies trimmed to form a hedge.
**CAUSE** The larvae of a small fly, *Phytomyza ilicis*, which lays eggs on young foliage in early summer. There is one generation a year.

**HOLLY LEAF MINER DAMAGE**
**CONTROL** Apart from creating discoloured patches on the foliage, this fly has little impact on the health and vigour of hollies. Mined leaves that turn yellow and drop in early summer are likely to be part of the normal loss of old foliage that occurs at that time. Holly leaf miner generally has to be tolerated, as insecticides do not penetrate the thick, waxy surface of holly leaves.

## HOLLYHOCK RUST

**SYMPTOMS** Bright yellow or orange spots with orange-red centres develop on the upper leaf surface. Bracts may also be affected. The lower leaf surface develops numerous raised orange pustules (*see p.17*), each about the size of a pin-head and corresponding with the spots on the upper surface. The spots may join together, killing large areas of leaf tissue and causing it to fall away, leaving the leaves with a ragged appearance. The pustules later turn buff-brown in colour.
**CAUSE** The fungus *Puccinia malvacearum*, which is very common, but most troublesome in damp seasons. The spores are ejected from the pustules and spread on air currents. The fungus overwinters on infected plant debris and on any infected leaves which may remain on the plant during mild winters.
**CONTROL** Inspect plants regularly and remove any infected leaves. Spray plants from an early stage at regular intervals with a fungicide such as penconazole, mancozeb, or bupirimate with triforine. Remove potential weed hosts such as wild mallow (*Malva* species) from the area. There are no resistant hollyhocks available, but the species *Althaea rosea* is said to be less susceptible.

## HOLM OAK LEAF MINER

**SYMPTOMS** Deciduous oaks can be attacked by this leaf miner, but it is most often seen on evergreen types, such as the holm oak, *Quercus ilex*. Affected leaves develop elongated, brownish-white, discoloured blotches on the upper leaf surface. The corresponding areas on the lower surface have a blistered appearance where the lower epidermis has become detached. Damage is particularly noticeable during winter.
**CAUSE** Caterpillars of a tiny moth, *Phyllonorycter messaniella*, which has three generations a year on holm oak.
**CONTROL** Treatment is difficult, as insecticides do not readily penetrate into the tough, shiny leaves of holm oak. Many trees may also be too large for effective spraying. Fortunately, oaks can support heavy infestations without adverse effects on their growth. Although the foliage may be unsightly in the winter when the year's accumulated damage is present, the tree's appearance will improve. In the spring the new growth will cover up the damaged foliage and some of the older leaves will be shed in early summer.

## HONEY FUNGUS

**SYMPTOMS** Affected trees, shrubs, woody climbers and occasionally woody herbaceous perennials start to die back. Occasionally other plants, including strawberries and potatoes, may become infected. Leaves may discolour and wilt, or fail to leaf up in the spring. Death may appear to be quite rapid, or an infected plant may take several years to die. A flowering or fruiting plant may put on a particularly good display or set an unusually good crop before dying. Resin may exude from around the base of the trunks of infected conifers. The roots and/or stem or trunk bases of infected plants develop a creamy white mycelium (fungal sheet) on them, sandwiched between the bark and the woody tissues beneath (*see p.34*). The mycelium has a characteristic mushroom-like smell. The fungus also produces tough, brown to black fungal strands called rhizomorphs. They may show branching and are hollow inside, the black outer "rind" being lined with white or faintly pink fungal growth. Rhizomorphs resemble old tree roots and are easily mistaken for them. They vary in length and may be plump or flattened, measuring from under a millimetre to several millimetres in diameter. The rhizomorphs feed on a woody host, or the remains of an old tree stump or dead plant. They may then grow through the soil and latch onto other food sources such as living trees or woody debris. The rhizomorphs may also grow beneath the bark of an infected plant, or be found growing over the surface of a root. Most are found within the top 20cm of the soil. Clumps of honey-coloured toadstools may appear around the base of infected plants in late summer or autumn. The toadstools may also appear around old tree stumps, following the root line of underground roots, or growing from underground rhizomorphs in the vicinity of an infected plant. The toadstools are very variable, but usually grow in clumps and have honey-coloured caps which may be marked with dark brown flecks. The spores are white and may accumulate on the caps of other fungi within the clump. The stipe, or stem, of the toadstool usually bears a distinct creamy-white ruff or collar.
**CAUSE** Various species and races of the fungus *Armillaria*. These vary in

their ability to infect potential hosts, some being far less pathogenic than others and living only on dead or very stressed plants. It is very hard to differentiate between the species, except on the basis of the damage they cause. Spread occurs either by rhizomorphs which grow through the soil at the rate of about 1m a year, or by root contact.

**CONTROL** Remove infected plants promptly, digging out stumps and as much of the root system as possible. Keep plants in good health by regular feeding, mulching and other relevant care, as this makes them less prone to infection. The spread of disease from the toadstools is minimal, so it is not necessary to clear them away. Certain plants, such as birches (*Betula*), cedars (*Cedrus*), most of the common hedging conifers, cotoneasters, currants (*Ribes*), forsythias, hydrangeas, lilac (*Syringa*), *Malus*, peonies, privet, *Prunus*, rhodo-dendrons, roses, willows (*Salix*) and wisterias, are known to be particularly susceptible and so should not be used as replacement plants in gardens which have honey fungus. Trees, shrubs and climbers that show a good degree of resistance should be used in preference to others. These include yews (*Taxus*), bamboos, beeches (*Fagus*), box (*Buxus*), mallows (*Malva*), oaks (*Quercus*), abutilons, actinidias, carpenterias, celastrus, cercis, catalpas, ceratostigma, chaenomeles, choisyas, clematis, cotinus, elaeagnus, fothergillas, hebes, kerrias, passifloras, phlomis, photinias, pieris, pittosporums, rhus, romneyas, sarcococcas and tamarix.

## HONEYDEW

This is the sugary excrement produced by some sap-feeding pests such as aphids, whiteflies, mealybugs and some scale insects. These pests suck sap from the phloem vessels in leaves and stems. The sap they imbibe is very rich in sugars, and is only partially digested by the insects, so their excrement is a sticky liquid due to its high sugar content. Although these pests usually live on lower leaf surfaces, it is upper leaf surfaces that becomes sticky. The insects flick honeydew droplets away from them and these then fall down onto leaves and other surfaces beneath. Under damp conditions non-parasitic fungi called sooty moulds grow on honeydew and form a black or greyish-green deposit on the foliage.

## HONEYSUCKLE APHID

**SYMPTOMS** Greyish-green aphids form dense colonies on the shoot tips and flower buds of honeysuckles in late spring and summer. This can result in the flower buds aborting and shoot tips drying up and browning.
**CAUSE** A sap-feeding insect, *Hyadaphis passerinii*.

ACTUAL SIZE

HONEYSUCKLE APHID

**CONTROL** Check the new growth and flower buds for the presence of aphids from mid-spring onwards. If necessary spray with pirimicarb, heptenophos with permethrin, malathion, pirimiphos-methyl, bifenthrin, permethrin, derris, pyrethrum or insecticidal soap.

## HORNBEAM WITCHES' BROOMS

**SYMPTOMS** Extensively branched twigs develop, closely clustered together, to form "witches' brooms" scattered throughout the crown of the tree. An off-white layer of fungal spores may develop on the surface of the leaves from late spring until mid-summer. During the winter the growths may, at first, be mistaken for birds' nests and can spoil the outline of the tree or hedge, but during the summer they appear as dense, leafy areas and so are less obtrusive.
**CAUSE** The fungus *Taphrina carpini*, spread by water splash and on air currents.
**CONTROL** Prune out the growths if desired. Spread seems to be very slow, but removal will help to limit the problem. No chemical control is available.

## HORSE CHESTNUT LEAF BLOTCH

**SYMPTOMS** Chestnut to tan-brown, irregular leaf blotches (*see p.20*), each with a bright yellow margin, develop from mid-summer onwards. The damage is often most severe on the leaf margins and leaf tips. Numerous pinprick-sized, black fungal fruiting bodies may develop on the upper leaf surface over the spots. Severely infected leaves turn brown, curl and fall prematurely.
**CAUSE** The fungus *Guignardia aesculi*, which although causing early leaf fall has little effect on the tree's overall health, because most of the damage occurs quite late in the season.

**CONTROL** Collect up and dispose of fallen leaves in the autumn to minimize disease carry-over.

## HORSE CHESTNUT MARGINAL LEAF SCORCH

**SYMPTOMS** Marginal leaf browning or scorching occurs, usually in early to mid-summer. Leaves curl slightly and scorched areas turn tan-brown. All or part of the tree may be affected. If the damage is only limited, the rest of the tree usually appears perfectly healthy.
**CAUSE** The precise cause is unknown, but it is believed that this disease may be due to a xylem-limited bacterial infection (a bacterial infection restricted to the xylem, or water-conducting elements, within the tree).
**CONTROL** None known, but symptoms may be less apparent in some years than others.

## HORSE CHESTNUT SCALE

**SYMPTOMS** Many trees and shrubs, including horse chestnuts (*Aesculus*), limes (*Tilia*), sycamore (*Acer pseudoplatanus*), maples (*Acer*), dogwoods (*Cornus*), magnolias, bay trees (*Laurus nobilis*), elms (*Ulmus*) and *Skimmia japonica,* are attacked. Infestations are most recognisable in early to mid-summer when the females deposit their eggs under white, waxy threads; these are secreted from the rear edge of the shell-like covering over the insects' bodies. Mature scales are brown, up to 5mm long and are found on the trunks and larger branches. After egg-laying, the scales die and fall away, leaving circular white egg patches on the bark (*see **p.40***). During the summer young scales – flat, oval, yellowish objects – live on the undersides of leaves. In the autumn they move onto the bark and overwinter as blackish-yellow immature nymphs. Little problem with honeydew and sooty mould occurs with this scale, except on *Skimmia japonica*.

**CAUSE** A sap-feeding insect, *Pulvinaria regalis*, which has one generation a year. The heaviest infestations occur on trees growing in streets and car parks where there is a warmer microclimate due to heat being reflected off concrete, tarmac and brickwork.

**CONTROL** On large trees, horse chestnut scale has to be tolerated. Fortunately, it causes little real harm apart from the unsightliness of the egg masses. On small trees and shrubs, scrape off scales and egg masses before eggs start hatching in mid-summer. If necessary, spray the undersides of leaves with malathion in mid-summer.

## HOVERFLIES

Many of these flies have black-and-yellow striped abdomens and can be seen sitting on flowers or hovering near them. Some have larvae that feed on greenfly and other aphids. They may consume as many as 400 before they are ready to pupate and are helpful in controlling aphids. Other types of hoverfly have larvae that feed in rotting vegetation, in wasp nests, in the mud in pools or in rot holes in trees. The only hoverfly to cause problems in gardens is the narcissus bulb fly (*see p.150*).

ACTUAL LENGTH: 6–15MM

**HOVERFLY**

## HYDRANGEA LEAF SPOT

**SYMPTOMS** Grey, brown or purple-brown spots of varying sizes are usually most apparent on the older foliage. Pinprick-sized fungal fruiting bodies may develop within the damaged areas. A yellow halo may be present around the edge of the leaf spot, often indicating that a bacterial infection is responsible. Although spoiling the appearance of the foliage, none of these leaf spots causes serious, long-term damage.

**CAUSE** A wide range of fungi and bacteria, most of which are weak pathogens, only able to attack hydrangeas which are suffering from other more fundamental problems.

**CONTROL** Remove severely infected leaves. Improve the vigour of the plant by regular watering, feeding and any other appropriate cultural measures.

## HYDRANGEA POWDERY MILDEW

**SYMPTOMS** Dense, white powdery deposits form on the upper leaf surfaces, and occasionally on petioles and new stem growth. The affected leaves may become distorted, particularly if infection occurs when they are still young. The leaves usually remain on the plant but may show patchy necrosis. They are rarely killed outright.

**CAUSE** The fungus *Microsphaera polonica*, which is encouraged by moist or humid air around the foliage and dry soil conditions around roots.

**CONTROL** Remove any severely affected leaves. Keep the plant well watered at the base, but avoid overhead irrigation. Spray with a suitable fungicide such as carbendazim or bupirimate with triforine, if the disease appears early in the year and threatens to become extensive.

## HYDRANGEA SCALE

**SYMPTOMS** This pest is most visible during late spring and early summer when the females deposit eggs amongst a mass of white, waxy fibres on the stems and foliage of host plants (*see **p.15***). Hydrangea is the most frequently attacked plant but this scale also occurs on *Prunus* and *Acer*. The egg masses are oval in shape and up to 5–6mm long. The mature scale is brownish-yellow but drops off when the insect dies, leaving the egg mass as the most obvious sign of infestation. Eggs hatch in mid-summer and the flat, oval, pale yellow scale nymphs live on the underside of leaves next to leaf veins during the rest of the summer. They move onto the bark in the autumn and overwinter as immature nymphs.

**CAUSE** A sap-feeding insect, *Pulvinaria hydrangeae*, which has one generation a year. Heavy infestations reduce the plant's vigour and result in shoots dying back.

**CONTROL** Spray with malathion or insecticidal soap in mid-summer against the more vulnerable newly

hatched scales. Hydrangeas can be damaged by pesticides, so avoid treatment when plants are exposed to bright sunlight or high temperatures. Water the plants first if they are dry at the roots.

## INONOTUS HISPIDUS

**SYMPTOMS** Annual, bracket-shaped, semi-circular fungal fruiting bodies develop on large limbs or the main trunk of trees. Each measures between 10 and 35cm across and is 2–10cm deep. They are covered in dense, hair-like growth and are an orangey-rust colour initially, turning dark brownish-black as they age. Their presence causes a yellowish-orange discoloration of the wood.

**CAUSE** The fungus *Inonotus hispidus*, which usually enters through wounds or stubs left by poor pruning.

**CONTROL** Infected trees may be rendered dangerous, their safety depending largely on where the decay has occurred. Employ a reputable tree surgeon to advise as to the best course of action.

## IRIS INK DISEASE

**SYMPTOMS** Blue-black streaking appears on the outer surface of the bulb. In extreme cases, or if left unchecked, the whole bulb turns black and rots. If infected bulbs produce foliage, it is streaked yellow initially and later develops a red discoloration, particularly towards the tips. Black necrotic spots may develop on the yellowed areas.

**CAUSE** The fungus *Drechslera iridis*, which produces spores that are spread on air currents. The fungus can overwinter on infected bulbs, bulb debris, and possibly in soil around infected bulbs.

**CONTROL** Dispose of any bulbs showing symptoms. Remove any plants which show foliar symptoms, and grow irises on a fresh site.

## IRIS LEAF SPOT

**SYMPTOMS** Water-soaked spots develop on the foliage, later turning brown with grey centres. The spots join together as they enlarge, so destroying the foliage. Affected irises are weakened by the loss of functional foliage and so are less vigorous and flower poorly.

**CAUSE** The fungus *Mycospharella macrospora*, which is encouraged by mechanical injury and wet weather. It produces spores which are spread on air currents and by rain and water splash. It overwinters on infected leaves and plant debris.

**CONTROL** If infection occurs after flowering, cut off badly affected areas. If it occurs early in the season, spray affected plants with a fungicide such as mancozeb. Keep plants well fed and ensure they are planted in a suitable, sunny site; they are less susceptible to infection if growing vigorously.

## IRIS RHIZOME ROT

**SYMPTOMS** Foliage yellows and withers, starting at the leaf tips or leaf margins. A soft, often foul-smelling rot develops at the base of the affected leaves and rapidly spreads down the rhizome (see **p.55**). As the infection spreads, the foliage

symptoms become more extreme, and leaves may yellow completely and topple over.

**CAUSE** The bacterium *Erwinia carotovora* subsp. *carotovora*, which is soil-borne and is particularly prevalent in heavy, wet soils and in wet seasons. The bacteria enter the iris via wounds.

**CONTROL** Plant irises on a well-drained soil. Choose only healthy rhizomes, and plant them shallowly. Avoid injuring the rhizomes and control pests such as slugs which may damage them. Remove and dispose of badly affected irises. It may be possible to save rhizomes which are only slightly affected by cutting off all damaged areas right back to sound, healthy tissue.

## IRIS RUST

**SYMPTOMS** Numerous small, raised, orange to yellow pustules develop on the leaves (see **p.15**). Each is lens-shaped and measures about 1mm in length. The pustules enlarge slightly and erupt to produce masses of yellow-orange spores. The infected leaves discolour, turning yellow at first and later, if the infection is severe, browning and dying off. Older leaves are generally the most badly affected.

**CAUSE** The fungus *Puccinia iridis*, which is encouraged by wet weather and excessively soft leaf growth.

**CONTROL** Cut off severely affected leaves. Infection usually occurs towards the end of the summer, so it does not cause too much loss of vigour. However, in severe cases or if the infection develops earlier in the year, spray with a

suitable fungicide such as penconazole, mancozeb or bupirimate with triforine.

## IRIS SAWFLY

**SYMPTOMS** V-shaped notches are eaten from the leaf margins initially (see **p.29**), but there is later extensive defoliation, particularly on the upper parts of the leaves. Only waterside irises are attacked, especially yellow flag iris, *Iris pseudacorus*.

**CAUSE** The caterpillar-like larvae of a sawfly, *Rhadinoceraea micans*. It is of local occurrence in Britain, but is numerous where it is found. The greyish-brown larvae, up to 15mm long, are present during early to mid-summer. The adult insects are black, with smokey-grey wings, and can be seen resting on waterside irises in late spring and early summer. There is one generation a year.

**CONTROL** The larvae can be controlled by spraying with pyrethrum, bifenthrin or permethrin, but insecticides should not be used where there is a risk of pesticides getting into the water. Fish, frogs and other pond wildlife are very vulnerable to insecticides. Removal of the larvae by hand would be the best option in most gardens.

## IRIS SCORCH

**SYMPTOMS** The upper parts of affected leaves wilt and bend over and become reddish-brown. The symptoms may initially be confused with those of iris rhizome rot, but the leaf discoloration is different and there is no deterioration of the rhizome. The roots die back; some rot and others disintegrate internally, leaving just a hollow outer "tube" with the

cotton-threadlike remains of the central cylinder inside.

**CAUSE** The precise cause is not known, but the problem seems to be caused by root injury in winter or early spring. Winter waterlogging may be responsible.

**CONTROL** Ensure good drainage. It may be possible to save slightly affected rhizomes by lifting, removing all deteriorating roots and leaves, washing thoroughly and replanting on a new, very well-drained site. Foliar feeding, in an attempt to stimulate the production of replacement roots, can help.

## IRON DEFICIENCY AND LIME-INDUCED CHLOROSIS

**SYMPTOMS** Usually seen in combination with manganese deficiency. Leaf yellowing, or chlorosis (see **p.24**), is often combined with the development of brown areas of discoloration which start at the leaf margins and then spread between the veins. Younger growth is usually affected earlier and more severely than older growth.

**CAUSE** Acid-loving plants and lime-hating plants have roots which are poorly adapted for the absorption of necessary trace elements from an alkaline soil. If conditions are too alkaline, they readily develop deficiency symptoms, in particular of iron and manganese.

**CONTROL** Avoid growing lime-hating plants on a soil which is not sufficiently acid. Treat affected shrubs with a chelated compound containing iron, manganese and other trace elements which will be available to the plant because they are in a form which does not become "locked up" in the alkaline soil. Use acidic mulches such as chopped, composted bracken or conifer bark. Incorporate acidic materials such as these into the planting hole at planting. Feed plants with a fertilizer formulated for use on acid-loving (lime-hating) plants. Before planting, consider acidifying soil using sulphur or aluminium sulphate and ferrous sulphate. Sulphur treatments can also be used around existing plants. Follow instructions carefully.

## IRREGULAR WATERING

**SYMPTOMS** Fruits are unusually small and may have a particularly thick skin. The surface may be ruptured, with splits caused when the fruit swells in response to the sudden availability of water (*see* **p.57**). Once the fruit's surface is broken it is open to infection by secondary organisms, such as grey mould, which may then cause further, rapid deterioration. Other fruit symptoms may include blossom end rot (*see p.108*). Leaves fail to grow at a regular rate and so develop an uneven, almost puckered surface (*see* **p.26**). They may also appear unusually small. Buds may fail to form, or form and later drop, and bark may split (*see* **p.36**).
**CAUSE** An irregular or erratic supply of water at the plant roots. Plants growing on a light, sandy soil with a low organic matter content are particularly prone, as are any plants growing in containers.
**CONTROL** Keep plants adequately watered at all times. Incorporate plenty of bulky organic matter into light soils before planting.

Apply a bulky mulch around the root areas to encourage soil moisture retention.
• *see also* BUD DROP *p.111;* BUD, DRY *p.112.*

## IVY LEAF SPOT

**SYMPTOMS** Brown or grey, near-circular spots develop on ivy foliage. On pale-leaved or variegated forms, the spots may be surrounded by a narrow band of reddish-purple discoloration. On variegated leaves, leaf-spotting is usually most severe on the pale or white areas of the leaf. Numerous, pinprick-sized, raised fungal fruiting bodies may develop on the spots, usually in concentric rings.
**CAUSE** Various fungi, including *Mycosphaerella hedericola, Colletotrichum trichellum* and *Phyllosticta hedericola.*
**CONTROL** Remove severely infected leaves to limit disease spread. Although unsightly, ivy leaf spot infections will not lower the overall vigour of the plant.

## IVY ON TREE TRUNKS

**SYMPTOMS** Ivy growing up the trunk and, if left unchecked, into the crown of trees. On young trees, growth may be slightly constricted, but this is unusual on larger trees.
**CAUSE** Ivy, with its climbing, self-clinging habit, moves onto and up suitable surfaces, including tree trunks. It is, however, rare for it to cause damage, but young or poorly growing trees may have their stem or trunk growth constricted. Dense ivy growth on trunks may also mask the development of other problems, such as decay-

causing bracket fungi, but it does provide a useful wildlife habitat.
**CONTROL** Rarely justifiable. If desired, the ivy should be removed regularly, before it develops a lot of growth. Established plants can be cut off at ground level and, once dead, their top growth pulled away from the trunk and branches of the tree.

## JUNIPER APHIDS

**SYMPTOMS** Leaves and stems become coated with black, sooty moulds that grow on the sugary excrement of aphids. Heavy attacks cause yellowing and dieback in mid- to late summer.
**CAUSE** Two species of aphid occur on junipers. *Cinara fresai* is 2–4mm long and brownish-black. Colonies on stems close to the soil often have a protective covering of soil provided by ants, which collect honeydew from the aphids. The other species, *C. juniperi*, is 2–3mm long, pale brown, and occurs mainly on younger growth. Both feed by sucking sap.
**CONTROL** Look for aphid infestations from mid-spring onwards and, if seen, spray with pirimicarb, malathion, pirimiphos-methyl, permethrin or bifenthrin.

## JUNIPER SCALE

**SYMPTOMS** Flat, rounded, or elongate and rectangular white objects with a small yellowish patch develop on the foliage of junipers, cypresses, Leyland cypress and thuyas. The rounded scales are protective covers for females and are up to

1.5mm long. The smaller rectangular scales cover males. Heavy infestations cause yellowing and dieback of green shoots (*see p.37*).
**CAUSE** A sap-feeding insect, *Carulaspis juniperi*, which has one generation a year.
**CONTROL** On plants small enough to be sprayed, use malathion in early summer when the more vulnerable newly-hatched scale nymphs are present.

## JUNIPER WEBBER MOTH

**SYMPTOMS** Patches of brown foliage bound together with silk threads develop. Dark brown caterpillars with paler stripes running along their bodies, up to 11mm long, live within the webbing (*see p.32*). Leaves damaged by the caterpillars dry up and die in early summer. Compact, upright junipers, such as 'Skyrocket', can be badly damaged.
**CAUSE** Caterpillars of a moth, *Dichomeris marginella*, which has one generation a year.
**CONTROL** Infestations can sometimes be dealt with by cutting out webbed shoots in late spring or early summer. Alternatively, use the biological control *Bacillus thuringiensis*, or spray forcibly with permethrin, bifenthrin, pirimiphos-methyl or fenitrothion to penetrate the webbing.

## KABATINA SHOOT BLIGHT

**SYMPTOMS** Cankers develop on the shoots of Leyland cypress, certain thuyas, cypresses and junipers. The cankers usually develop at the base of small branches and cause dieback and, occasionally, death.

**CAUSE** Fungi of *Kabatina* species. The fungal spores are carried on air currents and by rain splash.

**CONTROL** Remove affected shoots promptly to minimize spread of the disease.

## KEITHIA THUJINA NEEDLE BLIGHT

**SYMPTOMS** Small, black fungal fruiting bodies, each measuring about 1mm across, are embedded in the foliage of *Thuja plicata*. These later split as the spores are released, leaving small, brown sunken pits, or holes in the leaves. Foliage, particularly towards the base of the plant, may turn brown (*see* **p.28**) and die back if the infection is extensive.

**CAUSE** The fungus *Didymascella thujina*. The spores are air-borne and most infections take place in late summer or early autumn, especially in humid or damp weather. They may over-winter in mucilage attached to the damaged areas.

**CONTROL** Prune out infected areas promptly, before the spores are released. Remove severely infected stems.

## KNOPPER GALLS

*See* ACORN GALL WASP *p.98*.

## LABURNUM LEAF MINERS

**SYMPTOMS** Whitish-brown, dried-up patches develop on laburnum leaves (*see* **p.18**) between late spring and

early autumn. If the leaf is held up to the light it is often possible to see leaf-mining larvae feeding within the leaf.

**CAUSE** Two leaf miners commonly occur on laburnum. The larvae of a leaf-mining fly, *Agromyza demeijerei*, cause narrow linear mines along the leaf margin which later enlarge into irregular blotch mines extending from the margin. Caterpillars of a leaf-mining moth, *Leucoptera laburnella*, cause roughly circular mines in the centre of leaves, but in heavy attacks the mines join together and lose their distinctive shape.

**CONTROL** Attacks in late summer may be unsightly, but they have little effect on the tree's health and vigour and so can be tolerated. If damage is seen occurring in early summer, spray with pirimiphos-methyl or malathion.

## LACEBUG

*See* RHODODENDRON LACEBUG *p.168*.

## LACEWINGS

The larvae of lacewings are beneficial insects that feed on aphids and other small insects. Some lacewing larvae carry the sucked-out remains of earlier meals amongst clumps of bristles on their upper surface. A well-fed larva is up to 10mm long and, apart from a pair of sharp jaws protruding at the front, is hidden under a tangled mass of dead aphids. The harmless adult lacewings have slender, elongate bodies which are pale green, black or brown. They have two pairs of transparent wings, sometimes with darker markings, which have many

veins – hence the name lacewing. The adults have long, thin antennae and feed on pollen and nectar or honeydew. One species of green lacewing, *Chrysoperla carnea*, overwinters as adults and is often found indoors.

ACTUAL LENGTH: 2–20MM

**LACEWING**

## LACKEY MOTH

**SYMPTOMS** Gregarious, hairy caterpillars occur on a wide range of trees and shrubs from mid-spring to early summer. They are up to 48mm in length and have slate-grey heads with red, white, black and blue stripes on their bodies. The caterpillars cover their feeding area with silk webbing and can cause extensive defoliation (*see* **p.30**).

**CAUSE** Caterpillars of *Malacosoma neustria*, which hatch in the spring from eggs laid in late summer. These form a solid band of several hundred eggs glued together around a twig.

**CONTROL** Prune out egg bands or the webbed shoots when they are seen. For extensive infestations, use biological control with the bacterium *Bacillus thuringiensis* or spray with permethrin, bifenthrin, pirimiphos-methyl or fenitrothion.

## LADYBIRDS

Most of these beneficial insects feed on greenfly and other aphids, both as adult beetles and larvae; some ladybirds also prey on scale insects or red spider mites, while others feed on fungal

spores. Adult ladybirds are familiar insects, especially those with wing cases that are red with black spots (*see* **p.60**), but some species are yellow and black, or brown with white spots. The larval stages are generally slate-grey in colour with orange or white spots on the upper surface. One ladybird, *Cryptolaemus montrouzieri*, is commercially available for controlling mealybugs in greenhouses.

ACTUAL LENGTH: UP TO 15MM

**LADYBIRD LARVA**

## LAETIPORUS SULPHUREUS (CHICKEN OF THE WOODS)

**SYMPTOMS** Overlapping groups of annual, bright yellowish-orange fungal brackets, measuring between 10 and 40cm in total diameter, develop on the trunk or large limbs of host trees (*see* **p.38**). Each bracket is thin and has a wavy margin, with soft flesh resembling cooked chicken, and a yellow undersurface. Hardwoods, including oak, sweet chestnut, cherry (*Prunus*) and occasionally beech, and conifers, including larch, yew and pine, may become infected. The infected wood dis-colours, turning yellowish-orange at first and later breaking up into brown, cubical blocks. Off-white to yellow mycelium (fungal sheets) develop between the blocks of decaying wood.

**CAUSE** The fungus *Laetiporus sulphureus*, which produces numerous spores from its lower surface.

**CONTROL** Infected trees may become dangerous as the strength of wood is reduced. An arboriculturist should examine the tree and advise on its health and safety.

## LARCH ADELGID

**SYMPTOMS** Small patches of fluffy white wax appear on the foliage in early summer. Heavy infestations can cause leaf yellowing, and sooty moulds develop on the pests' sugary excrement (*see p.32*).
**CAUSE** Small, black sap-feeding insects, *Adelges laricis*, that resemble aphids and secrete waxy fibres.
**CONTROL** Large trees cannot be sprayed effectively but are usually able to tolerate infestations. Small trees can be sprayed with malathion or pirimiphos-methyl.

## LAVATERA LEAF AND STEM ROT

**SYMPTOMS** Red-brown necrotic spots develop on stems and occasionally on leaves and petioles of lavateras and hollyhocks. The spots enlarge and join together; they may girdle stems and so cause dieback or even death. Tiny, black, bristle-like spore-bearing structures may develop on the surface of the lesion.
**CAUSE** The fungus *Colletotrichum malvarum*, which can be seed-borne.
**CONTROL** Remove and dispose of infected plants promptly. Grow replacement plants on a fresh site and do not save seed from those showing infection.

## LAVENDER SHAB

**SYMPTOMS** The plant wilts, discolours and dies back, usually stem by stem. The whole plant may be killed.

Numerous tiny, blackish-brown, raised fungal fruiting bodies the size of pinpricks develop on the dead foliage and stems. The symptoms resemble those caused by wet soil or winter cold injury, but are unique in having the fungal fruiting bodies.
**CAUSE** The fungus *Phomopsis lavandulae*, which enters the plant through pruning wounds or other points of damage on the bark.
**CONTROL** Prune out infected areas immediately, cutting well back into healthy wood. It may not be possible to save infected plants, as symptoms develop rapidly.

## LEAF AND BUD EELWORMS

**SYMPTOMS** Brown islands or wedges develop in leaves (*see p.25*) where the internal spread of the eelworms is restricted by the larger leaf veins. Eventually all of the leaf will turn brown and dry up. Typically, the youngest leaves at the shoot tips are free of infestation, but the older leaves lower down the stem show progressively more severe damage.
**CAUSE** Two microscopic species, *Aphelenchoides ritzemabosi* and *A. fragariae*, attack a wide range of plants in gardens and greenhouses.
**CONTROL** These eelworms cannot be controlled with the pesticides available to amateur gardeners. Badly infested plants should be burnt. With indoor plants, avoid wetting the foliage, as eelworms can emerge from leaves and spread rapidly in a surface film of water.
• *see also* CHRYSANTHEMUM EELWORM *p.117*.

## LEAF CURL

*See* PEACH LEAF CURL *p.156*.

## LEAF-CUTTING BEES

**SYMPTOMS** Circular or elliptical pieces are cut from the leaf margins (*see p.30*) of roses, wisterias, epimediums and other plants. The smooth outline and regular shapes of the missing pieces readily distinguish leaf-cutting bee damage from caterpillar or other damage. The female bee uses the leaf pieces to make thimble-shaped cells in her nest. This may be in a hollow plant stem or in a tunnel dug in rotten wood or soil. Quite often a bee makes use of soil in flower pots and the cells, resembling old cigar stubs, come to light when the plant is repotted. When a cell has been built, the bee stocks it with nectar and pollen, lays a single egg, and completes the cell by capping it with circular leaf pieces. A nest may contain up to twenty cells.
**CAUSE** Leaf-cutting bees, *Megachile* species, which resemble honeybees but have ginger hairs beneath their abdomens. They are solitary bees, each female constructing her own nest during the summer. They often visit a plant repeatedly to collect material and by the time their nest is finished they may have removed several hundred pieces.
**CONTROL** Like other bees, these are useful pollinators and can be tolerated unless small plants are being badly defoliated. If necessary, watch the plant until the bee returns, then swat it.

## LEAFHOPPERS

Leafhoppers are sap-feeding insects, mostly 2–3mm long, which live on the undersides of leaves. The adults readily jump off leaves when disturbed, but their wingless nymphs are less active. Their

feeding usually causes a coarse, pale mottling of the upper leaf surface. Garden plants frequently attacked are blackberries, plums, phlomis, salvias, mints, beeches and many others. General control measures for leafhoppers are given in the entry for rose leafhopper (*p.171*).
• *see also* GLASSHOUSE LEAFHOPPER *p.132*; RHODODENDRON LEAFHOPPER *p.169*.

## LEAF MINERS

Leaf-mining is a habit adopted by many insects, mainly the larvae of some flies, moths, sawflies and beetles. The damaged parts of the leaf dry up and become white or brown. The mines may be straight or meandering lines, circular or irregular blotches, and the form of the mine is usually distinctive for any particular leaf-mining insect. If a mined leaf is held up to the light, the larva or pupa can often be seen at the end or edge of the mined area. Leaf miners on plants other than those listed below can be controlled by removing affected leaves or spraying with malathion or pirimiphos-methyl.
• *see also* APPLE LEAF MINER *p.101*; BEECH LEAF-MINING WEEVIL *p.106*; BEET LEAF MINER *p.106*; CELERY LEAF MINER *p.115*; CHRYSANTHEMUM LEAF MINER *p.117*; DELPHINIUM LEAF MINER *p.122*; GEUM SAWFLIES *p.131*; HOLLY LEAF MINER *p.136*; LABURNUM LEAF MINERS *p.141*; LEEK MOTH *p.144*; LILAC LEAF MINER *p.145*; PYRACANTHA LEAF MINER *p.166*.

## LEAF SCORCH

**SYMPTOMS** The plant grows well, but its foliage is spoiled by dry, brown scorching, which is usually first seen around the leaf edges. The whole leaf may ultimately be affected, or the central area may retain its normal colour until the end of the season. If the problem is severe, or occurs in several years in succession, plant growth is affected and dieback may occur. On maples and sycamore (*Acer*), stems which deteriorate as a result of scorch often develop a white surface discoloration. In extreme cases extensive dieback, and occasionally even death, may occur.

**CAUSE** Various factors may be involved. The most common are an over-wet or excessively dry soil, cold winds or hot, bright sun. Plants which are lacking in vigour due to some other cause, or those which are not yet properly established, are particularly prone to scorch.

**CONTROL** Attempt to determine which adverse condition or conditions are involved, and improve growing conditions accordingly. If poor drainage is responsible, improve this and apply a foliar feed throughout the growing period to encourage replacement root growth. If the soil tends to become too dry, water well and apply a bulky mulch around the root area. Consider erecting windbreaks or planting to provide wind protection. If necessary, and feasible, consider moving the plant to a more suitable site at the appropriate time of year.

## LEAF SPOTS (BACTERIAL)

**SYMPTOMS** Necrotic patches, usually circular or angular in outline, develop on foliage. Many bacterial leaf spots are surrounded by a bright

yellow "halo" (see **pp.13, 29**). There are no minute, raised fungal fruiting bodies (pycnidia) as seen on fungal leaf spots.

**CAUSE** Various bacteria may be involved, and these are usually spread by rain or watering splash from leaf to leaf or, occasionally, from stem lesions to leaves (for example, bacterial canker may cause leaf spotting on *Prunus*). In most cases the leaf spotting itself is not very damaging, but may indicate that the plant is suffering from another, more serious problem. Examples include *Pseudomonas maculicola* on brassicas, *Pseudomonas lachrymans* on cucumber, black blotch on delphiniums and *Pseudomonas mori* on mulberry.

**CONTROL** Remove infected leaves promptly and avoid overhead irrigation.

## LEAF SPOTS (FUNGAL)

**SYMPTOMS** Necrotic spots, usually circular and either grey or brown (see **p.13**), develop on the leaves and may join together, causing large areas of dead leaf tissue to develop (see **p.12**). Occasionally leaves may be killed, but in most cases leaf spots are not very damaging in their own right, though they may indicate that the host plant is suffering from some other, rather more fundamental problem. The spots often show concentric zones of discoloration, and may bear numerous, raised, pinprick-sized, black or brown fungal fruiting bodies. Many common examples have entries elsewhere under their host name; others include *Phyllosticta antirrhini* on antirrhinums; *Discula quercina* on oaks

and beeches; *Ramularia beticola* and *Cercospora beticola* on beet crops; *Septoria* on chrysanthemums, pinks and carnations (*Dianthus*), foxgloves (*Digitalis*), hebes, hawthorn and cyclamens; *Ascochyta cinerariae* on cinerarias; *Colletotrichum clivae* on clivias; *Marsonina daphnes* on daphnes; *Phyllosticta magnoliae* on magnolias; *Ramularia* on primroses and polyanthus, and *Phyllosticta wistariae* on wisterias.

**CAUSE** A wide range of fungi (see above, and references to specific entries below).

**CONTROL** Remove affected leaves if necessary. Rake up and dispose of fallen leaves at the end of the season to minimize infection carry-over. Fungicides such as carbendazim and mancozeb can be used to control certain leaf spots, but check product labels for details.

• see also ALTERNARIA LEAF SPOT *p.98*; ARBUTUS LEAF SPOT *p.102*; BRASSICA LIGHT LEAF SPOT *p.110*; CELERY LEAF SPOT *p.115*; HYDRANGEA LEAF SPOT *p.138*; IVY LEAF SPOT *p.140*; PANSY LEAF SPOT *p.154*; PEA LEAF AND POD SPOT *p.155*; RHODODENDRON LEAF SPOT *p.169*; STRAWBERRY LEAF SPOT *p.179*; WATER LILY LEAF SPOT *p.188*.

## LEAF WEEVILS

**SYMPTOMS** Ragged holes are eaten in the foliage of many deciduous trees, especially *Prunus*, *Malus*, sorbus, birches (*Betula*), oaks (*Quercus*) and beeches (*Fagus*), in late spring and early summer. Turquoise-green or golden-brown beetles, 3–7mm in length, may be present in large numbers.

**CAUSE** Several *Phyllobius* species occur in gardens and may cause extensive damage

to young foliage in mid-spring to early summer.

**CONTROL** Small trees can be sprayed with fenitrothion, pirimiphos-methyl or bifenthrin if leaf weevils are numerous enough to cause problems.

## LEAFY GALL

**SYMPTOMS** Numerous small, bunched, often somewhat distorted, thickened or fasciated leaves develop around stem bases (see **p.39**). The plant may also produce normal growth.

**CAUSE** The bacterium *Corynebacterium fascians* (syn. *Rhodococcus fascians*), which is soil-borne and enters wounded plant tissue.

**CONTROL** Remove and dispose of affected plants, together with the soil or compost in which they are growing. Avoid spreading the bacteria by washing hands and tools after handling infected plants. Thoroughly scrub pots, trays and green-house staging.

## LEATHERJACKETS

**SYMPTOMS** Legless, greyish-brown tubular larvae, up to 45mm long and without obvious heads, occur in the soil (see **p.53**). They are mainly a lawn pest, causing yellowish-brown patches in mid-summer where the grubs have severed the roots (see **p.58**). Leatherjackets can also cause problems in vegetable gardens and flower borders where they kill seedlings and young plants.

**CAUSE** Leatherjackets are the larvae of crane flies (daddy-long-legs). Many species occur in gardens, mainly *Tipula* and *Nephrotoma* species, and they lay eggs in the soil in late summer.

A-Z OF PESTS, DISEASES AND DISORDERS

ACTUAL LENGTH: UP TO 45MM

**LEATHERJACKET**

**CONTROL** By the time leatherjackets have grown large enough to cause noticeable damage, it is generally too late for insecticides to be effective. Young larvae can be controlled with lindane dust or pirimiphos-methyl; biological control with the pathogenic nematode *Steinernema carpocapsae* can be used against older larvae when the soil is moist and warm (at least 14°C/57°F). Leatherjackets can be removed from lawns by covering small areas with black polythene overnight after heavy rain or irrigation. The grubs will come up onto the lawn surface and can be removed when the polythene is lifted.

## LEEK MOTH

**SYMPTOMS** Onion and leek foliage develops whitish-brown patches (*see* **p.23**) where caterpillars up to 12mm long have been feeding as leaf miners. The older caterpillars also bore into the stems of leeks and bulbs of onions. There are two generations during the summer, with leaf mining occurring in late spring to mid-summer and late summer to mid-autumn. When fully fed, the caterpillars emerge and pupate in silk cocoons on the foliage.
**CAUSE** Caterpillars of a small moth, *Acrolepiopsis assectella*, which in Britain is mainly a problem in coastal south and east England.
**CONTROL** In places where leek moth is a regular problem, spray with malathion or pirimiphos-

methyl when signs of leaf mining are seen. Search for pupae in their net-like cocoons on the leaves, and crush them.

## LEEK RUST

**SYMPTOMS** The outer leaves develop numerous lens-shaped, bright orange, raised fungal pustules, each 1–2mm in length (*see* **p.17**). These erupt to produce conspicuous masses of bright orange spores. Affected leaves yellow and die back. The inner growth is rarely, or only mildly, affected.
**CAUSE** The fungus *Puccinia allii*, which may also affect related plants such as chives, onions, shallots, garlic and, occasionally, ornamental alliums. The disease is encouraged by damp conditions and wet weather.
**CONTROL** Remove and dispose of affected leaves at harvest. Dispose of all debris at the end of the season and grow leeks and other susceptible crops on a fresh site. Avoid excessive use of high-nitrogen fertilizers, as these encourage the production of soft growth which is then more readily infected. At planting, apply sulphate of potash at 15–20g per m² to harden growth slightly. Space plants widely to improve air circulation. Where available, choose leek cultivars resistant to this infection, such as 'Walton Mammoth', 'Titan', 'Poristo', 'Poribleu' and 'Splendid'.

## LEEK WHITE TIP

**SYMPTOMS** Leaf tips turn yellow at the end of summer. As the season progresses, the damage spreads down the leaves; they then either dry out and become pale brown

or bleached and papery in texture, or they rot.
**CAUSE** The fungus *Phytophthora porri*, which can persist in plant debris in the soil for several years after infection, even in the absence of any new leek crops during this time.
**CONTROL** Remove affected leeks promptly and do not grow them or other susceptible plants, such as onions or ornamental alliums, on the contaminated soil for at least five years.

## LEOPARD MOTH

**SYMPTOMS** Apple trees (*Malus*) are the principal host plant, but the caterpillars also attack *Pyrus*, sorbus, maples and sycamores (*Acer*), birches (*Betula*), hawthorn (*Crataegus*), oaks (*Quercus*) and many others. The creamy-yellow or white caterpillar is up to 55mm long and has many black spots on its body. It tunnels in trunks or branches which have a diameter of up to 10cm, and is often detected when the weakened branch or trunk snaps on a windy day. Another sign of infestation is pellets of compacted sawdust (the caterpillar's excrement) coming from a hole in the caterpillar's tunnel.
**CAUSE** Caterpillars of a moth, *Zeuzera pyrina*, which take two years to complete their feeding. Often the infestation consists of a single caterpillar, but this can have disastrous consequences if it is present in the trunk of a young tree.
**CONTROL** Preventive measures are not possible since attacks are sporadic, and spraying against the adult moths during summer is unlikely to be effective. If caterpillars are detected, damaged branches can be

pruned out or a piece of wire inserted into the tunnel to skewer the caterpillar.

## LETTUCE DOWNY MILDEW

**SYMPTOMS** Angular yellow areas of discoloration develop on the leaves (*see* **p.22**), the older, outer leaves being affected first. The infected areas turn pale brown and either papery or soft. Off-white, fluffy fungal growth, occasionally with a slightly purple tinge, develops on the lower leaf surface, corresponding with the patches of discoloration. Severely affected leaves die off. Seedlings and mature plants may be affected.
**CAUSE** The fungus *Bremia lactucae*, which is encouraged by humid or damp growing conditions.
**CONTROL** Remove infected leaves promptly, or the entire plant if the infection is severe. Spray with a fungicide containing mancozeb. Improve ventilation and/or air circulation. Grow resistant lettuces such as 'Avoncrisp', 'Avondefiance', 'Beatrice', 'Debby', 'Lakeland', 'Musette' or 'Saladin'.

## LETTUCE GREY MOULD

**SYMPTOMS** Fluffy grey fungal growth develops on lettuce leaves (*see* **p.28**). If the grey mould infection takes place towards the base of the plant, the stem rots, often together with the internal tissues there, causing a yellowish-orange or brown slimy rot. The whole plant may then die off. Seedlings are also susceptible.
**CAUSE** The fungus *Botrytis cinerea*, which is encouraged by damp growing conditions and often enters the plant

through wounds. It may also be seed-borne in lettuce. The spores are carried on air currents. Black fungal resting bodies (sclerotia) may form on the infected tissues, but will not yet have developed if the plants are removed promptly.

**CONTROL** Remove all plant debris which may harbour the fungus. Avoid injury to the plant. Control likely sources of injury, such as downy mildew or slugs. Improve ventilation, and water early in the evening to allow time for the foliage to dry off before nightfall. Remove leaves infected with the fungus. Spray with a fungicide containing carbendazim.

## LETTUCE ROOT APHID

**SYMPTOMS** Lettuce plants tend to wilt and make slow growth in sunny weather. When plants are dug up, a white, powdery wax can be seen on the roots and soil particles. Creamy-yellow aphids, up to 2mm long, will be found on the roots.

**CAUSE** Sap-feeding insects, *Pemphigus bursarius*, which feed on the roots of outdoor lettuce during mid- to late summer.

**CONTROL** Some lettuce cultivars, such as 'Avoncrisp', 'Avondefiance', 'Debby', 'Lakeland' and 'Sigmaball', are resistant to lettuce root aphid. Root aphids are more difficult to control than leaf-feeding types. Infestations may be checked by watering plants with an insecticide containing heptenophos with permethrin or pirimiphos-methyl, made up to spray strength. Keep the plants well watered and grow lettuce crops on a new site each year.

## LETTUCE VIRUSES

**SYMPTOMS** Leaves show yellow mottling, mosaic patterning or vein clearing (yellowing immediately adjacent to the veins). Plants are stunted and the foliage may be distorted.

**CAUSE** Various viruses, including lettuce big vein, lettuce mosaic, cucumber mosaic and beet western yellows virus.

**CONTROL** Remove plants showing symptoms promptly. Control pests such as aphids, which may spread the viruses. Wash hands thoroughly after handling infected plants. Grow lettuces on a fresh site. Choose resistant cultivars, where available. Those showing resistance to lettuce mosaic virus include 'Action', 'Corsair', 'Debby' and 'Musette'.

• *see also* CUCUMBER MOSAIC VIRUS *p.120*.

## LILAC BLIGHT

**SYMPTOMS** Angular brown spots develop on lilac leaves. Affected leaves die back. Buds blacken and die, and young shoots may be girdled and wilt and die (*see **p.33***). Large, elongated cankers may form on older stems. Secondary infection by grey mould is common and may mask the symptoms.

**CAUSE** The bacterium *Pseudomonas syringae*, which is readily spread on air currents and by rain and water splash.

**CONTROL** Prune out affected areas of the plant promptly, cutting well back into perfectly healthy tissue.

## LILAC LEAF MINER

**SYMPTOMS** The leaves of lilac, privet and ash develop brown shrivelled areas (*see **p.19***) where the internal

tissues have been eaten by leaf-mining caterpillars. When half-grown, the whitish-green caterpillars emerge from their mine and roll up the leaf tip to form a tube held in place by silk threads. There are two generations, with damaged leaves being seen from late spring to early summer and from late summer to early autumn.

**CAUSE** Caterpillars of a small moth, *Caloptilia syringella*.

**CONTROL** Pick off affected leaves; clipping privet hedges will remove many of the mined or rolled leaves. If necessary, spray with malathion or pirimiphos-methyl when mining commences.

## LILY BEETLE

**SYMPTOMS** Bright red beetles, 6–7mm long with black heads, appear on lilies (*Lilium*) and fritillaries from early spring onwards. In mid-summer the larval stage also occurs on the foliage. The grubs have rotund bodies, up to 8–9mm long, and are orange-red with black heads. They cover themselves with their own wet, black excrement. Both grubs and adults eat leaves, flowers and seed pods, and can ruin plants.

**CAUSE** A leaf beetle, *Lilioceris lilii*, which is active as an adult from spring until autumn. In Britain these beetles are mainly found in south-east England, but may occur elsewhere.

ACTUAL LENGTH: 6–7MM

**LILY BEETLE**

**CONTROL** The long period during which the adults and larvae are active makes this a difficult pest to control. With small numbers of lilies and fritillaries it is possible to protect them by picking the adults and grubs off by hand. Spraying with bifenthrin, heptenophos with permethrin or pirimiphos-methyl can help; the grubs are more vulnerable than adult beetles.

## LILY DISEASE

**SYMPTOMS** Dark green, water-soaked spots develop on the foliage. Affected leaves soon turn brown or bleached and die off. Leaves wither from the base of the stem upwards, wilt and remain hanging on the stem. Flower buds are killed or damaged, so that they become very distorted, and plants fail to flower. If infection spreads into the stem, the plant topples over. The bulb usually remains healthy but may start to rot.

**CAUSE** The fungus *Botrytis elliptica*, which produces numerous spores that are spread on air currents and by rain and water splash. The fungus may persist from year to year as black sclerotia (fungal resting bodies) which develop on infected plant debris. Wet weather encourages disease development and spread.

**CONTROL** Remove and dispose of infected leaves, or the whole of the top growth if necessary. Bulbs from infected plants can be kept, provided they show no rotting. Remove all debris at the end of the season. Some species, notably *Lilium hansonii*, *L. martagon*, *L. pyrenaicum* and *L. davidii* var. *wilmottiae*, seem to show a degree of resistance.

## LILY VIRUSES

**SYMPTOMS** Foliage shows yellow mottling, streaking and mosaicing. Plants are usually stunted, and leaves show twisting or other deformity. Flower buds may not develop, or turn dry and brown and die off.

**CAUSE** Various viruses, including tulip-breaking virus, lily mosaic and lily mottle. Tulip-breaking virus also causes flower breaking symptoms on tulips and so lilies and tulips should not be grown in close proximity.

**CONTROL** None available. Remove infected plants, and wash hands thoroughly after handling them. Spray against aphids, as these may be vectors of the viruses.

## LIME-INDUCED CHLOROSIS

*See* IRON DEFICIENCY AND LIME-INDUCED CHLOROSIS *p.139*.

## LIME NAIL GALL MITE

**SYMPTOMS** From late spring onwards, slender, tubular, red or yellowish-green projections, up to 5mm long, develop on the surface of lime tree leaves (*see p.14*).

**CAUSE** Microscopic gall mites, *Eriophyes tiliae*, which induce the galls' growth by secreting chemicals as they feed on the underside of the leaves in early summer. The galls are hollow and the mites live within them during the summer months.

**CONTROL** Apart from creating the galls, the mites have no harmful effects on the tree and can be tolerated.

## LUPIN APHID

**SYMPTOMS** Dense colonies of whitish-grey aphids, which are up to 4.5mm long, occur on herbaceous and tree lupins during late spring and summer (*see p.41*). They live on the underside of leaves and on flower spikes which become sticky with honey-dew excreted by the pests. Infestations can be so heavy that plants wilt and die.

**CAUSE** A sap-feeding insect, *Macrosiphum albifrons*, which is a North American species that arrived in Britain in 1981 and is now widespread in Europe.

**CONTROL** Look for aphids on leaves from mid-spring; spray with heptenophos with permethrin, dimethoate, pirimicarb, malathion, pirimiphos-methyl, pyrethrum or insecticidal soap before a damaging infestation has developed. Use pirimicarb or insecticidal soap if it is necessary to spray during flowering.

## MAGNESIUM DEFICIENCY

**SYMPTOMS** Yellowing develops between leaf veins (interveinal chlorosis) and around the leaf margins, leaving clear green bands immediately adjacent to the veins (*see p.22*). As the green pigment deteriorates, other pigments may become visible, so instead of yellowing, red, purple or brown discoloration develops between veins. Camellias affected by this deficiency often develop distinct, raised, brown interveinal patches on the lower leaf surface. When magnesium is in short supply, it is transported out of the older leaves into new, developing leaves; older, lower leaves are therefore most readily affected.

**CAUSE** A deficiency of magnesium, which often occurs in plants growing in very acid soils or composts, or after periods of heavy rain or watering. Magnesium is readily leached through the soil when soil water levels are high. The use of high potash fertilizers or sulphate of potash may also increase the occurrence of this deficiency, as the magnesium may be rendered unavailable because of the high levels of potassium in the soil.

**CONTROL** For a rapid response, apply magnesium as a foliar spray. Use Epsom salts at the rate of 200g in 10 litres (8oz in 2½ gallons) of water to which is added a wetting agent such as a few drops of mild liquid detergent or soft soap. Several applications are usually needed, at weekly to fortnightly intervals. Alternatively, apply Epsom salts to the soil at 40g per m² or Kieserite at 70–140g per m².

## MAHONIA RUST

**SYMPTOMS** Bright orange spots develop on upper leaf surfaces (*see p.24*), with corresponding dark brown spore-filled fungal pustules beneath (*see p.12*). The leaf develops bright, often red or orange discoloration, and falls prematurely.

**CAUSE** The fungus *Cumminsiella mirabilissima*, which is encouraged by moist or humid air.

**CONTROL** Clear up and dispose of fallen leaves to minimize overwintering of the fungus. If feasible, improve air circulation around the plant. Infection is rarely damaging enough to merit treatment, but if necessary, spray with a suitable fungicide such as penconazole, mancozeb or bupirimate with triforine.

## MANGANESE DEFICIENCY

*See* IRON DEFICIENCY AND LIME-INDUCED CHLOROSIS *p.139*.

## MEALY CABBAGE APHID

**SYMPTOMS** All brassicas are susceptible to this pest between mid-spring and mid-autumn. Yellow patches appear on the foliage, and the underside of these areas will have dense colonies of whitish-grey aphids (*see p.32*). During early summer the aphids infest leaves developing at the shoot tips of young plants. These leaves become distorted and have a pale, mottled appearance. Heavy attacks can kill shoot tips, causing secondary buds to develop side-shoots, resulting in multi-headed plants.

**CAUSE** Sap-feeding insects, *Brevicoryne brassicae*, which sometimes also attack swede and radish.

**CONTROL** Spray with pirimicarb, heptenophos with permethrin, malathion, pirimiphos-methyl, bifenthrin or insecticidal soap, especially if young plants are under attack.

## MEALY PLUM APHID

**SYMPTOMS** The shoot tips and undersides of leaves on plums have dense colonies of whitish-green aphids on them during late spring to mid- or late summer. The foliage and fruits become sticky with honeydew excreted by the aphids. A black, sooty mould often grows on this substance.

**CAUSE** A sap-feeding insect, *Hyalopterus pruni*.

**CONTROL** Look for the aphid from late spring onwards, and if necessary spray with pirimicarb, heptenophos with permethrin, dimethoate, pirimiphos-methyl or fenitrothion. Sooty mould can be wiped off fruits with a damp cloth.

## MEALYBUGS

**SYMPTOMS** Cacti and succulents are the main host plants, but many other house plants and those grown in greenhouses are attacked. Mealybugs are soft-bodied, greyish-white or pink insects up to 4mm long. They frequently infest leaf axils (*see p.32*) and other inaccessible places, such as between twining stems (*see p.37*). A white, fluffy wax is secreted from the insects' bodies and this also conceals the eggs. Infested plants become sticky with honeydew excreted by the pest. Sooty moulds may grow on the honeydew.

**CAUSE** Several sap-feeding mealybugs (*Pseudococcus* and *Planococcus* species) occurring in greenhouses.

**CONTROL** During mid-summer, when temperatures are high, biological control with a ladybird predator, *Cryptolaemus montrouzieri*, can reduce infestations. At other times spray thoroughly with malathion or insecticidal soap.

• *see also* PHORMIUM MEALYBUG *p.159*; ROOT MEALYBUGS *p.170*.

## MELON COTTON APHID

**SYMPTOMS** Small, dark green or black aphids form dense colonies on the foliage and flowers of chrysanthemums (*see p.43*). Other plants

likely to be attacked include cucumbers, melons, begonias, dahlias and cinerarias. Heavy infestations cause stunted growth and the early death of blooms. Plants are soiled, with cast aphid skins, a sticky excrement called honeydew and sooty moulds.

**CAUSE** A sap-feeding insect, *Aphis gossypii*, which can also damage plants by transmitting tomato spotted wilt virus. It is mainly a greenhouse problem, where it can remain active throughout the year.

**CONTROL** This aphid usually has a high degree of pesticide resistance, so the usual chemicals available to amateur gardeners fail to work. Biological control with parasitic wasps, *Aphidius colemani*, or predatory midge larvae, *Aphidoletes aphidomyza*, can help during the warmer summer months. Frequent, thorough spraying with insecticidal soap can reduce infestations.

## MERIPILUS GIGANTEUS

**SYMPTOMS** Large, fleshy, golden-brown fungal brackets develop around the base of affected trees or in close proximity to the trunk or roots (*see p.39*). The lower surface is creamy-white (but bruises dark when handled). The fruiting bodies are conspicuous for their large size (up to 30cm across) and the number that are produced (they may total up to about 12kg in weight). They usually appear in late summer or autumn, and deteriorate into a smelly, black mass when frosted. The fungus causes root decay, often attacking the lower roots first, so the damage may not be apparent

on initial investigation. The roots become hollow and so make the tree unsafe. Infected trees often show dieback of branches; they produce small, sparse leaves, and may be blown over in heavy winds.

**CAUSE** The fungus *Meripilus giganteus*, occasionally known as the giant polypore. Mature or over-mature trees of beech (*Fagus*), oak (*Quercus*) and occasionally robinia may be attacked.

**CONTROL** There is no control available, and prompt inspection by a professional arboriculturist is advisable, as trees may be rendered dangerous. Removal of the fungal growth will not prevent further damage to the tree but will prevent spores being spread to other potential hosts.

• *see also* GRIFOLA FRONDOSA *p.134*.

## MICE

Several types of mice, such as *Apodemus* and *Mus* species, cause problems in gardens and buildings. They dig holes in the soil where peas, beans and sweetcorn have been sown, or where crocus corms have been planted. Green shoots are left lying on the soil surface but the corms and seeds are eaten (*see p.51*). Mice also cause damage in the autumn to spring period, when they come indoors and feed on stored fruits, vegetables and seeds.

**CONTROL** Set mouse traps in places where mice are causing damage. In the garden, put traps under the cover of logs or bricks, away from birds and pets. After planting crocuses, press the ground down firmly to make it harder for mice to locate the corms.

## MICHAELMAS DAISY MITE

**SYMPTOMS** Infested plants are stunted and have a brownish scarring on the stems. The flowers are reduced in size and number, with many of them being converted into rosettes of small green leaves (*see p.45*). *Aster novi-belgii* cultivars are particularly susceptible.

**CAUSE** Microscopic sap-feeding mites, *Phytonemus pallidus*, which live at the tips of shoots and in the flower buds.

**CONTROL** None of the pesticides available to amateur gardeners is effective against this mite. Infested plants should be scrapped, as their condition will not improve. Other forms of Michaelmas daisy, such as *Aster novae-angliae* and *A. amellus*, are less susceptible and should flower satisfactorily.

## MICHAELMAS DAISY WILT

**SYMPTOMS** The foliage wilts and later withers. The plant is usually affected in sections. Dark brown discoloration of the vascular tissue (conducting elements) within the stem may be apparent if the bark is peeled back. The whole plant is usually killed.

**CAUSE** The fungus *Phialophora asteris*, which is soil-borne. It may enter via damaged roots, or via pruning tools which have been used on an infected plant and then on a healthy plant without disinfecting them.

**CONTROL** No control is available. Remove infected plants and the soil in the vicinity of their roots. Do not plant Michaelmas daisies on the same site.

## MIDGES

*See* CHIRONOMID MIDGES *p.116*; GALL MIDGES *p.130*.

## MILLIPEDES

**SYMPTOMS** Millipedes feed mainly on rotting organic matter, but will sometimes damage seedlings and other soft plant tissues, such as strawberry fruits. They can also extend damage initiated by other pests, such as slugs on potato tubers. Millipedes have long, segmented bodies which may be cylindrical or flat; each body segment has two pairs of legs, unlike the superficially similar predatory animals called centipedes (*see p.116*), which have one pair of legs per segment.
**CAUSE** Many species can be found in garden soils. The one most often associated with plant damage is the spotted snake millipede, *Blaniulus guttulata* (*see p.54*). It is cream-coloured, with a row of red dots on either side of a cylindrical body up to 25mm long.

ACTUAL LENGTH: 15–40MM

**SPOTTED SNAKE MILLIPEDE**

**CONTROL** Protect strawberry fruits by using straw to lift the berries clear of the soil. Damage to potato tubers is generally initiated by slugs, so it is necessary to control them first (*see p.176*). Millipedes are encouraged by organic fertilizers and these should be avoided if millipedes are numerous enough to cause problems. They are not easily controlled with insecticides, but lindane dust has some effect.

## MINT BEETLE

**SYMPTOMS** Emerald-green beetles, up to 10mm long, occur in mid-summer on wild and cultivated mints (*see p.31*). In late summer, rotund, matt-black grubs also appear. Both adults and larvae eat holes in the foliage.
**CAUSE** A leaf-feeding beetle, *Chrysolina menthrastri*, which has one generation a year.
**CONTROL** Hand-removal is often sufficient, as the adult beetles are easy to see.

## MINT RUST

**SYMPTOMS** Leaves and stems become flecked with yellow and distorted. The stem may be conspicuously curled, often shortly after growth has begun in spring. Numerous orange-grey, cup-like fungal bodies develop on the lower leaf surface and the affected areas of stem. These become yellowish-orange and later blackish-brown. Mint is the most common host, but several other herbs, including marjoram and savory, may also be affected.
**CAUSE** The fungus *Puccinia menthae*, which survives the winter on the remains of the plants or as resistant, resting spores in the soil.
**CONTROL** It may be possible to kill the fungus by burning off all of the top growth of mint plants and scorching the soil surface at the end of the season. This is, however, difficult to do safely and reliably; it may be more advisable to dispose of all infected plants, replanting fresh stock on a new site.

## MITES

Mites are allied to spiders and most have four pairs of legs, although gall mites have only two pairs. All the plant pest species of mites are less than 1mm in length. They feed by sucking sap from their hosts. Gall mites secrete chemicals as they feed that cause the plant to produce abnormal growths that enclose the mites. Most mites capable of damaging plants belong to the red spider mite group. These cause fine, pale mottling of foliage and this can lead to defoliation. Heavily infested plants may become festooned with silk webbing. Tarsonemid mites infest the shoot tips and flower buds, causing stunted and distorted growth. Many mites are harmless, since they feed on fungal growth on rotting plants or on algae (*see* oribatid mites, *p.154*). Some mites are predators of pest insects or of other mites, and can be used to give biological control in greenhouses. Examples are *Phytoseiulus persimilis* for the control of glasshouse red spider mite, and *Amblyseius* species for western flower thrips.

• see also **Gall mites** – ACER GALL MITES *p.98*; BIG BUD MITES *p.107*; BROOM GALL MITE *p.110*; ELM GALL MITE *p.125*; FELT GALL MITES *p.126*; LIME NAIL GALL MITE *p.146*; MOUNTAIN ASH BLISTER MITE *p.149*; PEAR LEAF BLISTER MITE *p.157*; PLUM LEAF GALL MITE *p.161*; RASPBERRY LEAF AND BUD MITE *p.167*; RED BERRY MITE *p.168*; VINE ERINOSE MITE *p.186*; WALNUT BLISTER MITE *p.187*.
**Red spider mites** – BRYOBIA MITES *p.111*; CONIFER RED SPIDER MITE *p.119*; FRUIT TREE RED SPIDER MITE *p.129*; GLASSHOUSE RED SPIDER MITE *p.132*.
**Tarsonemid mites** – BULB SCALE MITE *p.112*; MICHAELMAS DAISY MITE *p.147*; STRAWBERRY MITE *p.179*; TARSONEMID MITES *p.181*.

## MOLES

Moles, *Talpa europaea*, are infrequently seen, as they spend most of their lives below soil level. They create a system of underground tunnels and chambers in which they live and feed on earthworms and other soil-dwelling invertebrates. The soil excavated during tunnel construction is deposited in heaps on the surface of lawns and flower-beds (*see p.59*). All of the mole activity in a small garden could be the work of a single animal. Moles also make shallow tunnels above which the soil is pushed up as a ridge.
**CONTROL** A variety of controls and deterrents is available. The cheapest and often most effective method is to use a mole trap. The position of a mole's tunnel is located by probing the soil around a recent molehill with a stick. When found, open up the tunnel with a trowel at least 30cm away from the molehill. Insert the trap, taking care to align it with the direction and depth of the tunnel. Cover the trap with turf and an inverted bucket to stop light entering the tunnel. If the mole keeps pushing soil into the trap, reset it else-where in the tunnel system. Alternatives to trapping are mole smokes and electronic devices. Smokes are ignited and placed in a tunnel. They give off toxic gases and may kill the mole if the gases do not escape too readily. Electronic devices emit noises said to drive moles away, but are expensive and give variable results.

## MOTHS

Moths, like butterflies, are characterized by having wings that are covered by many small, overlapping scales. These scales are often coloured, and their arrangement on the wings gives the distinctive colour patterns by which the different moth species can be recognized. Moths vary considerably in size; some of the smallest, like some of those with leaf-mining larvae, have a wing-span of about 5mm, while the biggest hawk moths have a wing-span of up to 80mm. In all cases it is the caterpillars which cause the damage to plants, rather than the adults, which take nectar from flowers. Most moths are not pests, as their caterpillars do not feed on cultivated plants, but those listed below can cause damage to garden plants.

• *see also* APPLE LEAF MINER *p.101;* BROWN-TAIL MOTH *p.111;* BUFF-TIP MOTH *p.112;* CABBAGE MOTH *p.112;* CARNATION TORTRIX MOTH *p.114;* CHINA MARK MOTH *p.116;* CODLING MOTH *p.118;* COTONEASTER WEBBER MOTH *p.119;* CURRANT CLEARWING MOTH *p.121;* DELPHINIUM MOTH *p.123;* ELEPHANT HAWK MOTH *p.125;* GYPSY MOTH *p.134;* JUNIPER WEBBER MOTH *p.140;* LACKEY MOTH *p.141;* LEEK MOTH *p.144;* LEOPARD MOTH *p.144;* LILAC LEAF MINER *p.145;* MULLEIN MOTH *p.149;* PEA MOTH *p.155;* PLUM MOTH *p.161;* PYRACANTHA LEAF MINER *p.166;* SMALL ERMINE MOTHS *p.176;* SWIFT MOTHS *p.180;* TOMATO MOTH *p.182;* TORTRIX MOTHS *p.183;* VAPOURER MOTH *p.185;* WINTER MOTHS *p.191;* YELLOW-TAIL MOTH *p.192.*

## MOUNTAIN ASH BLISTER MITE

**SYMPTOMS** The foliage of rowan or mountain ash and other *Sorbus* species develop a whitish-green blistered appearance during late spring or early summer. Later in the summer, the pale areas darken and become brown. Sometimes nearly every leaf is affected, and there may be some early leaf fall.

**CAUSE** A microscopic gall mite, *Eriophyes pyri*. This causes similar damage to pears (*see p.157*), although it is believed that different sub-species are involved which are specific to *Sorbus* and *Pyrus* species.

**CONTROL** None of the pesticides available to amateur gardeners gives good control. Light infestations on small trees can be dealt with by removing affected leaves. Heavier infestations should be tolerated, since despite the unhealthy appearance of the foliage, the mites have little effect on the tree's vigour. The degree of infestation can vary greatly from year to year.

## MULBERRY BACTERIAL BLIGHT

**SYMPTOMS** Small, black, angular spots, each with a yellow "halo", develop all over the leaf surface. As the infection progresses, the entire leaf may become yellowed and, if infection is severe, the tree's overall vigour may be reduced. Roughened areas may develop on the bark of shoots and can result in dieback.

**CAUSE** The bacterium *Pseudomonas mori*, which overwinters in the stem lesions.

**CONTROL** No control measures are available. Remove affected leaves and prune out any stems showing lesions.

## MULBERRY CANKER

**SYMPTOMS** Shoots die back due to the presence of small cankers which girdle the affected stems (*see p.38*). Tiny, reddish-brown pustules develop around the cankered area. Symptoms are usually first noticed in the summer.

**CAUSE** The fungus *Gibberella baccata*.

**CONTROL** Prune out cankered shoots and areas of dieback promptly in an attempt to limit spread.

• *see also* NECTRIA CANKER *p.151.*

## MULLEIN MOTH

**SYMPTOMS** Caterpillars up to 50mm long, greyish-white with black and yellow markings, are found during summer on mulleins, figwort (*Scrophularia*) and buddlejas. They feed on the foliage and flower spikes.

**CAUSE** Caterpillars of a moth, *Cucullia verbasci*.

**CONTROL** The brightly coloured caterpillars feed in exposed positions and can be removed by hand. Where this is not feasible, spray with pyrethrum, permethrin or bifenthrin before extensive damage is caused.

## MUSSEL SCALE

**SYMPTOMS** This pest lives on the older stems of woody trees and shrubs, including apple (*Malus*), cotoneaster, box (*Buxus*), dogwoods (*Cornus*), heathers and ceanothus. The scales are greyish-brown, 2–3mm long, and in shape and appearance resemble the shellfish of the same name. Heavy infestations can occur to the extent that the bark is entirely covered (*see p.36*), and scales may spread onto fruits and berries. Such plants lack vigour and shoots may die back.

**CAUSE** Sap-feeding insects, *Lepidosaphes ulmi*, which have one generation a year.

**CONTROL** On fruit trees, control overwintering scales by spraying with a tar oil wash in winter while the tree is dormant. Otherwise spray with malathion in early summer to kill the more vulnerable, newly-hatched scale nymphs. If the plant is in flower, spray at dusk to avoid harming bees.

## MYCOPLASMA INFECTION

**SYMPTOMS** Symptoms vary from host to host and with the precise identity of the organism. They may resemble those produced by viruses, and may include greening of flowers and yellow markings on leaves, often combined with distortion and stunting.

**CAUSE** Various organisms, usually described as "virus-like", thought to be similar in structure and activity to both bacteria and viruses.

**CONTROL** None is available. Remove any affected plants promptly.

## NARCISSUS BASAL ROT

**SYMPTOMS** Infection occurs at the basal plate (the flattened area at the base of the bulb), causing it to discolour and rot. Discoloration and rotting spread upwards through the inner scales, turning them chocolate-brown. A pale pink fungal growth may develop on the basal plate and amongst the affected scales. Bulbs in store turn brown all over, dry out and become mummified. Those in the ground usually rot off completely, and by the end of the season there may be no trace of them. Leaf symptoms on growing bulbs include yellowing, stunting and sparse foliage. **CAUSE** The fungus *Fusarium oxysporum* f. sp. *narcissi*, which is soil-borne. It persists in soil on infected bulbs that are not sufficiently affected for the symptoms to have been noticed. The disease is most damaging during or following very hot summer weather. **CONTROL** None available, but it may be possible to prevent infection by dipping the bulbs in a fungicide solution. This should be done before summer temperatures are at their highest, and within 48 hours of the bulbs being lifted. Dip for half an hour in a solution of carbendazim, following the manufacturer's instructions, and then dry off before storing. Dusting with sulphur, and early planting in autumn, may also reduce the risk of infection developing.

## NARCISSUS BULB FLY

**SYMPTOMS** Bulbs fail to grow or produce only a few thin leaves in the spring. If the bulb is cut in half vertically, it can be seen that the centre has been eaten away and filled with a fly larva's muddy excrement (*see* **p.54**). The plump maggot is a dirty cream colour and is up to 16mm long. There is usually only a single maggot in each bulb. Several maggots, each up to 8mm long, indicate that the bulb has been colonized by small bulb fly larvae (*Eumerus* species). These flies lay eggs on bulbs that are already damaged by other pests or diseases, whereas narcissus bulb fly attacks sound bulbs.

**CAUSE** Narcissus bulb fly, a hoverfly (*Merodon equestris*), resembles a small bumblebee and lays eggs on the necks of bulbs in early summer, when the foliage is dying down. It also attacks other bulbs, including hippeastrum and snowdrops. **CONTROL** Chemical controls against this pest are not effective, but cultivation techniques can reduce losses. The adult flies prefer warm, sunny places, so plant narcissus bulbs in shady or exposed sites. Mound up soil around the stem bases when plants start to die down, to make them less attractive as egg-laying sites. Valuable bulbs can be protected by growing them under horticultural fleece. This needs to be in position from late spring to mid-summer. When small bulb fly larvae are found, investigate whether the bulbs are suffering from other pests such as eelworm or narcissus fly, or from fungal diseases.

## NARCISSUS EELWORM

**SYMPTOMS** Plants produce stunted and distorted foliage and flower stems. Some viruses cause similar symptoms, so the presence of eelworms needs to be confirmed by cutting the bulb in half transversely. Eelworm activity in the bulb causes concentric brown rings or arcs where the pest has damaged the tissues (*see* **p.54**). In large plantings of daffodils, the pest gradually spreads, so that each year the area of killed and damaged plants gets larger. New bulbs planted as replacements in infested soil quickly succumb to eelworm.

**CAUSE** Microscopic nematodes that live within the tissues of the bulb, foliage and flower stems. **CONTROL** There are no effective pesticides available to amateur gardeners. Dig up and burn daffodils showing symptoms; others growing within a radius of 1m may be infested and should also be removed. Avoid introducing narcissus eelworm into a garden by planting sound bulbs purchased from a reputable supplier. Hot water treatment of bulbs can be attempted, although it is difficult to do this properly without access to equipment that will keep the temperature constant. In late summer, dormant bulbs must be immersed for three hours in water heated to 44.5°C (112°F), after which they are cooled and planted in fresh soil. Too much heat will damage the bulbs, while too little allows eelworms to survive.

## NARCISSUS LEAF SCORCH

**SYMPTOMS** As the leaves emerge, they become red-brown and scorched, and the symptoms then spread down the leaves (*see* **p.29**). Brown spots appear on the foliage and numerous tiny, pinprick-sized, raised fungal bodies may develop on damaged areas. Occasionally flower stems and spathes may also be attacked. The bulbs do not show rotting or discoloration. In addition to daffodils and narcissi, amaryllis, crinums, hippeastrums, nerines, snowdrops and sprekelias may be infected. **CAUSE** The fungus *Stagonospora curtisii*, which may survive in the dry, papery outer scales of the bulb.

**CONTROL** Spray emerging foliage with carbendazim or mancozeb. Remove and dispose of infected leaf tips. Dip bulbs for half an hour in a solution of carbendazim, according to the manufacturer's instructions, and dry off before storing.

## NARCISSUS SMOULDER

**SYMPTOMS** Infected bulbs produce deformed, chlorotic shoots with blackish-brown, withered leaf tips. Only a few leaves may be affected. In wet weather they may become covered with masses of grey spores. Infected leaves may rot off, and the deterioration then spreads down into the centre of the bulb. Numerous tiny, black sclerotia (fungal resting bodies) develop on the outer scales of the bulb and on infected leaves and debris. **CAUSE** The fungus *Sclerotinia narcissicola*, which may persist in the soil as sclerotia on infected bulbs and debris. **CONTROL** As soon as symptoms appear on leaves, spray with a fungicide containing carbendazim. Inspect bulbs carefully and plant only those which appear perfectly healthy and free from sclerotia. Dip bulbs for half an hour in a solution of carbendazim, according to the manufacturer's instructions, and then dry off before storing.

## NARCISSUS VIRUSES

**SYMPTOMS** Many different viruses may infect daffodils and narcissi, alone or in combination. The symptoms depend on the cultivar, the virus and growing conditions. Faint yellow mottling may occur in mild cases, particularly towards the leaf base.

Flowers may show bleached streaks (see **p.23**), and occasionally yellow streaking on the flower stem. Certain viruses, in particular narcissus yellow stripe virus, cause bright yellow leaf stripes, often combined with distortion and stunting, and the plant may fail to flower.
**CAUSE** Various viruses, transmitted by aphids, soil-living eelworms or handling. The vector varies depending on the virus.
**CONTROL** Mild symptoms are quite common and may never develop seriously. All infected plants are, however, a source of infection, so disposal is usually the best option, particularly for those showing symptoms of narcissus yellow stripe virus.

## NECTAR ROBBING

Many flowers produce nectar in order to attract bees and other pollinating insects. As these insects push their mouthparts into the nectaries at the base of a flower, their bodies become dusted with pollen from the anthers. When they visit other flowers of the same type, some of this pollen rubs off onto the stamens and pollination is achieved. Sometimes, however, bees cheat by biting a hole in the base of the flower (see **p.46**), enabling them to reach the nectar without touching the anthers and stamens. It is mostly short-tongued bumblebees that do this where they cannot reach the nectaries from the front of the flower. Once a short-cut to the nectar exists, other insects will take advantage. Nectar robbing is most frequently encountered on beans and may contribute to poor flower set. It is not possible to prevent nectar

robbing, but flower set is usually adequate if beans are kept well-watered and do not suffer from high temperatures.

## NECTRIA CANKER

**SYMPTOMS** Woody plants show roughened, flaked areas of bark on branches, twigs and stems. The cankers enlarge and may girdle the stem, causing dieback. During the summer, white fungal pustules may develop on cankered areas; in winter these are replaced by many small, red fruiting bodies.
**CAUSE** Various species of the fungus *Nectria*.
**CONTROL** Prune out affected areas as soon as symptoms develop. Spray with a fungicide containing copper; on fruit trees, the sprays containing carbendazim used to control mildew and scab should also give incidental control of this disease.

## NEEDLE BLIGHTS AND CASTS OF PINES

**SYMPTOMS** On young pines, the needles discolour, often showing diffuse brown or black lines. Infected needles may be shed prematurely. Damage to mature trees is insignificant, but young pines may be weakened. Another blight may cause distinct bands of discoloration to develop on the needles.
**CAUSE** *Lophodermium* species and *Dorthistroma pini*.
**CONTROL** None available. Rake up and dispose of infected needles as they fall.

## NEEDLE RUSTS OF CONIFERS

**SYMPTOMS** On spruce (*Picea*), needles become discoloured. If the fungus *Chrysomyxa*

*abietis* has attacked them, they may develop orange fungal pustules but remain attached to the tree through winter; or, if attacked by *Chrysomyxa rhododendri*, they may become covered with orange pustules, followed by white spore tendrils, and fall from the tree in autumn. This latter fungus is less common and spends part of its life cycle on rhododendrons. Norway and Sitka spruce are most frequently affected, but most spruces can be attacked. On pines another fungus causes small yellowish spots on the needles, and resin-like exudations may occur. Later, off-white, column-like growths develop on the needles and erupt to produce bright orange spores. This fungus may also attack various herbaceous hosts, including campanulas and some weeds, in particular sow thistle and groundsel.
**CAUSE** The fungi *Chrysomyxa abietis* and *C. rhododendri* on spruce; the fungus *Coleosporium tussilaginis* on pine and other plants.
**CONTROL** None available. Dispose of infected needles.

## NEMATODES

See EELWORMS *p.125*.

## NITROGEN DEFICIENCY

**SYMPTOMS** Leaves are small and chlorotic, and general growth is reduced. On some plants, red or purple leaf discoloration may develop as the green pigment levels drop. The lower, older leaves are affected first, but if the deficiency progresses, all parts of the plant may be similarly affected. Flowering, fruiting and tuber or root formation are also reduced.

**CAUSE** A deficiency of nitrogen, which can develop in any soil, particularly if it is light and has a low organic matter content, or if it has been heavily cropped. The use of large quantities of wood chippings may also result in nitrogen deficiency, as nitrogen is removed from the soil as the lignin in the wood is broken down.
**CONTROL** Apply nitrogen-rich fertilizers, composts and mulches to the soil regularly. Grow legumes, which are capable of fixing nitrogen (see *p.68*) using bacteria in their root nodules.

## NUT WEEVIL

**SYMPTOMS** In late summer, hazelnuts and cob nuts have a round hole, 1–2mm in diameter, in their shells (see **p.52**) where beetle grubs have bored their way out before pupating in the soil. The kernel inside the nut shell is eaten by a white maggot with a pale brown head.
**CAUSE** A brown weevil, *Curculio nucum*, which is about 10mm long. Almost half the body length is due to a slender snout or rostrum on the front of the head.

ACTUAL LENGTH: APPROX. 10MM

**ADULT NUT WEEVIL**

**CONTROL** Usually only a small proportion of the crop is attacked, so routine control measures are not required. If a heavy attack has occurred, spray against adult weevils with fenitrothion or pirimiphos-methyl in early summer when the young nuts are forming.

## OAK GALL WASPS

**SYMPTOMS** English oaks (*Quercus robur* and *Q. petraea*) are the host plant of more than 30 species of gall wasps. Many of these have complex life cycles, involving alternating sexual and asexual (female only) generations which create different forms of gall (*see p.17*) on different parts of the tree. Some examples are the oak apple gall wasp and the common spangle gall wasp. The former lays eggs in buds in the winter and this causes them to develop into pithy, creamy-pink, bun-shaped objects in the spring. The adults that emerge in mid-summer go into the soil and lay eggs on the roots, giving rise to spherical swellings which contain the wasps' grubs. When adults emerge they crawl up the tree and complete the two-generation cycle by laying eggs in the buds. The common spangle gall wasp creates flat, yellowish-brown, disc-like galls (*see p.17*) on the underside of oak leaves in late summer. The next generation causes spherical yellowish-red galls on the male catkins in the spring, with adults emerging to lay eggs in the foliage in mid-summer. With all gall wasps it is the larval stage that secretes chemicals that induce the plant to produce gall tissues.

**CAUSE** Oak apple gall wasp, *Biorhiza pallida*, and common spangle gall wasp, *Neuroterus quercusbaccarum*. Other gall wasps frequently encountered on oaks are the marble gall wasp, *Andricus kollari*, which makes woody spherical galls up to 25mm across on the stems, and the artichoke gall wasp, *Andricus fecundator*, which makes greatly enlarged buds at the shoot tips. The silk button gall wasp, *Neuroterus numismalis*, has galls shaped like ring doughnuts, 3mm across, on the underside of leaves in late summer (*see p.17*). The cherry gall wasp, *Cynips quercusfolii*, produces reddish-yellow spherical galls up to 18mm across under leaves in late summer.

**CONTROL** Control measures are unnecessary. None of the gall wasps has any serious effect on the growth of oak trees, even when numerous.
• *see also* ACORN GALL WASP *p.98.*

## OAK PHYLLOXERA

**SYMPTOMS** In late spring or early summer, yellow spots develop on the upper surface of oak (*Quercus*) leaves. Underneath these spots on the lower leaf surface are orange aphid-like insects up to 1mm long. The females are often surrounded by a ring of yellow eggs. Later in the summer, the spots on the leaves tend to join together and damaged parts of the leaf dry up and become whitish-brown.

**CAUSE** A sap-feeding insect, *Phylloxera glabra*, which is a specific oak pest and over-winters on the tree as an egg. The heaviest infestations tend to be on saplings rather than mature trees.

**CONTROL** Encourage good growth by watering young trees in dry summers. On small trees, phylloxera can be controlled by spraying with heptenophos with permethrin, dimethoate, malathion or pirimiphos-methyl in early summer if damage is seen.

## OAK POWDERY MILDEW

**SYMPTOMS** Tiny, almost imperceptible cinnamon-coloured spots develop on young leaves (*see p.24*) from late spring onwards. These are soon followed by a layer of white fungal spores and mycelium. Severely affected leaves may be killed, but the overall vigour of the tree is rarely affected.

**CAUSE** The fungus *Microsphaera alphitoides*, which overwinters in buds.

**CONTROL** Only really necessary on very young trees which are more likely to suffer serious effects. Keep the plants well watered, avoiding overhead irrigation. Spray with a suitable fungicide such as one containing carbendazim or bupirimate with triforine.

## OEDEMA

**SYMPTOMS** Raised, wart-like outgrowths develop, most frequently on the lower leaf surfaces (*see p.19*). These are the same colour as the leaf initially but may later become corky and brown. Affected leaves may sometimes be distorted but do not always die.

**CAUSE** The plant takes up more water than it can lose through the leaves by the process of transpiration. As a result, small groups of leaf cells swell up as they become over-filled with water, producing pale green warty growths. If the growing conditions do not improve, cells rupture and die off and so turn brown and corky. Oedema usually develops when humidity is high and water levels around the roots are excessive.

**CONTROL** Do not remove leaves showing the symptoms. This disorder is not infectious and removal of leaves further reduces the rate of moisture loss and so worsens the problem. Reduce watering and/or improve drainage. Increase air circulation around the plants by spacing them further apart and improving ventilation, as appropriate.

## OLEANDER SCALE

**SYMPTOMS** Flat, round, whitish-brown objects up to 2mm across develop on the stems and foliage of many greenhouse plants. These include oleander, jasmine, asparagus ferns, azalea, palms and ivies; it also occasionally occurs on spotted laurel (*Aucuba japonica*) growing in sheltered places in gardens. Heavy attacks cause loss of vigour and dieback.

**CAUSE** A sap-feeding insect, *Aspidiotus nerii*, which breeds continuously throughout the year in greenhouses. Eggs are laid under the protection of the females' bodies and shell-like wax scales. No honeydew is excreted, so there is no sooty mould on the foliage.

**CONTROL** Prune out heavily infested shoots or leaves. Spray with malathion or an insecticidal soap, particularly on the lower leaf surface. Two or three applications at fortnightly intervals may be needed to control the young nymphs as they hatch from the eggs.

## ONION BOLTING

**SYMPTOMS** Onions produce flower stems before reaching the normal stage of maturity, so the bulb size is smaller than it should be.

**CAUSE** Exposure to cold temperatures earlier in the year promotes early flowering. This is usually brought about by a late, cold spring and is particularly common on early cultivars. Very hot, dry conditions and day length may also induce bolting.

**CONTROL** Little can be done, but the incidence will be reduced by use of late cultivars which are less likely to be subjected to early cold weather.

## ONION DOWNY MILDEW

**SYMPTOMS** Affected leaves turn grey, wither and collapse, often with a "boiled" appearance. Close inspection may reveal dense, slightly fluffy, off-white fungal growth. However, secondary infections, particularly of the fungus *Alternaria*, may rapidly mask this. If left on the plant, the rotting of the leaves may spread down into the bulb and so cause further losses. The earlier in the season that the disease appears, the greater the reduction in bulb size.

**CAUSE** The fungus *Peronospora destructor*, which is particularly common in mild, wet weather and is encouraged by humid growing conditions.

**CONTROL** Remove infected leaves promptly and spray with a suitable fungicide such as mancozeb. If the disease is recurrent, preventive spraying may even prove necessary. Improve air circulation around the plants by increased spacing and regular weeding. Clear up and dispose of all debris at the end of the season. Grow the cultivar 'Norstar', which shows a degree of resistance.

## ONION EELWORM

**SYMPTOMS** Young onion plants are stunted and abnormally swollen. The tissues have a soft mealy texture and readily succumb to secondary rots (*see* **p.52**), so the plants usually die before reaching maturity. Onions attacked later in the summer do produce bulbs, but these will also be infested and will rot in store.

**CAUSE** Microscopic worm-like eelworms or nematodes, *Ditylenchus dipsaci*, that infest the foliage, bulbs and seeds of onion and allied plants. It can also attack some other vegetables and common weeds, without necessarily causing obvious symptoms, but these alternative host plants allow onion eelworm to survive when onions are not being grown.

**CONTROL** There are no effective insecticides available to amateur gardeners, so remove infested plants as soon as they are spotted. Buy onion sets and seeds from reputable suppliers in order to reduce the risk of introducing the pest into a garden. When onion eelworm has occurred, crop rotation can be effective in eliminating it. Good weed control and growing vegetables not affected by onion eelworm, such as lettuce, turnip, swede and any of the brassicas, for one or two years will clear the soil.

## ONION FLY

**SYMPTOMS** Onion is the most susceptible plant but leeks, shallots and garlic can also be attacked. In early summer young plants may collapse as the roots are eaten by white maggots up to 8mm long. Larvae of the second generation in late summer feed on roots and also bore into onion bulbs, allowing secondary rots to gain entry.

**CAUSE** Maggots of a fly, *Delia antiqua*, which is similar to a housefly in appearance.

ACTUAL LENGTH: UP TO 8MM

**ONION FLY MAGGOT**

**CONTROL** Onions grown from sets are less susceptible to first generation maggots than seedlings. Some protection can be given by dusting the soil along seed rows or around sets with pirimiphos-methyl. Maggots feeding inside onion bulbs cannot be reached with insecticides. Infested plants should be lifted carefully and destroyed before the maggots go into the soil to pupate.

## ONION NECK ROT

**SYMPTOMS** Scales of infected onions become soft, pale brown and semi-transparent (*see* **p.52**). A dense, grey fungal growth develops over the affected areas and the tissues start to dry out, so that the onion becomes mummified. Sclerotia (resilient black fungal resting bodies), which may be several millimetres in length, develop on the affected areas, particularly around the neck end of the bulb.

**CAUSE** The fungus *Botrytis allii*, which can persist in the form of sclerotia on onion debris or, having fallen from the bulb, loose in the soil. Infections usually spread by spores or by sclerotia. The symptoms are usually first noticed in store.

**CONTROL** Dust seeds or sets with carbendazim. Buy seed and sets only from reputable sources, as the fungus can easily be brought in on infected material. Do not grow onions on the same site for more than two years in succession. Improve cultural conditions so that hard, well-ripened bulbs are produced. Do not apply fertilizers after mid-summer and avoid excessive use of high nitrogen fertilizers. Keep crops evenly and adequately watered. Grow red or yellow bulbs instead of white, as these are generally less susceptible. The cultivar 'Norstar' is known to show a degree of resistance. Keep onions dry as they are drying off.

## ONION THRIPS

**SYMPTOMS** The foliage of onions and leeks develops a fine, white mottling during the summer. Narrow-bodied insects, up to 2mm long, and black or pale yellow in colour, occur on the leaves. Thrips are more troublesome in hot, dry summers and heavy infestations will check the plant's growth.

**CAUSE** The adults and nymphs of a sap-sucking insect, *Thrips tabaci*, which can also feed on many ornamental plants.

**CONTROL** Light infestations can be tolerated, but if much of the green colour is being lost before late summer, spray with malathion, permethrin, pirimiphos-methyl or pyrethrum.

## ONION WHITE ROT

**SYMPTOMS** Dense, white, slightly fluffy fungal growth develops around the base of the bulb and the roots (*see* **p.52**). Numerous small, black sclerotia then develop in amongst this growth. The sclerotia may fall off and into the soil where they may remain viable for more than seven years. The foliage on affected bulbs becomes chlorotic and wilts.

**CAUSE** The fungus *Sclerotium cepivorum*.

**CONTROL** There is no chemical control available. Remove and burn infected plants as soon as they are noticed and do not grow onions or related plants on the same site for at least eight years. Grow the cultivar 'Norstar' which shows a degree of resistance.

## ORCHID VIRUSES

**SYMPTOMS** Various symptoms, depending on the orchid cultivar and the virus or viruses involved. Yellow or very dark brownish-black streaking of the foliage may appear (*see p.12*). Yellow or brownish-black spots or rings may also develop. Flowering may be affected and the plant's overall vigour declines.

**CAUSE** Various viruses, which may infect alone or in combination. The mode of transmission varies but handling, pruning tools and sap-feeding insect pests may all be involved.

**CONTROL** None is available. Remove and dispose of infected plants promptly and control potential means of spread.

## ORIBATID MITES

Oribatid mites, such as *Humerobates* species, also known as beetle mites, commonly occur on the bark of many trees and shrubs, and sometimes on fences. They have shiny black or reddish-brown bodies, about 1mm long, and during the day tend to cluster together in crevices where they can be mistaken for eggs. They are often noticed on dead or

dying plants, but are not the cause of the plant's decline. Oribatid mites feed on algae on the bark and are mainly active after dark.

**CONTROL** No control measures are required, as these mites do not harm plants.

## PANSY DOWNY MILDEW

**SYMPTOMS** Purple-brown spots appear on the upper leaf surface, often combined with a yellow halo. These may be confused with the symptoms caused by pansy leaf spot (*see below*). The lower leaf surface develops small, grey or off-white patches of fuzzy fungal growth which correspond with the spots. Affected leaves become chlorotic and may wither and die off. In severe cases the whole plant may become so weakened that it dies.

**CAUSE** The fungus *Peronospora violae*, which is encouraged by humid air and moist growing conditions.

**CONTROL** Remove affected leaves. Spray with a suitable fungicide such as mancozeb.

## PANSY LEAF SPOT

**SYMPTOMS** Brown, purple-brown or very pale buff-coloured spots develop on the foliage (*see p.13*). The older leaves are affected first and most severely. If the affected leaf tissue is killed, the whole area may fall away, producing shothole symptoms.

**CAUSE** Various fungi, including species of *Ramularia*, *Cercospora* and *Phyllosticta violae*.

**CONTROL** Rarely necessary; remove any affected leaves. If plants are growing very poorly, spraying with mancozeb may be successful.

## PANSY POWDERY MILDEW

**SYMPTOMS** Powdery, white fungal growth develops on the upper leaf surface. This may cause distortion and leaf yellowing, occasionally followed by leaf death.

**CAUSE** *Oidium* species, which may be more troublesome if soil conditions are dry and the plants are growing in overcrowded conditions.

**CONTROL** Ensure that the soil or compost is adequately watered during dry weather, but avoid overhead watering. Pick off severely affected leaves and, if necessary, spray with a fungicide such as carbendazim or bupirimate with triforine.

## PANSY SICKNESS

**SYMPTOMS** The whole plant wilts and does not recover, even if watered. The foliage yellows and dies off and the plant is killed. The symptoms may take several weeks to develop fully.

**CAUSE** The soil-borne fungus *Pythium violae*, and occasionally other fungi associated with root and foot rots. The fungus builds up in the soil, particularly when pansies or violas are grown in the same site for several years in succession. The fungus seems to be able to survive in soil for several years, even in the absence of suitable host plants.

**CONTROL** Remove and dispose of affected plants. Do not incorporate them into the compost heap. Plant new stocks of pansies or violas on a fresh site.

## PAPERY BARK

**SYMPTOMS** The bark peels off as a thin, papery, brown sheet (*see p.40*). All woody parts of the tree may be affected but small- to medium-sized shoots more frequently show these symptoms than do large branches or tree trunks. Shoots may also die back.

**CAUSE** Problems associated with inadequate functioning of the root system of the tree, most commonly waterlogging.

**CONTROL** Improve growing conditions, in particular soil drainage. Prune off all dead stems and those showing the symptoms. Prune back to a healthy bud growing from perfectly healthy-looking wood. The removal of the dead wood is necessary not because the problem is infectious, but because once dead or injured, the wood is prone to invasion by other organisms which may subsequently cause further dieback.

## PARSNIP CANKER

**SYMPTOMS** Roughened cankered areas develop on the root, usually reddish-brown, orange-brown or black in colour (*see p.57*). The shoulder of the root is the area most commonly affected, but lesions may occur anywhere.

**CAUSE** The fungus *Itersonilia pastinacae* or, less commonly, *Mycocentrospora aceria*. The fungus may be spread into the soil from leaf spots and usually infects via injured fine roots. Other sources of injury, such as that caused by carrot fly, may also provide the initial point of entry.

**CONTROL** No control is available. Reduce the risk of infection by growing resistant cultivars such as 'Avonresister'. Improve soil drainage and avoid or control any source of injury to the roots. Sowing late (in late spring) and in closely spaced rows produces parsnips with smaller roots which seem to be less susceptible than larger-rooted ones.

## PARSNIP VIRUSES

**SYMPTOMS** Yellow markings on the leaves, usually as yellow bands adjacent to the leaves or mosaic-like patterns or flecking. The foliage may be slightly stunted and distorted and overall growth retarded.
**CAUSE** Various viruses, in particular parsnip yellow fleck virus which is spread by aphids.
**CONTROL** Remove infected plants to prevent further spread of the infection. Try to control aphids to decrease the risk of infection.

## PEA AND BEAN WEEVIL

**SYMPTOMS** Peas and broad beans have U-shaped notches eaten in their leaf margins (*see **p.31***) by greyish-brown beetles which are 3–4mm long.
**CAUSE** Adult beetles, *Sitona lineatus*; the larval stage lives in the soil and feeds on the nitrogen-fixing nodules on the roots of peas and beans.

ACTUAL LENGTH: 3–4MM

**PEA AND BEAN WEEVIL**

**CONTROL** Only a small proportion of the foliage is eaten, and plants can usually

tolerate the damage. Control measures are only required where small plants are being heavily attacked. If necessary, spray plants with pirimiphos-methyl, malathion or permethrin, but protect bees by not spraying while broad beans are in flower.

## PEA DOWNY MILDEW

**SYMPTOMS** The foliage develops yellow blotches of discoloration, each of which corresponds with patches of fluffy, off-white fungal growth on the lower surface. Occasionally, yellowish patches of discoloration develop on the pods, but this symptom is less common. Infection early in the season may cause serious stunting and little, if any, pod set.
**CAUSE** The fungus *Peronospora viciae*, which is only likely to cause serious damage in wet seasons.
**CONTROL** Remove any signs of early infection and spray with a fungicide containing mancozeb. The fungus forms spore stages which can persist in the soil, so it is important to remove all debris at the end of the season and grow peas on a fresh site. Grow resistant cultivars such as 'Hurst Green Shaft' and 'Kelvedon Wonder'.

## PEA LEAF AND POD SPOT

**SYMPTOMS** Brown or yellow, often sunken spots develop on the leaves (*see **p.14***), stems, pods (*see **p.51***) and flower stems. Numerous tiny, pinprick-sized, raised fungal fruiting bodies may develop on the lesions.
**CAUSE** Various fungi, including *Ascochyta pisi* and *A. pinodes*, which most

frequently attack fully grown peas. They may occasionally attack seedlings, which may be killed. The fungus carries over from year to year in infected plant debris and, if seeds are collected from infected pods, the resultant seedlings will succumb, too.
**CONTROL** Clear up all diseased plant material at the end of the season. Plant fresh seed on a new site. Do not save seed from infected plants.

## PEA MARSH SPOT

**SYMPTOMS** Circular brown areas of discoloration develop within the seed. A brown cavity later develops. This spoils the eating quality of the crop and also renders them useless for saving as seed.
**CAUSE** A deficiency of manganese. This is most likely to develop on poorly drained soils, particularly those which are alkaline and have a high organic matter content.
**CONTROL** Improve soil conditions and apply lime to lower the pH.

## PEA MOTH

**SYMPTOMS** Caterpillars, up to 6mm long, live inside pea pods and feed on the developing seeds (*see **p.51***). The caterpillars have black heads and small dark spots on creamy-white bodies.
**CAUSE** Caterpillars of a moth, *Cydia nigricana*. Eggs are laid on peas that are in flower, so early or late sown peas that flower outside the moth's flight period from early to late summer will escape damage.
**CONTROL** Spray peas that flower during the moth's flight period with

permethrin, bifenthrin, pirimiphos-methyl or fenitrothion about seven days after the onset of flowering.

## PEA POWDERY MILDEW

**SYMPTOMS** Powdery white fungal growth develops on the leaves and occasionally the leaf stalks and the stems. Leaves may become slightly distorted and yellowed.
**CAUSE** The fungus *Erysiphe polygoni*, which is encouraged by dry conditions at the plant roots, and by poor air circulation around the foliage.
**CONTROL** Control is only necessary if the infection occurs early in the season. More usually it develops towards the end of summer and so has little if any effect on productivity. Grow resistant cultivars including 'Kelvedon Wonder', 'Oregon Sugar Pod II', 'Sugar Gem' and 'Top Pod'.

## PEA THRIPS

**SYMPTOMS** Pea pods are discoloured with a silvery or brownish scarring and the pods contain only a few seeds at the stalk end. The foliage is also discoloured. Adult thrips are black and have narrow elongate bodies 2mm long. The immature nymphs have a similar appearance but are pale yellow.
**CAUSE** Sap-feeding insects, *Kakothrips robustus*, which can be troublesome in hot, dry summers.
**CONTROL** Watch for signs of thrips' activity as the pods develop and, if necessary, spray with heptenophos with permethrin, dimethoate, bifenthrin, pirimiphos-methyl or fenitrothion.

## PEA VIRUSES

**SYMPTOMS** Yellow mosaicing, streaking, spotting or flecking occurs, mainly on the leaves; occasionally symptoms may also be seen on the stems. Plants are stunted and distorted and may fail to flower or may flower but fail to set much of a crop. Pods formed may be discoloured and distorted.

**CAUSE** Various viruses, including pea enation virus, pea mosaic virus and bean leaf roll virus, the first being spread by several different species of aphid.

**CONTROL** Remove and burn affected plants promptly to minimize the risk of the infection spreading. Keep pests such as aphids under control. Avoid handling healthy-looking plants after handling those which are showing symptoms. If necessary, scrub your hands thoroughly. Do not save seed. Grow the cultivars 'Ambassador' and 'Sugar Pod II', which show resistance to pea enation virus.

## PEA WILT

**SYMPTOMS** Rapid wilting of the plant, usually in association with yellowing and dieback of the leaves (*see p.28*). Infection can occur on plants of any age, but is most common towards the middle of the season. Dark brown discoloration of the vascular tissue (the conducting elements) in the stem develops but is only visible if the outer parts of the stem are gently stripped away.

**CAUSE** The fungus *Fusarium oxysporum* f. sp. *pisi*, which is soil-borne and so particularly troublesome if peas or related plants are grown on the same site for several years in succession.

**CONTROL** No control is available. Remove infected plants and grow any replacements or future crops on a fresh site. Grow cultivars showing a degree of resistance, such as 'Greenshaft', 'Kelvedon Wonder', 'Sugar Snap', 'Sugarbon' and 'Onward'.

## PEACH APHIDS

**SYMPTOMS** Peach, nectarine and almond foliage becomes crinkled and curled, with a yellowish-green colour, during late spring to early summer. Green or black aphids suck sap from the underside of the leaves, although they may die out later in the summer, leaving only whitish cast skins.

**CAUSE** Several aphids occur on peach and related plants. The more important ones are the peach potato aphid, *Myzus persicae*, which is yellowish-green in colour, and a black species, the peach aphid, *Brachycaudus schwartzi*.

**CONTROL** Look for aphids on the foliage from mid-spring onwards and spray with pirimicarb, heptenophos with permethrin, dimethoate or pyrethrum if they are seen.

## PEACH LEAF CURL

**SYMPTOMS** Leaves may become infected before or as soon as they unfurl in the spring. They develop symptoms of puckering and blistering (*see p.26*), being pale green at first and soon turning bright red or purple. A white, powdery spore layer then develops over the leaf surface. Affected leaves drop prematurely. A second flush of foliage is usually produced later in the season and this almost invariably remains healthy. The overall vigour and cropping of the tree is only likely to be affected if the infection occurs several years in succession. On nectarines, but not on peaches, the smooth skin on the fruit may develop slightly raised, rough patches as a result of infection spreading to the fruits. A very similar disease may affect poplars.

**CAUSE** The fungus *Taphrina deformans*, which produces spores that are carried by wind and rain, and lodge in cracks and crevices in the bark and in bud scales. This overwintering allows them to cause such early infections.

**CONTROL** Remove infected leaves promptly, preferably before the spore layer develops. Keep the tree well watered and fed to encourage the development of plenty of new replacement growth. Peaches and nectarines which grow under glass are rarely attacked because the glass acts as a shelter, preventing the spores from landing. Temporary protection, provided by the erection of an open-sided polythene shelter over fan-trained trees, works well. To be effective, the shelter must be in position from mid-winter until late spring. Spray with a copper fungicide several times between mid- and late winter to try to protect the developing foliage from attack. The spraying must be completed before the flower buds start to open. Repeat the spray in the autumn, just before leaf fall.

## PEAR AND APPLE CANKER

**SYMPTOMS** Areas of bark sink inwards and become pale in colour. As the canker enlarges, it becomes elliptical and the bark flakes in concentric rings. The branch may swell around the cankered area. The growth above the canker is poor and the foliage may be sparse and discoloured. Cankers often enlarge sufficiently to girdle the stem or branch, and the growth above this point then dies completely. The canker usually starts near a bud, or in a wound such as a leaf scar or insect injury. During the summer months numerous small, raised, white, fungal spots develop on the cankered area and in the winter these are replaced by similar but red fruiting bodies.

**CAUSE** The fungus *Nectria galligena*, which produces most of its spores in the spring. The spores are wind-borne. Poor growing conditions, such as a very heavy soil or inadequate feeding may increase the likelihood of canker being troublesome.

**CONTROL** Remove cankered stems and dispose of them well away from nearby apple, beech, ash, willow, *Sorbus* and pear trees. On large branches or trunks, carefully cut out the cankered area using a sharp knife, so that all diseased bark and wood is removed. Treat wounds with a canker paint containing cresylic acid. Sprays of carbendazim used to control scab or powdery mildew will give incidental suppression of canker. Some apple cultivars tend to show a degree of resistance under certain growing conditions: these include 'Bramley's Seedling', 'Newton Wonder', 'Laxton's Superb' and 'Lane's Prince Albert'. Avoid cultivars which are notably susceptible, such as 'Cox's Orange Pippin', 'James Grieve', 'Worcester Pearmain' and 'Spartan'.

## PEAR AND CHERRY SLUGWORM

**SYMPTOMS** Pale yellow, caterpillar-like larvae, which secrete a black, slimy mucilage over their bodies, giving them a slug-like appearance, feed on pear and cherry foliage from late spring to mid-autumn. Other plants attacked include hawthorn, japanese quince, mountain ash and plums. The larvae are up to 10mm long and are swollen at the front end, making them club-shaped. They feed by grazing away the upper leaf surface, causing the remaining damaged tissues to dry up and turn brown (*see p.19*).
**CAUSE** Larvae of a sawfly, *Caliroa cerasi*, which has two or three generations during the summer.
**CONTROL** On trees small enough to be sprayed, slugworms are easy to control with most contact insecticides, including pyrethrum, permethrin, pirimiphos-methyl and fenitrothion.

## PEAR BEDSTRAW APHID

**SYMPTOMS** Leaves at the shoot tips on pear become yellowish-green and curled during the late spring. Whitish-grey aphids infest the undersides of the leaves and are sometimes present in such large numbers in late spring to early summer that they swarm over the branches and trunk. The aphids excrete a sugary honeydew that can allow sooty moulds to grow on the foliage.
**CAUSE** A sap-feeding insect, *Dysaphis pyri*, which in mid-summer migrates to wild flowers known as bedstraws (*Galium* species).

**CONTROL** A tar oil wash in early to mid-winter will control overwintering eggs. In the spring, inspect the foliage and spray with pirimicarb, heptenophos with permethrin, dimethoate, pirimiphos-methyl, fenitrothion or pyrethrum if aphids are seen.

## PEAR BORON DEFICIENCY

**SYMPTOMS** Distorted fruits are marked internally with numerous small, brown spots. Most fruits are affected. The fruit's texture is unpleasant because of the patches of dead, brown cells. Some stem dieback may occur and overall vigour may be reduced. The symptoms may be confused with those of the far more serious pear stony pit virus (*see p.157*).
**CAUSE** A deficiency of boron.
**CONTROL** *See* BORON DEFICIENCY, *p.109*.

## PEAR LEAF BLISTER MITE

**SYMPTOMS** In mid- to late spring, the foliage of pears develops yellowish-green or pink blisters or raised blotches. On expanded leaves these form broad bands on either side of the midrib (*see p.16*). By mid-summer the blisters have darkened and become black. Less than 3mm in diameter, they should not be confused with the larger black lesions caused by pear scab disease.
**CAUSE** Microscopic gall mites, *Eriophyes pyri*, which live within the leaves. The blistered appearance of the foliage is induced by chemicals secreted by the mites as they feed.
**CONTROL** None of the pesticides available to amateur gardeners gives

good control. Fortunately, although blister mite gives a tree an unhealthy appearance, it has little impact on the tree's ability to produce fruit. On small trees, affected leaves can be picked off if the infestation is light, otherwise this pest has to be tolerated.

## PEAR MIDGE

**SYMPTOMS** Pear fruitlets make rapid initial growth but, a few weeks after petal fall, start to turn black at the eye end of the fruits (*see p.48*). In the centre of the fruitlets are whitish-orange maggots up to 2mm long. Affected fruitlets become extensively blackened and fall from the tree in late spring and early summer. Sometimes nearly all potential fruit is lost.
**CAUSE** Larvae of a small fly or gall midge, *Contarinia pyrivora*, which go into the soil when the fruitlets drop. They pupate inside silk cocoons and emerge as adults the next spring.
**CONTROL** On small trees, pick off and destroy infested fruitlets before the maggots complete their feeding. Prevent the adult flies from laying eggs by spraying with pirimiphos-methyl or fenitrothion when the blossom is at the white bud stage – when the petal colour can be seen, but before the flowers open.

## PEAR SCAB

**SYMPTOMS** Blackish-brown scabby patches develop on the fruits (*see p.48*), and similar but more greenish-grey spots develop on the leaves. In severe cases, fruit may be almost entirely covered with scabby patches and is very small, and misshapen. Infected fruits

may crack or split and so become infected with secondary organisms such as brown rot. Affected leaves yellow and fall early.
**CAUSE** The fungus *Venturia pirina*, which overwinters on scabby patches on the young stems and also on fallen infected leaves. The disease is usually most prevalent on trees with crowded branches, and in damp seasons.
**CONTROL** Rake up and dispose of affected leaves as they fall. Prune out cracked and scabby shoots to limit the overwintering of the fungus. Spray with a suitable fungicide such as carbendazim, mancozeb or triforine with bupirimate. Grow cultivars showing some resistance, including 'Jargonelle' and 'Catillac'.

## PEAR STONY PIT VIRUS

**SYMPTOMS** The fruits become knobbly and pitted (*see p.48*). Within the flesh are many dead sclerenchyma (stone cells), which make the fruits unpleasant to eat. The symptoms may be confused with those of boron deficiency. Affected fruits rarely appear all over the tree, but are frequently borne on just one branch.
**CAUSE** Stony pit virus, which is most common on old trees. The virus particles are only known to be transmitted by grafting.
**CONTROL** There is no known cure. Avoid cultivars known to be particularly susceptible, including 'Doyenné du Comice', 'Laxton's Superb', 'Winter Nelis' and 'Anjou'.

## PEAR SUCKER

**SYMPTOMS** Pear leaves become very sticky with honeydew excreted by blackish-green insects up to 2mm long. The immature nymphs have flattened bodies, while the adults resemble winged aphids. Sooty moulds frequently develop where honeydew has accumulated on the upper surface of leaves.

**CAUSE** A sap-feeding sucker or psyllid, *Psylla pyricola*, which has several generations during summer and overwinters as an adult.

ACTUAL SIZE ➛

**PEAR SUCKER**

**CONTROL** Pear sucker is less frequent on pears in gardens than in commercial orchards. Routine control measures are not needed, but look for the pest during late spring to late summer and, if necessary, spray with pirimiphos-methyl, fenitrothion or permethrin.

## PEDICEL NECROSIS

**SYMPTOMS** Flower buds develop normally, but before or just after opening the pedicel (area of flower stem immediately behind the bud) discolours, shrinks inwards and flops over with the weight of the bud. The flower bud then turns brown and fails to open, or opens and dies off.

**CAUSE** Extremely rapid or soft growth, such as that produced by excessive use of high-nitrogen fertilizers. A sudden check in growth, for example caused by drought, inadequate fertilizer use, or the wrong amount of potassium, may also be involved. Plants with very large or double flowerheads may be more prone to this than single varieties.

**CONTROL** Remove damaged flowerheads. Feed the plant with sulphate of potash at about 20g per m² in spring and autumn. Avoid excessive use of high nitrogen fertilizers. Mulch the soil if necessary to improve its ability to retain moisture.

## PELARGONIUM RUST

**SYMPTOMS** Rings of dark brown fungal pustules develop on the lower leaf surface, often in concentric rings (*see p.15*). Corresponding yellow blotches develop on the upper surface. Occasionally, the pustules appear on the upper surface, too. The infection spreads rapidly and infected leaves become extensively discoloured, wither and die. If the disease spreads unchecked, the plant may be killed.

**CAUSE** The fungus *Puccinia pelargonii-zonalis*, which produces spores that are spread on air currents and by water splash. Spread is encouraged by humid atmospheres, and so is usually most severe during the winter and on plants growing in crowded conditions under glass.

**CONTROL** Remove infected leaves. Spray with a fungicide containing mancozeb, penconazole or triforine with bupirimate. Improve air circulation to reduce humidity. Space plants further apart, or if under cover, open vents, windows and doors, as appropriate.

## PELARGONIUM VIRUSES

**SYMPTOMS** Yellow markings, such as flecks, ring spots and streaks (*see p.21*), on the foliage. Leaves may, occasionally, be distorted and flowers may show colour-breaking symptoms (pale streaks on the petals). The overall vigour of the plant is rarely affected, but the appearance of foliage and flowers may be spoiled.

**CAUSE** Various viruses, either alone or in combination, some of which may be transmitted by aphids, others by handling, or by eelworms in the soil.

**CONTROL** There is no control available, but it is always advisable to inspect plants very carefully before purchase and to dispose of any showing severe symptoms.

## PEONY WILT

**SYMPTOMS** Shoots of herbaceous peonies wilt and turn brown at the stem base. In damp conditions, fuzzy grey fungal growth may develop at the base of the stem. The withered stem dies, shortly followed by other stems on the plant. Sclerotia (shiny black fungal resting bodies) form on and within the infected stem. Occasionally the fungus attacks higher up the stem, causing similar symptoms (*see p.43*). Wilt is most common on herbaceous peonies, but tree species may also be affected.

**CAUSE** The fungus *Botrytis paeoniae*, which produces numerous spores that are carried on air currents and by water splash. The sclerotia are usually responsible for wilt infections, which appear early in the season; subsequent infections may be due to spore spread.

**CONTROL** Cut out infected stems promptly, cutting right back into healthy growth and below soil level if necessary. If the stems are not removed, the sclerotia fall into the soil

where they can persist for several years. Remove all debris at the end of the season, as this may harbour sclerotia. If a plant has been severely damaged by the infections, scrape off the uppermost few millimetres of soil in the autumn in case it contains sclerotia – this must be done with care so as to avoid injuring the roots. Spray emerging shoots with a suitable fungicide such as carbendazim.

## PERIWINKLE RUST

**SYMPTOMS** Dark brown spore pustules develop beneath the leaves (*see p.15*). The foliage becomes pitted and distorted and the whole plant also appears distorted and may fail to flower.

**CAUSE** The fungus *Puccinia vincae*, which produces a fungal mycelium that invades all parts of the plant, including the roots.

**CONTROL** As the fungus is present throughout the plant, there is little point in attempting control, and all infected or obviously distorted plants should be removed and disposed of promptly.

## PESTALOTIOPSIS

**SYMPTOMS** Grey-brown necrotic patches develop on the foliage and may also appear on shoots. The tree or shrub appears scorched. Leaves drop prematurely and stem infection may cause extensive dieback.

**CAUSE** The fungus *Pestalotiopsis*; most frequently *P. guepini*, which is encouraged by high humidity.

**CONTROL** Remove affected leaves and prune out infected stems. Spray with mancozeb.

## PETUNIA FOOT ROT

**SYMPTOMS** The plant wilts and dies, often quite rapidly. The stem base may become blackened and atrophied.
**CAUSE** Various species of the soil-borne fungus *Phytophthora*, which produces resilient resting bodies that enable it to persist in the soil for several years.
**CONTROL** No control is available. Remove infected plants promptly, together with the soil in the immediate vicinity of the roots. Do not grow petunias on that site for several years; instead choose other bedding plants such as zonal pelargoniums, French and African marigolds, begonias and heliotrope, since these are not susceptible.

## PETUNIA VIRUSES

**SYMPTOMS** Yellow flecking, spotting, mosaicing and streaking on the foliage. The leaves may also be stunted and generally chlorotic and may be deformed. The whole plant appears stunted and the flowers may show colour breaking (pale streaking) in the petals.
**CAUSE** Various viruses, some of which may be transmitted by aphids and other sap-feeding pests; others may be spread by handling or when propagating from cuttings.
**CONTROL** Remove and dispose of affected plants promptly. Always wash hands thoroughly after handling infected plants. Control pests which are potential vectors.

## PHELLINUS IGNARIUS

**SYMPTOMS** Pale brown to buff-coloured fungal bracket, with a rounded outline, develops on the branches or trunk of the tree. There may be no apparent loss of vigour of the tree but internally the wood is weakened.
**CAUSE** The polypore fungus *Phellinus ignarius.*
**CONTROL** No control is available. Infected trees may stand safely for many years, but they may be rendered dangerous, so it is better to employ a reputable tree surgeon to examine the tree and advise as to the best course of action.

## PHELLINUS POMACEOUS

*See* FOMES POMACEOUS *p.128.*

## PHLOX EELWORM

**SYMPTOMS** In early to mid-summer, phlox stems become abnormally swollen and are liable to split. Growth is stunted and leaves at the shoot tips are greatly reduced in width, sometimes being little wider than the midrib (*see* **p.33**). Badly infested plants rot and die.
**CAUSE** Microscopic eelworms or nematodes, *Ditylenchus dipsaci*, that feed within the stems and foliage. They overwinter in dormant buds at the base of the plant.
**CONTROL** Phlox eelworm does not attack the roots, so clean plants can be propagated by taking root cuttings. There are no effective chemicals available to gardeners, so affected plants should be removed before the pest spreads.

## PHLOX POWDERY MILDEW

**SYMPTOMS** White, powdery fungal growth develops on the upper and occasionally lower leaf surfaces. Infected leaves may be distorted and puckered and turn yellow and die off. The stems may show similar symptoms.
**CAUSE** The fungus *Sphaerotheca fuliginea*; this produces numerous spores which are spread by water splash or on air currents; it is encouraged by dry soil conditions and humid air.
**CONTROL** Spray affected plants with a suitable fungicide such as carbendazim, triforine with bupirimate, penconazole or sulphur. Water more regularly, but avoid splashing water over the foliage.

## PHOMOPSIS CANKER

**SYMPTOMS** Dieback of shoots, often accompanied by girdling of the stem by cankers. Girdled areas become constricted, leaving wider areas of stem above and below. On some species the cankers are accompanied by resin production. Many conifers may be attacked – in particular, Douglas fir, and occasionally various species of cedar, larch and other firs.
**CAUSE** The fungus *Phacidiopycnis pseudotsugae*, which enters stems and branches through wounds.
**CONTROL** Prune out and burn affected areas where feasible. Prune carefully, avoiding leaving stumps, and carry out any extensive pruning during the active growing season.

## PHORMIUM MEALYBUG

**SYMPTOMS** A white, waxy substance accumulates at the ensheathing leaf bases on New Zealand flax plants. This is secreted by greyish-white, soft-bodied insects that are up to 4mm long. Heavy infestations cause a loss of vigour and plants may eventually be killed.
**CAUSE** A sap-feeding insect, *Trionymus diminutus*. Most mealybugs are greenhouse pests, but this species is not harmed by winter frosts. It breeds continuously throughout the warmer months of the year.
**CONTROL** Malathion will reduce but not eliminate infestations. When purchasing a phormium, inspect the leaf bases carefully to ensure that it is free of this imported pest.

## PHOSPHORUS DEFICIENCY

**SYMPTOMS** Growth is weak and reduced and leaves are small and may fall early. Leaves develop a blue-green or purplish discoloration, or scorching around the edges. Plants may show delayed or poor flowering and fruiting.
**CAUSE** A deficiency of phosphorus which is most likely to occur on acid soils or after periods of heavy rain or watering.
**CONTROL** Apply fertilizers containing a high proportion of phosphorus, or apply super-phosphate at the rate and in the manner specified by the manufacturer.

## PHYTOPHTHORA

**SYMPTOMS** Rotting at the roots and base (collar) of woody plants causes poor growth. Foliage is sparse and may become discoloured. Stems show signs of dieback and the whole plant may be killed. Inspection of the roots reveals that the finer ones are killed, and larger roots show a blackish-brown discoloration. If the bark is removed at the base of the main stem or trunk, a reddish- or blackish-brown discoloration is visible.

**CAUSE** Various species of the soil- or water-borne fungus *Phytophthora*, in particular *P. cinnamomi*.

**CONTROL** There is no cure available to gardeners. Remove infected plants together with the soil in the immediate vicinity of the roots. Improve drainage, as a poorly drained soil may encourage the development and spread of the disease.

## PHYTOTOXICITY

Plants are sometimes damaged by insecticides or fungicides and this is known as phytotoxicity. The manufacturer's instructions should be read and followed carefully to ensure that the correct dilution is being applied. Plants known to be harmed by a chemical will be listed by the manufacturer, but this information is not known for all potential plant/chemical interactions. The risk of spray damage occurring can be minimized by not treating plants that are already under stress as a result of exposure to bright sunlight, extremes of temperature or dryness at the roots. Seedlings will be more susceptible than mature plants, and petals are more likely to be scorched than foliage. Where there is doubt and there are a number of plants to be sprayed, it is sensible to treat just a few and watch for adverse reactions over the next few days. Plants that have been damaged may suffer a significant check in growth, but usually make a recovery.

## PIGEONS

The wood pigeon, *Colomba palumbus*, can cause damage at any time of year, but they are particularly troublesome in the winter when snow or frost encourages them to feed in gardens. Brassicas, such as cabbages, Brussels sprouts and cauliflowers, can have their leaves stripped to the midribs (*see p.30*).

**CONTROL** Shooting is usually not a feasible proposition in gardens, so other means are needed to protect brassicas. Netting is the only certain way of keeping birds away. Various scaring devices, such as scarecrows, humming tapes, glitter strips and models of cats or birds of prey, are generally not very successful. Birds soon get used to scarers, particularly if they are hungry and other food supplies are scarce.

## PINE ADELGID

**SYMPTOMS** The new shoots of pines, especially scots pine, develop a fluffy, white, waxy coating in mid- to late spring (*see p.39*). Black aphid-like insects can be found underneath the wax. The insects and wax persist through the summer but gradually become less obvious.

**CAUSE** Sap-feeding insects, *Pineus pini*, which overwinter on the tree as immature nymphs and have several generations during the summer.

**CONTROL** Spraying is usually unnecessary; although the waxy secretions may be unsightly, the adelgids have little obvious effect on the tree's vigour once it has become established. Young trees could be sprayed with malathion or pirimiphos-methyl on a dry, mild day in late winter against the overwintering nymphs before they produce their waxy covering. Treatment later in the spring is less effective, since the wax acts as a protective barrier that is very difficult to penetrate with pesticides.

## PINE, NEEDLE BLIGHTS AND CASTS

*See* NEEDLE BLIGHTS AND CASTS OF PINES *p.151*.

## PINEAPPLE GALL ADELGID

**SYMPTOMS** In early summer, spruce trees (*Picea* species), especially the Norway spruce or traditional Christmas tree, develop swellings at the shoot tips that resemble green pineapples (*see p.38*). These galls contain many small aphid-like insects. In late summer the galls begin to dry up and turn brown. Slits open up in the gall as it dries, allowing the winged adult insects to emerge. Old galls can persist on trees for many years and spoil their appearance, particularly if they are to be used as Christmas trees.

**CAUSE** Sap-feeding insects, *Adelges abietis*, which initiate the galls by secreting chemicals as they feed on the buds in the spring. These chemicals prevent normal shoot development and, instead, the bases of the leaves become swollen and fused together, while leaving cavities within where the adelgid nymphs continue their feeding and growth.

**CONTROL** Pineapple gall adelgids have to be tolerated on trees that are too tall to be sprayed. Apart from the visual effect of the old galls, little real damage is done. On small trees, galls can be picked off or, as a preventive measure, trees can be sprayed on a mild, dry day in late winter with malathion or pirimiphos-methyl. This treatment is aimed at overwintering nymphs before they mature and lay eggs on the buds.

## PLANE ANTHRACNOSE

**SYMPTOMS** A brown, discoloured patch develops at the base of the leaf bud in winter. This may not be easy to see, but in spring affected buds die without producing a leaf. Adjacent areas of stem turn brown and die back and a small cankered area may appear on the stem. Buds which do open into leaves may show more obvious symptoms as the newly-produced growth is killed rapidly. Very dark brown, necrotic areas develop on leaves, primarily adjacent to the main veins. Throughout the summer, leaves which started off perfectly healthy may develop symptoms due to the leaf blight stage of the disease and this is the stage most frequently noticed.

**CAUSE** The fungus *Apiognomonia veneta* syn. *Gnomonia platani*, which is encouraged by cool weather conditions in the spring and which may overwinter on fallen infected leaves.

**CONTROL** None available. If the infection is mild and the tree well-sized, it may be possible to prune out all infected areas. Rake up and dispose of fallen leaves.

## PLUM BLOSSOM WILT

**SYMPTOMS** Blossom withers and dies shortly after appearing. Trusses of dead blossom remain hanging on the tree. Infection may then spread to adjacent leaves, causing them to wilt, turn brown, die and remain hanging on the branch. Tiny, pinprick-sized, raised, buff-coloured fungal pustules develop on infected areas. Localized dieback may occur.

**CAUSE** The fungus *Sclerotinia laxa*, which may overwinter as cankers on infected stems, or pustules on infected flowers and foliage. The spores are wind-borne and spread is most rapid during damp weather.

**CONTROL** Spray with copper oxychloride just before flowering. Prune out infected flower trusses, preferably before the infection spreads to leaves or into the spur.

## PLUM LEAF GALL MITE

**SYMPTOMS** Whitish-green, pouch-like swellings or galls develop on the leaves of plums, damsons and gages during the early summer, mainly on the upper leaf surface, especially around leaf edges (*see **p.16***).

**CAUSE** Microscopic gall mites, *Phytoptus similis*, which suck sap from the underside of the leaves during the spring. They also secrete chemicals into the leaf, causing the growth of the galls, which are hollow, enclosing the mites.

**CONTROL** Nothing available to gardeners is effective against gall mites. Fortunately, the mites have little harmful effect, even when galls are numerous, so their presence can be tolerated.

## PLUM LEAF-CURLING APHID

**SYMPTOMS** The foliage on plums, damsons and gages becomes tightly curled and crinkled during the spring (*see **p.27***). Green insects, up to 2mm long, and their whitish cast skins can be found on the underside of the leaves. Winged aphids

develop during late spring and early summer and fly away to spend the rest of the summer on various herbaceous plants. After that time, the infestation on plums dies out and trees start to produce normal foliage again.

**CAUSE** A sap-feeding aphid, *Brachycaudus helichrysi*, which migrates back to plums in late summer to lay over-wintering eggs near the buds.

**CONTROL** Overwintering eggs can be reduced by using a tar oil wash on a mild, dry day in early to mid-winter. Newly-hatched aphids are present shortly after bud burst, so spray then with pirimicarb, dimethoate or heptenophos with permethrin. These all have some systemic action and will reach aphids hidden in curled leaves, although the aim is to control the aphids before they cause extensive leaf distortion.

## PLUM MOTH

**SYMPTOMS** A pinkish-white caterpillar, up to 12mm long, with a brown head feeds inside a ripening plum fruit. Most feeding occurs around the stone and the damaged area becomes filled with the caterpillar's brown excrement pellets (*see **p.48***). Damaged fruits tend to ripen prematurely and often have a depressed area on the surface where the fully-fed larva has eaten its way out before seeking an over-wintering site under a loose flake of bark.

**CAUSE** Caterpillars of a moth, *Cydia funebrana*, which usually has one generation a year, but may have two in hot summers.

**CONTROL** Spray with permethrin, bifenthrin, pirimiphos-methyl or

fenitrothion in early summer, with a second application three weeks later to control the young caterpillars as they hatch. Pheromone traps (*see p.86*) are available for detecting the presence of this pest and indicating the best time to spray. On isolated plum trees, pheromone traps may disrupt the females' mating success sufficiently to reduce the number of maggoty fruits.

## PLUM POX (SHARKA)

**SYMPTOMS** The fruit develops discoloured, usually dark brown streaks or rings which are easiest to see on red-fleshed plums. On darker-coloured or yellow fruits there is clear pitting on the surface. Affected fruits may fall early and have an unpleasant and very acidic taste. Pale green blotches or rings develop on the foliage. Although very damaging, this disease is much less common in Britain than the rest of Europe. It may be confused with physical injury caused by very heavy rainfall or hail.

**CAUSE** The plum pox virus which is transmitted by aphids and which may be introduced on plum rootstocks.

**CONTROL** None available. In Britain, any suspected outbreaks must be reported to the Ministry of Agriculture, Fisheries and Food.

## PLUM RUST

**SYMPTOMS** The upper leaf surface of plums or anemones is covered with bright orange-yellow spots from mid-summer onwards. In severe cases the whole leaf may appear yellow. Numerous rust-brown, fungal

pustules full of brown spores develop in corresponding areas beneath the leaf (*see **p.22***). Towards late summer and early autumn spores darken and appear black. Affected leaves fall prematurely and if severe infections occur for several years in succession the tree may be weakened.

**CAUSE** The fungus *Tranzschelia pruni-spinosae*, which is encouraged by damp weather. Despite the link between the disease on plums and on anemones, it may appear in the same garden on one without any sign of it on the other.

**CONTROL** Rarely necessary on plums; rake up and dispose of fallen infected leaves. Feed the tree to maintain vigour. Remove and destroy any infected anemones nearby. Avoid growing the plum 'Victoria', which seems particularly susceptible.

## PLUM SAWFLY

**SYMPTOMS** Unlike plum moth, this pest has caterpillar-like larvae that attack plum fruitlets, rather than the mature fruits. The larvae are creamy-white with brown heads and are up to 10mm long. Damaged fruitlets have a round hole bored in them, from which black excrement pellets may protrude. These fruits drop from the tree in late spring and early summer.

**CAUSE** Larvae of a sawfly, *Hoplocampa flava*, which lays eggs on the blossoms during mid-spring.

**PLUM SAWFLY DAMAGE**

**CONTROL** Plum sawfly is of local occurrence and so preventive measures are only worthwhile if this pest has been troublesome in previous years. If control is necessary, spray seven to ten days after petal fall with fenitrothion, pirimiphos-methyl or permethrin. On small trees, remove damaged fruitlets before the larvae leave to go into the soil where they will overwinter. Some plums, such as 'Victoria' and 'Czar', are more susceptible than others.

## POCKET PLUM

**SYMPTOMS** As infected fruits develop, they become deformed, twisted and often banana-shaped. They contain no stone, are hollow like pockets, and pale green and smooth. A white spore layer then develops over the surface and the plums start to shrivel.
**CAUSE** The fungus *Taphrina pruni*, which appears to be capable of overwintering in twigs. The spores are carried on air currents or by water splash.
**CONTROL** None available. Remove affected plums promptly, preferably before they turn white. Although causing a loss of crop, this disease rarely affects all the fruit in one year, and may not re-appear for several years.

## POLLEN BEETLES

**SYMPTOMS** Black or bronze-green beetles, 1–2mm long, cluster in flowers (*see p.43*) in the spring and again in mid-summer. Yellow flowers, such as daffodils, are affected in the spring but many others, including sweet peas, runner beans, marrows, roses, dahlia and shasta

daisy, attract these beetles during the summer. They can be a nuisance, especially when flowers are cut for the house, but generally little damage is caused. When abundant, the beetles will sometimes cause damage by eating their way into unopened rose buds, and they may discolour cauliflowers by feeding on the developing flowerhead.
**CAUSE** There are many pollen beetles, *Meligethes* species, most of which are entirely harmless, as they feed on wild flowers. The adult beetles feed on pollen, but do not eat enough to interfere with pollination; they may even assist with pollination if they move from one flower to another and carry pollen on their bodies. The species that cause concern in gardens are usually those that develop as larvae in the flower buds of brassica seed crops. These include the widely grown agricultural crop, oil seed rape. In mid-summer, adult pollen beetles that have bred in oil seed rape fields disperse into the countryside and gardens, where they seek out flowers. They will not breed in garden flower buds, but they can be a temporary nuisance during the summer, and again in the spring when the over-wintered adults re-emerge.
**CONTROL** Pollen beetles on flowers in the garden have to be tolerated. Insecticides should not be used, as they will also kill bees, butterflies and other desirable flower visitors. Spraying would in any case need to be done at frequent intervals in order to cope with the continual migration of beetles into gardens. When cut flowers are taken indoors, pollen beetles have an annoying

habit of leaving the flowers and crawling about the room. By placing cut flowers in a shed or garage for a few hours during the day before they are brought into the house, the beetles can be persuaded to leave the flowers without being a nuisance.

## POOR POLLINATION OF RASPBERRIES AND STRAWBERRIES

**SYMPTOMS** Fruits develop but are distorted and may bear dry areas. On raspberry, individual drupelets or small groups of them may be dry and brown. Adjacent drupelets may be perfectly normal. The overall growth of the plant is normal and it may bear other, perfect fruits.
**CAUSE** Poor pollination, which results in uneven swelling and development of the fruit. A lack of bees, or weather conditions which prevent their flying and pollinating activities.
**CONTROL** Grow susceptible plants in an area which is provided with shelter. Grow plenty of other plants and flowers to attract bees and other pollinating insects.

## POPLAR CANKER

**SYMPTOMS** In spring a dense, off-white slime is exuded from cracks in young branches. The damaged areas may be girdled and so die back. On older branches, the slime may also be apparent, and roughened areas of bark develop, measuring up to 3cm in diameter. Larger, more "open" cankered areas may develop and stand out clearly from the branch as they become surrounded by raised callus.
**CAUSE** The bacterium *Xanthomonas populi*, which

is believed to enter the tree via wounds, such as as those around bud scales, leaf scars, frost crack and damage caused by insects feeding. The precise means of spread is not known, but rain splash and insects are probably the main ones.
**CONTROL** Prune out infected areas where feasible, in an attempt to limit the spread of the bacteria.

## POPLAR YELLOW BLISTER

**SYMPTOMS** In late spring or early summer raised, rounded blisters develop on upper, and occasionally lower, leaf surfaces. The concave, lower surface of each blister turns bright yellow. In damp weather young shoots may also be attacked. Although leaves are disfigured, the overall vigour of the tree is not affected.
**CAUSE** The fungus *Taphrina populina*, which overwinters as spores on the bud scales.
**CONTROL** No control measure is available, and none is justified.

## POTASSIUM DEFICIENCY

**SYMPTOMS** Most frequently poor flowering and/or poor fruit set. Flowers may be undersized. On tomatoes this deficiency is at least partially responsible for the disorder known as blotchy ripening. Potassium is also important in the ripening of wood and in the structure of the stem, and too little may increase the risk of frost or similar damage. Leaf tips may appear scorched around the edges and show brownish-purple spotting beneath (*see p.24*).
**CAUSE** A deficiency of potassium in the soil, which

is most common on light, sandy soils and those with a low clay content.

**CONTROL** Apply sulphate of potash in spring and autumn.

## POTATO BLACK LEG

**SYMPTOMS** Foliage is chlorotic and stunted; the leaves are small and slightly incurled. These symptoms are easily confused with other problems, but with black leg the stem base is blackened and rotted at ground level. The vascular strands within the stem are similarly discoloured, showing as distinct black spots when the stem is cut across. The parent tuber is completely rotted. The plant may be killed before a crop is produced.

**CAUSE** The bacterium *Erwinia carotovora* var. *atroseptica*, which is encouraged by wet soil conditions. It is often introduced on symptomless but mildly affected seed tubers. When these are planted they cause affected plants to appear scattered through the crop of other perfectly healthy plants. The bacteria may enter tubers via wounds either whilst in the soil, or at lifting.

**CONTROL** Take great care at lifting time to keep damage to a minimum. If possible, do not lift crops during wet weather as infection is more likely then. Store only those tubers which appear quite healthy. Avoid growing 'Majestic', 'Arran Pilot', 'Maris Bard', 'Desirée' and 'Estima', as these cultivars seem to be particularly susceptible.

## POTATO BLIGHT

**SYMPTOMS** Necrotic brown patches develop on the leaves, largely on the leaf tips and around the edges (*see* **p.28**). As the spots enlarge, the leaves wither and die. In wet weather or under humid conditions slightly fluffy, white fungal growth may be visible around the edges of the spots. This is most common on lower leaf surfaces. The haulms of infected plants develop brownish-black patches and may collapse. Infected tubers develop slightly sunken, dark patches on the skin with a reddish-brown discoloration beneath. The discoloration may then spread into the flesh (*see* **p.56**). Secondary organisms often invade, turning the dry, brown areas into unpleasant-smelling, slimy wet rot. The whole tuber is then rotted.

**CAUSE** The fungus *Phytophthora infestans*, also responsible for blight on tomatoes. Spores are produced from the leaf lesions and carried by rain splash or on air currents. For infection to occur, the fungus needs two consecutive 24-hour periods, each with a minimum temperature of 10°C (50°F) and each with a minimum of at least 11 hours where the relative humidity is 89% or more. Spores from the infected top growth are washed down through the soil and may infect tubers.

**CONTROL** Spray the foliage with a fungicide such as Bordeaux mixture, copper oxychloride or mancozeb. In seasons where weather conditions are favourable to blight, spray before the symptoms are seen. If a "Blight Infection Period" is mentioned on the radio for your area, it is advisable to spray. Reduce the chances of the infection developing on the tubers by earthing up deeply. If the haulms show signs of infection, remove them as soon as they start to die down. Grow cultivars which show some resistance, such as 'Cara', 'Estima', 'Kondor', 'Maris Peer', 'Pentland Crown', 'Record' and 'Romano'.

## POTATO COMMON SCAB

**SYMPTOMS** Raised, roughened scabby patches develop on the skin of the tuber (*see* **p.56**). The skin ruptures, leaving the scabs with ragged edges. Damage may be superficial or result in cracking. The flesh is usually undamaged, although just beneath the scabby patches it may become discoloured.

**CAUSE** *Streptomyces scabies*, which is particularly common on sandy, light soils with a low organic matter content. It is present naturally in most soils and potatoes planted on ground that was previously grassland are particularly prone to attack. Soil which has been limed is also likely to have more *Streptomyces* in it, whereas acid soil has less.

**CONTROL** Improve the organic matter content of the soil. Water regularly, as there appears to be a link between this disease and dry soil conditions. Do not lime soil prior to planting potatoes. Use acidic materials such as sulphate of ammonia and superphosphate to reduce scabbing. No fungicide treatment is available. Grow resistant cultivars such as 'Arran Comet', 'Arran Pilot', 'Golden Wonder', 'King Edward' and 'Maris Peer'.

## POTATO CYST EELWORM

**SYMPTOMS** Potato plants die prematurely and hence produce a poor crop of small potatoes. Affected plants die during mid- to late summer, with leaves progressively yellowing and drying up from the bottom of the stems upwards. When cyst eelworms first occur in a garden, small patches within the potato rows show symptoms. These patches gradually enlarge each time potatoes are subsequently grown in that part of the garden until eventually it is impossible to grow a worthwhile crop of potatoes. Tomato plants are also attacked. If infested plants are carefully dug up, spherical eelworm cysts, up to 1mm in diameter and white, yellow or brown in colour can be seen on the roots (*see* **p.55**). These are the bodies of mature female eelworms and each one can contain up to 600 eggs.

**CAUSE** Two species of cyst eelworm attack potato and tomato. The golden cyst eelworm, *Heterodera rostochiensis*, has female cysts which are white at first, but pass through a yellow phase before becoming brown when mature. The white cyst eelworm, *H. pallida*, has cysts that change from white to brown without an intermediate stage. Both species develop inside the roots, where their feeding disrupts the uptake of water and nutrients; when mature, the females become swollen and burst through the root wall.

**CONTROL** There are no effective pesticides available to amateur gardeners for controlling eelworms. Potato cyst eelworms are particularly difficult because their eggs remain viable inside the cysts for many years. The eggs are stimulated to hatch by chemicals that are secreted into the soil by the roots of suitable host plants. Adopting a system of crop rotation can delay the build-up of damaging infestations, but once a serious eelworm problem has arisen, the usual three or four-year rotation is not long enough to eliminate the pest. Some potato cultivars have some resistance to the golden cyst eelworm. These include 'Accent', 'Nadine', 'Pentland Javelin', 'Rocket', 'Ukama' (earlies); and 'Alhambra', 'Cara', 'Kingston', 'Maris Piper', 'Montana' and 'Stemster' (main crop). In addition there are the maincrop cultivars 'Cromwell' and 'Sante' which are resistant to golden cyst eelworm and have some tolerance of the white cyst eelworm. The latter will, however, breed successfully in the roots, so infestations will increase.

## POTATO DRY ROT

**SYMPTOMS** The skin of the potato tuber becomes wrinkled at one end, followed by rapid shrinking, the development of concentric rings of wrinkles, and discoloration (*see **p.57***). The affected tissues become black, grey or brown. Pink, white or blue-green pustules of fungal spores develop on the shrivelled tissues. The symptoms are likely to be found on stored tubers only, though it may also cause poor growth after planting. **CAUSE** A species of the fungus *Fusarium*, which infects the tubers whilst they are still in the ground or as they are being lifted. The fungus enters through the lenticels (breathing pores), eyes, or wounds caused by abrasions, pests or scab infection.

**CONTROL** Handle potatoes very carefully at lifting and when storing. Minimize other problems such as wireworm, slugs and scab infections. Ensure that the potatoes are properly mature at harvest time and that they are stored in a frost-free but cool and dry place. Do not grow 'Arran Pilot', which is notably susceptible, but consider 'King Edward' or 'Pentland Crown' instead, which are notably resistant.

## POTATO EARLY BLIGHT

**SYMPTOMS** Despite its name, this disease usually occurs late in the season in Britain and Europe, though it is an early season infection in North America. Dark brownish-black spots, often angular in outline and showing concentric ringing, appear on the foliage of both potato and tomato. Sunken spots may develop on the stems and a dark rot and associated furry fungal growth may also occur at the sepal end of the fruits. Inadequately fed plants may be more susceptible. However, although causing some damage, this disease rarely affects cropping. **CAUSE** The fungus *Alternaria solani*, which can carry over from year to year on infected plant debris. **CONTROL** Rarely necessary, but the treatments applied to control potato or tomato late blight (*Phytophthora infestans*) should give incidental control.

## POTATO GANGRENE

**SYMPTOMS** Small areas of damage, slightly sunken with clearly defined edges, develop on the tubers after lifting (*see **p.57***). The skin

may become slightly wrinkled. The affected tubers rot, turning wet and pale pink, and then darken. There is no visible fungal growth. If planted, a gangrene-infected potato produces a poor, gappy crop or no crop at all. **CAUSE** The fungus *Phoma exigua* var. *foveata*. **CONTROL** Avoid damaging tubers on lifting and inspect all tubers carefully before planting and throughout storage. Dispose of any tubers showing symptoms.

## POTATO HOLLOW HEART

**SYMPTOMS** The tubers may look perfectly normal on the outside, as does the plant itself. When cut open, the tuber has an internal cavity – this may be simple and surrounded by corky, somewhat darkened tissue; it may be star-shaped, or it may consist of two or more interconnecting cavities. Large potatoes are particularly susceptible; small ones are rarely affected. **CAUSE** This disorder occurs when potatoes have been growing rapidly (perhaps due to overwatering, or excessive fertilizer use) and are then subjected to a sudden check in growth, usually caused by lack of water. Conversely, it may also occur if the plants are subjected to an over-dry soil, followed by a sudden increase in the soil moisture due to heavy rain or watering. In some instances it appears to be induced by excessively high temperatures. **CONTROL** Grow plants in uniform stands and ensure regular and adequate but not excessive watering, particularly during dry weather.

## POTATO INTERNAL RUST SPOT

**SYMPTOMS** The tubers and the whole plant look perfectly normal, but when cut open, the flesh of the tuber is marked with many irregular, small rust-brown spots. **CAUSE** The precise cause is unknown, but unsuitable soil conditions are probably largely responsible. It may occur on sandy and clay soils but is usually most severe on light soils or loam soils with a low organic matter content. Low potassium and lime levels may also be responsible. **CONTROL** Improve the condition of the soil prior to planting – incorporate plenty of humus and water regularly as the crop grows. The soil could also be limed but this must be done well in advance of planting. Grow 'King Edward' or 'Arran Consul', which appear to show a degree of resistance.
• *see also* POTATO COMMON SCAB *p.163*.

## POTATO POWDERY SCAB

**SYMPTOMS** Small, nearly circular scabby patches develop on the tuber. Each has a raised margin. The scabs later burst open and produce masses of brown spores which are released into the soil (*see **p.56***). Occasionally a canker form of this disease occurs when affected tubers are seriously deformed. They may then be incorrectly assumed to have potato wart disease. **CAUSE** The fungus *Spongospora subterranea*, which is particularly common in wet seasons and on heavy soils, especially those which have grown many crops of potatoes.

**CONTROL** Dispose of infected tubers; do not incorporate them into the compost heap. Do not grow potatoes on that site for at least three years. Improve soil aeration before planting potatoes. Avoid growing 'Pentland Crown', as this is particularly susceptible.

• *see also* POTATO WART DISEASE *below.*

## POTATO SILVER SCURF

**SYMPTOMS** Inconspicuous silvery markings, usually lines, develop on the skin of the tuber (*see **p.56***). The flesh of the tuber is not affected and this disease is of little practical consequence unless the tubers are to be used for showing. The condition usually only becomes apparent when the tubers are stored, particularly if the storage conditions are too humid. The markings may turn black if the fungus produces spores.
**CAUSE** The fungus *Helminthosporium solani*.
**CONTROL** None necessary.

## POTATO SPRAING

**SYMPTOMS** The flesh of tubers is marked by tan-brown or red-brown rings or arcs of discoloration (*see **p.56***). There may be development of some corky tissue associated with the discoloration, too. Tubers may, sometimes, be distorted and the stems and foliage may be mottled yellow.
**CAUSE** Tobacco rattle virus and occasionally potato mop top virus, which are spread by free-living nematodes (eelworms) in the soil. Infection may be started if infected tubers are planted.
**CONTROL** No chemical control available. Remove affected

plants, as tobacco rattle virus has a wide host range and may affect sweet pepper, hyacinth, gladiolus, tulip, China aster, tobacco plants and many common weeds. Practise good weed control. Grow potatoes on a fresh site. 'Pentland Dell' appears to be particularly susceptible to this problem, so other cultivars should be grown in preference.

## POTATO VIRUSES

**SYMPTOMS** These vary with the precise combination of viruses involved, the potato variety and the growing conditions. Leaf symptoms include yellow flecking, streaking or mosaicing, dark spotting and leaf distortion, upward rolling and stiffening of the leaflets. If home-saved seed potatoes are used for many years in succession, there may be a gradual but noticeable decline in the plant vigour and cropping and this is often due to the build-up of viral diseases.
**CAUSE** Many different viruses, including potato virus X, potato virus Y and potato leaf roll virus.
**CONTROL** None is available. Always grow potatoes on a fresh site and do not save tubers for seed. Buy tubers certified as virus-free.

## POTATO WART DISEASE

**SYMPTOMS** The tubers are covered in dense, warty outgrowths which are at first off-white, and later black. Affected tubers may rot. Tubers may only start to show symptoms in store. Occasionally small, pale green growths rather like those of crown gall may develop on the stem at soil level. This disease is rare,

but very damaging. The symptoms may be confused with the distorted form of powdery scab, or even with fusarium rot of potato.
**CAUSE** *Synchytrium endobioticum*, which is usually found in wet years or on heavy soils, and infects tubers through the eyes. Spores are produced by the wart-like growths. These spores are very resilient and can remain viable in the soil for about thirty years, even in the absence of a potato crop. They are easily spread from one area to another in particles of soil adhering to tools and footwear.
**CONTROL** Infected tubers should be reported to the local Ministry of Agriculture, Fisheries and Food. Potatoes must not be grown on infested land.

• *see also* POTATO POWDERY SCAB *p.164.*

## POWDERY MILDEWS

**SYMPTOMS** White, powdery fungal growth develops on the leaf surface (*see **p.25***). Upper leaf surfaces are usually affected first and most severely but the symptoms may spread to the lower leaf. The mildew may also develop on all other above ground parts of the plant; the precise location of its growth depends on the host plant and the species of mildew involved. Affected leaves or other plant parts may yellow and become distorted. Distortion is particularly common if infection occurs on young foliage. Affected fruits may crack and split because they are unable to expand normally (*see **p.49***). Sometimes buff or even pale brown fungal growth may be seen; this is most common on rhododendron (where the

mildew is often restricted to the lower leaf surface), gooseberry and laurel. The mildew may kill off small areas of leaf tissue which then drop away, causing a shot-hole effect. Growth may be poor and in extreme cases dieback or even death may occur following premature leaf fall.
**CAUSE** Various fungi, in particular many species of *Oidium*, *Microsphaera*, *Podosphaera*, *Uncinula*, *Erysiphe* and *Phyllactinea*. Each usually only infects a single genus or closely related host plants. These fungi are often encouraged by plants growing in dry soils and by humid or damp air around the top growth.
**CONTROL** Remove infected leaves promptly. Spray with a suitable fungicide such as carbendazim, mancozeb, triforine with bupirimate or sulphur. Keep plants adequately watered but avoid overhead watering. Where available, grow resistant varieties.

• *see also* APPLE POWDERY MILDEW *p.101;* BEGONIA POWDERY MILDEW *p.107;* GERANIUM DOWNY MILDEW *p.131;* HYDRANGEA POWDERY MILDEW *p.138;* OAK POWDERY MILDEW *p.152;* PANSY POWDERY MILDEW *p.154;* PEA POWDERY MILDEW *p.155;* PHLOX POWDERY MILDEW *p.159;* QUINCE POWDERY MILDEW *p.166;* RHODODENDRON POWDERY MILDEW *p.169;* ROSE POWDERY MILDEW *p.171.*

## PRIVET THRIPS

**SYMPTOMS** During the summer, privet and lilac leaves develop a dull green appearance, becoming increasingly silvery brown on the upper surface (*see **p.22***). Elongate, narrow-bodied insects up to 2mm long occur on leaves. The adults

are black with three white bands across wings that are folded back over the body. The immature nymphs are similar in shape but are creamy-yellow and wingless.
**CAUSE** Sap-feeding insects, *Dendrothrips ornatus*, which are more troublesome in hot, dry summers.
**CONTROL** Light infestations have little effect on the plant, but if damage is becoming noticeable, especially before late summer, spray with dimethoate, malathion, pirimiphos-methyl, permethrin or pyrethrum.

## PROLIFERATION

**SYMPTOMS** Only the flower or bud is affected. Buds develop within the centre of developing buds (*see p.44*). These may never mature further or may produce fully grown flowers which grow out from the centre of the original flower. Occasionally a "triple-decker" effect is seen, with three flowers on top of each other. Stems and foliage may also appear between flowers. On fruiting plants, each flower may develop into a fruit.
**CAUSE** Most commonly, physical injury such as that caused by a late frost which damages developing buds. Occasionally virus infection is responsible. If symptoms occur all over the plant and for several successive years, then virus would seem a likely cause and the affected plant should be removed.
**CONTROL** Rarely necessary, though affected shoots could be pruned out.

## PYRACANTHA LEAF MINER

**SYMPTOMS** An oval, silvery-brown patch develops on the upper leaf surface of

*Pyracantha*, or firethorn, where the moth's caterpillar has eaten the internal tissues. As the caterpillar matures, it spins silk threads that cause the leaf to fold in half upwards so that the mine is no longer visible. Leaves at the bottom of the plant are often more heavily attacked than elsewhere on the shrub.
**CAUSE** Caterpillars of a tiny moth, *Phyllonorycter leucographella*, which can have three generations a year, with leaf damage being most obvious over the winter period. It was first recorded in Britain in 1989, but is now widespread in England and is continuing to spread. It originates from southern Europe, but occurs in gardens in many northern European countries.
**CONTROL** Pyracanthas can tolerate heavy infestations, so control measures are often unnecessary. Small plants with extensive leaf mining could be sprayed with malathion or pirimiphos-methyl when mines are seen.

## PYRACANTHA SCAB

**SYMPTOMS** Greyish-black or dark khaki-coloured, scabby patches develop on the leaves and berries (*see p.49*). Infected leaves become chlorotic and may fall prematurely. Infected berries may remain small, start to crack or split and fall.
**CAUSE** The fungus *Spilocaea pyracanthae*, which is encouraged by damp weather. It does not appear to overwinter on fallen leaves (unlike apple and pear scab), but overwinters on infected leaves which remain on the plant, and as tiny pustules on the shoots.
**CONTROL** Prune out severely infected areas, and twigs bearing the pustules. Spray

with carbendazim. Some cultivars show a degree of resistance to this infection (and also to fireblight) and these include 'Orange Charmer', 'Shawnee' and 'Golden Charmer'.

## PYTHIUM ROOT ROT

**SYMPTOMS** The whole plant wilts and dies back. Seedlings are particularly prone, but older plants may succumb, particularly if otherwise lacking in vigour. Roots are killed and may disintegrate, giving the plant the appearance of having only a few roots.
**CAUSE** The fungus *Pythium*, which is soil- or water-borne.
**CONTROL** Remove infected plants promptly, together with the soil or compost in the immediate vicinity of the roots. Water seeds and seedlings with a copper-based fungicide such as Bordeaux mixture, Cheshunt compound or copper oxychloride. Ensure strict hygiene and do not water seedlings with water from a water butt or use unsterilized compost or soil for raising seedlings.

## QUINCE LEAF BLIGHT

**SYMPTOMS** Numerous irregular, small spots appear on the foliage. These are dark red at first but soon turn black and may join together (*see p.48*). The leaves turn yellow and then brown and fall prematurely. Occasionally similar spotting develops on the fruits and these may show some

distortion. Shoot tips may also be infected and then develop dieback.
**CAUSE** The fungus *Diplocarpon mespili* (syn. *Fabrea maculata*), which may overwinter on infected shoots.
**CONTROL** Rake up and dispose of fallen leaves. Prune out infected stems. Spray the tree with a copper-based fungicide such as copper oxychloride and repeat this the following spring as the leaves break.

## QUINCE POWDERY MILDEW

**SYMPTOMS** Powdery, white deposits on the upper leaf surface and occasionally on the lower surface.
**CAUSE** The fungus *Podosphaera leucotricha*, which may also affect apple.
**CONTROL** Keep quinces well watered, as dry soil seems to encourage mildew development. Apply a mulch to the moist soil to preserve soil moisture. Prune out severely infected stems. Prune established trees to improve air circulation within the crown. Spray with a suitable fungicide, such as carbendazim or bupirimate with triforine.
• *see also* APPLE POWDERY MILDEW *p.101*.

## RABBITS

Rabbits (*Oryctolagus cuniculus*) can feed on a very wide range of plants. Herbaceous plants may be grazed down to ground level, while the foliage and soft shoots of woody plants are eaten up to a height of about

50cm. Rabbits also gnaw the bark from the base of trunks, especially on young apple trees, and the tree dies if bark is lost from most or all of the trunk's circumference. Bark feeding occurs at any time of year, but trees are particularly at risk when snow or frost makes other food items unavailable. Rabbits are inquisitive animals that often feed on newly planted plants, even if they have previously ignored similar established plants in the same garden. They were greatly reduced by a viral disease, myxomatosis, during the 1950s and 60s, but they now have greater resistance and are often numerous in rural areas.

**CONTROL** Direct control measures, such as shooting, trapping and gassing of rabbit warrens, are usually not suitable in gardens. A better solution is to keep rabbits away from plants by netting. Fences around the perimeter of a garden need to be at least 120–140cm tall with a further 30cm sunk below soil level, and angled outwards to discourage rabbits from burrowing underneath. The wire netting needs a maximum mesh size of 25mm, otherwise young rabbits may squeeze through. Where individual plants or small flower beds are being protected, it is usually sufficient to enclose them with netting 90cm high without the need to bury part of the netting. Gates to gardens or enclosed flower beds also need to be rabbit-proof and kept closed when not in use. Tree bark can be protected by placing wire netting or spiral tree guards around the base of the trunks. Animal repellent products cannot be relied on to give long-term protection.

Although most plants can be eaten, there are some which are usually left alone; these feature in various published lists of "rabbit-proof plants". Looking at other nearby gardens will also provide suggestions for plants that local rabbits ignore.

## RASPBERRY BEETLE

**SYMPTOMS** All cane fruits, including raspberry, blackberry, loganberry and tayberry, are susceptible. The ripe fruits have dried-up patches at the stalk end of the berries (*see p.50*). When fully grown, the larvae are up to 8mm long and are creamy-white with pale brown markings on their upper surface. The grubs feed initially at the base of the berries but later feed in the inner core or plug. They are often seen crawling around the bowl after the fruits have been picked.

**CAUSE** The larval stage of a small greyish-brown beetle, *Byturus tomentosus*. Eggs are laid on the flowers during early and mid-summer.

**CONTROL** Timing of control measures is important in order to control the newly-hatched grubs before extensive damage has occurred. On raspberries, spray when the first pink fruits are seen; on loganberries and other hybrid berries, at 80% petal fall; and on blackberries, when the first flowers open. Spray with fenitrothion or derris at dusk to minimize the risk to bees and other pollinating insects.

## RASPBERRY CANE BLIGHT

**SYMPTOMS** The shoots or the whole cane start to die back during the summer. This is due to an infection at the base of the canes, which causes them to become brown and the bark to rupture. They also become very brittle and may snap off readily at or just above ground level. Minute, black, pinprick-sized, fungal fruiting bodies develop on dead areas and exude spores. Other spore forms develop later and these forcibly eject their spores into the air.

**CAUSE** The fungus *Leptosphaeria coniothyrium*, which is carried both by rain or watering splash and on air currents. Infection usually takes place through wounds, such as those caused by cane midge attack, late spring frosts, or following pruning. The fungus can persist in the soil on infected plant debris.

**CONTROL** Prune and train the plant carefully to avoid infection. Protect from frost where feasible and control any pests which may cause wounds through which the fungus can gain entry. If infection occurs, prune out affected canes and cut away any discoloured wood from the crown. Then spray the crowns and any new growth with a copper-based fungicide.

## RASPBERRY CANE SPOT

**SYMPTOMS** Purple spots with silvery-white central areas develop on the stems of raspberries and hybrid cane fruits. They may also appear on the leaves and flower stalks (*see p.34*). On loganberries, the fruits may be infected. The bark may split as the spots enlarge and whole canes may be killed.

**CAUSE** The fungus *Elsinoe veneta*, which usually appears first on the canes in early summer, and may then spread to the foliage and the fruits.

**CONTROL** Prune out infected canes. Avoid growing the raspberry cultivars 'Norfolk Giant' and 'Lloyd George' as these are particularly prone to this infection.

## RASPBERRY LEAF AND BUD MITE

**SYMPTOMS** Pale yellow, rounded blotches develop on the upper surface of raspberry leaves (*see p.18*) from late spring onwards. On the underside of the leaves, corresponding with the yellow blotches, are slightly darker areas, compared with the usual silvery-green colour. By mid-summer the foliage may be extensively discoloured and leaves at the shoot tips may be distorted. These symptoms can be confused with those of some virus diseases. Canes affected by the mite normally grow to their usual height and produce an adequate crop, whereas virus-infected plants lack vigour and crop badly.

**CAUSE** Microscopic, sausage-shaped mites, *Phyllocoptes gracilis*, which suck sap from the underside of leaves. In the autumn they hide in or near buds but cause no damage in the winter.

**CONTROL** None of the insecticides available to amateur gardeners is effective against this mite. Fortunately, its impact on the plant's health and cropping ability is not as great as the symptoms might suggest. Raspberry cultivars vary in their susceptibility; 'Malling Jewel' is often attacked, while 'Malling Promise' has some resistance. Plants growing in warm, sheltered places are also likely to have heavier infestations.

## RASPBERRY RUST

**SYMPTOMS** Bright orange, raised pustules full of fungal spores develop on leaves of raspberries and blackberries in spring or early summer. They develop first on the upper leaf surface. Towards the end of summer or in early autumn, dark brown-black spores develop on the lower surface. Premature leaf fall occurs (see *p.40*).
**CAUSE** The fungus *Phragmidium rubi-idaei*, which can overwinter on infected fallen leaves and on the canes.
**CONTROL** Rake up and dispose of infected leaves. The raspberry cultivar 'Glen Prosen' shows a degree of resistance. Avoid growing 'Glen Moy', 'Glen Clova, 'Malling Delight', 'Malling Jewel' and 'Malling Joy'.

## RASPBERRY SPUR BLIGHT

**SYMPTOMS** Dark purplish patches develop around the buds of new canes. These increase in size and spread up and down the canes, causing extensive discoloration. In autumn and winter they change colour, become greyish-silver and are covered in numerous minute raised, black, pin-prick-sized, fungal fruiting bodies. In spring, infected canes bear few viable buds and produce weak shoots which wither and die, or persist but crop very poorly.
**CAUSE** The fungus *Didymella applanata*, which is usually most troublesome in wet weather, but occurs in most seasons regardless of the weather. Plants that have been fed excessive quantities of nitrogen or which are overcrowded are more prone to infection and encourage rapid disease spread.

**CONTROL** Prune out any badly infected canes. Spray with a suitable fungicide such as carbendazim or one based on copper.

## RASPBERRY VIRUSES

**SYMPTOMS** Yellow patterning of the foliage (see *p.18*), most commonly mosaic patterns often combined with slight distortion, stunting and reduction in vigour and cropping. There may also be downward curling of the leaves. The foliage symptoms may be confused with those caused by raspberry leaf and bud mite, but the other symptoms are not present.
**CAUSE** Various viruses, either alone or in combination with each other. The most common are raspberry mosaic, curly dwarf and raspberry yellow dwarf.
**CONTROL** Remove and dispose of infected plants promptly. Control aphids and other sap-feeding pests as these may spread the virus. Do not grow replacement raspberries on the same site again, as some viruses may be transmitted by soil-living nematodes.

## RATS

Rats cause damage to stored fruits and vegetables, and can also damage growing root crops. Rats also spoil foodstuffs by contaminating it with their excrement and urine. Rats often carry in their urine the bacterium that causes Weil's disease; this can cause a serious illness in humans. The brown rat, *Rattus norwegicus*, is of widespread occurrence in Britain and the rest of Europe. Local authorities will often undertake rat control in homes and gardens. Otherwise they

can be controlled with rat traps or poisoned baits based on coumatetralyl or difenacoum. Poisoned baits need to be placed where rats are active but covered up so that children and pets do not have access to them.

## RED BERRY MITE

**SYMPTOMS** Blackberry fruits ripen unevenly, with some parts of the berries remaining red and hard (see *p.49*). This problem is more frequently seen in hot summers. The first fruits to develop usually ripen normally, but later fruits show progressively more severe symptoms.
**CAUSE** Microscopic gall mites, *Acalitus essigi*, that feed by sucking sap from the flowers, foliage and fruits. Only the fruits suffer damage, which is due to a toxic saliva secreted by the mites as they feed.
**CONTROL** None of the insecticides available to amateur gardeners is effective against this mite. Some of the overwintering mites can be eliminated by pruning out to the ground and burning the old fruiting canes.

## RED SPIDER MITE

*See* CONIFER RED SPIDER MITE *p.119*; FRUIT TREE RED SPIDER MITE *p.129*; GLASSHOUSE RED SPIDER MITE *p.132*.

## RED THREAD

*See* TURF RED THREAD *p.184*.

## REPLANT PROBLEMS

*See* ROSE SICKNESS *p.172*.

## REVERSION

*See* CURRANT REVERSION *p.121*.

## RHODODENDRON BUD BLAST

**SYMPTOMS** Flower buds fail to open, and turn brown and dry (see *p.41*). They may remain on the plant for several years. The surface of the bud is covered in numerous tiny, black, bristle-like fungal outgrowths. These are the spore-bearing parts of the fungus, and it is the presence of these which differentiates between this and dry bud with which it is often confused. Quite often infected buds and healthy ones are seen growing side by side. Leaf, stem and general growth are unaffected.
**CAUSE** The fungus *Pycnostysanus azaleae*, spread by rhododendron leafhoppers as they lay their eggs in the bud scales.
**CONTROL** The leafhoppers first appear in mid- to late summer, so pick off and dispose of infected buds before this time of year. Control of the leafhopper is possible but rarely fully effective in controlling bud blast spread, because the pests simply re-invade.
• *see also* BUD DROP *p.111*; BUD, DRY *p.112*.

## RHODODENDRON LACEBUG

**SYMPTOMS** During the summer the upper leaf surface of rhododendrons, especially those growing in sunny positions, develop a coarse yellowish mottling. The underside of the leaves is covered in rusty-brown excrement spots deposited by the lacebugs (see *p.21*). These are present between late spring and early autumn. The wingless, immature nymphs are yellowish-brown, while the adults are 4mm long, brownish-black and with wings that are folded flat over their backs. The transparent wings have many veins, giving them a lace-like appearance – hence the name lacebug.

**CAUSE** Sap-feeding insects, *Stephanitis rhododendri*, which have one generation a year. In late summer the adult lacebugs lay overwintering eggs in the midribs of the leaves.

ACTUAL LENGTH: 4MM

**RHODODENDRON LACEBUG**

**CONTROL** Avoid planting rhododendrons in warm, sunny spots. If the pest does occur, spray the foliage with permethrin, pirimiphos-methyl, dimethoate or pyrethrum. Damaged leaves do not regain their green colour and can persist on the shrub for more than a year; new growth develops normally once the pest has been dealt with.

# RHODODENDRON LEAF SPOT

**SYMPTOMS** Spots with clearly defined edges develop on the leaves. The older leaves are usually worst affected and, if severely damaged, some may fall prematurely. The spots are brownish-purple, often with a distinct black or very dark purple ring around the edge (*see p.14*). The spots often have concentric ringing with numerous pinprick-sized, raised, black fungal fruiting bodies on the dead areas.

**CAUSE** The fungus *Gleosporium rhododendri*, which, although causing

unsightly damage to leaves, is only likely to reach serious levels on a plant lacking vigour or in poor health.

**CONTROL** Remove severely affected leaves. If necessary, spray with a fungicide containing mancozeb. Take steps to improve the general vigour of the plant by regular feeding, watering and mulching.

# RHODODENDRON LEAFHOPPER

**SYMPTOMS** The creamy-yellow immature nymphs are present on the underside of rhododendron leaves from late spring to mid-summer, but cause no noticeable damage. The adult insects are more readily seen, as they often sun themselves on the upper leaf surface between mid-summer and mid-autumn. They are 8–9mm long and turquoise-green with two orange stripes on each forewing. When disturbed, they jump off the leaf and fly a short distance before returning to the plant.

**CAUSE** A sap-feeding insect, *Graphocephala fennahi*, which causes no direct damage to rhododendrons. The females, however, insert overwintering eggs into the flower buds and this makes entry wounds for the fungal disease rhododendron bud blast (*see p.168*).

**CONTROL** Reduce the incidence of bud blast by spraying against the adult leafhoppers during late summer and early autumn. Dimethoate, pirimiphos-methyl, permethrin, malathion or fenitrothion will control leafhoppers, but more may fly in from nearby gardens, especially during hot weather, which increases leafhopper activity.

# RHODODENDRON PETAL BLIGHT

**SYMPTOMS** Small spots develop on the petals – these are white on coloured flowers and pale brown on white blooms (*see p.42*). The spots increase in size and appear water-soaked. The entire flower rapidly collapses and becomes slimy and shapeless. Affected flowers do not fall but remain on the plant. Dry weather may then cause the dead flowers to dry out and they may remain on the plant in this state until the following spring.

**CAUSE** The fungus *Ovulinia azaleae*, which is encouraged by mild, humid growing conditions. The spores develop on the petals as they deteriorate and are spread by insects and on air currents.

**CONTROL** Remove and dispose of infected flowers promptly to minimize spread.

# RHODODENDRON POWDERY MILDEW

**SYMPTOMS** Yellow blotches develop on the upper leaf surface. Beneath each is a corresponding patch of felty, buff-coloured fungal growth (*see p.15*). Premature leaf fall occurs. Occasionally the fungal growth occurs on the upper surface and is then more off-white and powdery than buff and felty.

**CAUSE** The precise identity of the fungus or fungi is as yet undetermined, but the most important is probably a species of *Erysiphe*. Dry soil and wet conditions around the foliage in late summer and autumn encourage its development and spread.

**CONTROL** Pick off severely affected leaves. Spray with a suitable fungicide such as

carbendazim, bupirimate with triforine, or mancozeb. Keep plants well watered and mulched but avoid overhead watering. Avoid growing *R. cinnabarinum* and its hybrids, and hybrids with species from the subsections *Campylocarpa* or *Fortunea* as their parentage.

# RHODODENDRON RUST

**SYMPTOMS** The upper leaf surface develops bright orange-yellow spots with corresponding yellow and brown fungal pustules appearing beneath. The disease is not very common and may be confused with the damage caused by rhododendron lacebug (*see p.168*).

**CAUSE** The fungus *Chrysomyxa ledi* var. *rhododendri*, which persists within the affected leaves as long as they remain on the plant. It is encouraged by damp conditions.

**CONTROL** Remove infected leaves promptly, pruning out entire shoots if necessary. Some success may be achieved by spraying with a fungicide such as mancozeb or penconazole, though this treatment does not work as reliably as it does on most other rust infections.

# RHUBARB GUMMING

**SYMPTOMS** The sticks develop normally but some are spoiled by drops of resin-like exudation which appear along their length. Sticks may crack around the areas of gumming, allowing the entry of secondary organisms which then cause rotting.

**CAUSE** The precise cause is not known but poor growing conditions, in particular an erratic supply of moisture, are believed to be involved. Inadequate feeding may also play a part.

**CONTROL** Improve growing conditions and pay particular attention to watering and feeding the plant.

## ROBIN'S PIN CUSHION

**SYMPTOMS** Wild hedgerow roses, sucker growth on grafted roses and some species roses develop roughly spherical galls, up to 60mm across, on stems (see *p.39*) and occasionally leaves in mid- to late summer. The gall consists of a hard central core in which many white grubs live and feed. The exterior of the gall is covered in a dense mass of moss-like growths which are reddish or yellowish-green in colour. In the autumn the mossy covering dries up, leaving the inner core in which the grubs overwinter before pupating in the spring.

**CAUSE** A small gall wasp, *Diplolepis rosae*, which lays eggs in buds in mid-summer. The galls that subsequently develop are also sometimes known as bedeguar galls.

**CONTROL** Apart from creating the galls, this insect has little adverse effect on the plant and so control measures are unnecessary. If the galls are considered unsightly they can be pruned out.

## ROOT APHIDS

**SYMPTOMS** Infested plants tend to have reduced vigour and are more likely to wilt during sunny weather. The aphids, up to 2–3mm long, usually a dirty cream colour but sometimes bluish-green, live amongst the roots or at the stem bases of host plants (see *p.55*). Root aphids often

secrete a white, powdery or fluffy wax from their bodies and this gives infested roots and nearby soil particles a white coating.

**CAUSE** Several different species of root aphids occur in gardens. These include *Smynthurodes betae* on runner and French beans, *Pemphigus bursarius* on lettuce (see LETTUCE ROOT APHID *p.145*), *Thecabius auriculae* on auriculas, *Trama troglodytes* on Jerusalem artichoke, *Maculolachnus submacula* on roses (see ROSE ROOT APHID *p.172*), *Aphis sambuci* on pinks, *Dysaphis crataegi* on carrots and parsnips.

ACTUAL LENGTH: 2–3MM

**ROOT APHID**

**CONTROL** Root aphids are more difficult to control than those that feed on the foliage. Drenching the soil with spray-strength dimethoate, heptenophos with permethrin, pirimicarb or pirimiphos-methyl will reduce infestations. Crop rotation helps avoid infestation by aphids that have overwintered in the soil on the remains of the previous year's crops.

## ROOT KNOT EELWORMS

**SYMPTOMS** Many plants can be attacked by root knot eelworms. These microscopic animals live within the roots and cause knobbly swellings (not to be confused with the nitrogen-fixing nodules normally found on the roots of legumes). Root knot eelworms disrupt the uptake

of water and nutrients, so that plants lack vigour and have poor leaf colour. In northern Europe this pest is mainly encountered in greenhouses, but can attack garden plants on light sandy soils.

**CAUSE** Several *Meloidogyne* species may infest a wide range of plants.

**CONTROL** There are no effective chemical controls available to gardeners. Infested plants should be destroyed, together with the soil around the roots.

## ROOT MEALYBUGS

**SYMPTOMS** Root mealybugs are mainly a problem on pot plants, particularly cacti, succulents, pelargonium, ferns, African violet and fuchsia, causing them to lack vigour. The mealybugs are up to 2mm long and they cover themselves and the roots with a white, waxy powder (see *p.55*). They have flatter, more elongate bodies than root aphids (see *p.170*), which have a globular body shape.

**CAUSE** Several *Rhizoecus* species occur in greenhouses and on the roots of house plants. They tend to be more troublesome on plants which are normally grown in a dryish potting compost.

**CONTROL** Root mealybugs are difficult to eliminate. Drenching the potting compost with spray-strength malathion can reduce infestations. With many plants it is better to discard the old infested plants.

## ROSE APHIDS

**SYMPTOMS** The younger leaves, shoot tips and flower buds are often covered in dense colonies of green, yellowish-green or pink insects up to 2–3mm long

(see *p.43*). Heavy infestations cause stunted growth and poor quality blooms, while the foliage becomes sticky with a sugary honeydew excreted by the aphids. Sooty mould may develop on the honeydew, which also traps the white cast skins shed by the aphids as they grow.

**CAUSE** Several species of aphid occur on roses; the most troublesome is usually *Macrosiphum rosae*.

**CONTROL** Look for aphids from mid-spring onwards. Spray with pirimicarb, heptenophos with permethrin, dimethoate, fenitrothion, pirimiphos-methyl or malathion before heavy infestations develop. Organic sprays, such as pyrethrum, derris and insecticidal soaps, are also effective but may need to be used more frequently.

• see also ROSE ROOT APHID, *p.172*.

## ROSE BALLING

**SYMPTOMS** Flower buds fail to open or start to do so but never open fully (see *p.44*), generally in damp weather. The outer petals become pale brown, dry and papery. The inner petals are perfectly normal but hidden beneath the outer casing of dead ones. If damp weather persists, the bud rots off, usually becoming covered in grey mould growth. The rose's general growth is not affected; healthy buds and flowers may appear among those showing balling.

**CAUSE** Most commonly rain followed by bright or hot sun on the petals. The petals become slightly scorched and form an outer casing which prevents undamaged petals from appearing.

**CONTROL** Little can be done to prevent this, but do not water roses on hot, bright days. Water during the evening and directly at the base of the plant. Prune out blooms showing the symptoms before they develop grey mould, as this secondary infection may then result in dieback.

## ROSE BLACK SPOT

**SYMPTOMS** Diffuse purple-black spots or blotches develop on the leaves (*see* **p.13**), shortly followed by yellowing, and then premature leaf drop. The spots do not have sharply defined edges and can be seen to be made up of tightly packed strands of fungal growth growing through the surface of the leaf. The spots enlarge and may sometimes join together. If black spot occurs several years in succession, particularly if infection occurs early in the summer, the rose may be severely weakened. Much smaller purple-black spots (usually only a millimetre or two across) may also develop on the stems, particularly on species roses.
**CAUSE** The fungus *Diplocarpon rosae*, which overwinters on stem lesions, and also on the bud scales and on fallen leaves.
**CONTROL** Dispose of infected leaves promptly. In spring, prune out stems showing the lesions associated with black spot. Spray with a suitable fungicide immediately after spring pruning, or before the leaves start to break, whichever is appropriate to the pruning regime. Repeat applications of fungicide according to the instructions. Many applications are usually

necessary to achieve reasonable control. Fungicides could include triforine with bupirimate, penconazole, mancozeb, carbendazim or copper with ammonium hydroxide. Winter treatments of soil are not usually effective, because of the way in which the leaves are curled and packed on the ground. Affected leaves should, however, be raked up and disposed of promptly. Check current catalogues and choose those roses described as showing good disease resistance.

## ROSE CANKER AND DIEBACK

**SYMPTOMS** Stems discolour and die back or fail to produce growth in the spring. They may have a purple or blackish patch of discoloration which becomes uniform if the stem is allowed to die off totally (*see* **p.40**). In extreme cases, the rose may be killed. During damp weather fuzzy grey fungal growth may develop from areas damaged by *Botrytis* (grey mould) and minute raised, fungal bodies may develop on the areas killed by the other fungi.
**CAUSE** Various fungi, in particular *Leptosphaeria Coniothyrium* and *Botrytis cinerea*. These gain entry to the stems, usually at the base, via wounds or other points of injury. These diseases are all encouraged by poor growing conditions such as over-wet soil, too deep planting, and excessively high mulching (when the mulch is placed too close to the stems and at some height up them.)
**CONTROL** Prune out affected shoots completely, cutting right back into healthy tissue. Clear mulch or excess soil

away from the base of the plant. Provide good growing conditions, water, feeding and so forth to ensure good, sturdy replacement growth.

## ROSE GREY MOULD

*See* GREY MOULD *p.134*; ROSE CANKER (*above*).

## ROSE LEAF-ROLLING SAWFLY

**SYMPTOMS** During late spring and early summer, rose leaflets become tightly rolled in a downwards direction (*see* **p.27**). The rolled leaflets often hang down compared with unaffected parts of the leaf. In the following weeks the leaflets are eaten by pale green caterpillars, up to 10mm long, hidden within them.
**CAUSE** The leaf curling is brought about by chemicals injected into the leaves by female sawflies, *Blennocampa phyllocolpa*, as they insert their eggs. There is one generation a year, with egg laying occurring in early summer.
**CONTROL** It can be difficult to prevent damage occurring, especially when hot weather coincides with the sawfly's flight period. This increases their activity and the number of eggs laid. Light infestations can be dealt with by picking off rolled leaflets. If many leaves are affected, it is better to leave them on the plant and spray with an insecticide such as heptenophos with permethrin or pirimiphos-methyl.

## ROSE LEAFHOPPER

**SYMPTOMS** Rose leaves develop a coarse, whitish mottling on the upper leaf surface (*see* **p.21**) during the summer. Roses growing

against a wall or in other sheltered positions may be heavily infested and by late summer much of the green colour will have gone from the leaves. Pale yellow insects, up to 3mm long, live on the underside of the leaves and readily jump off when disturbed. They have narrow bodies which are broadest at the head end and taper to the rear. The immature nymphs are creamy-white and crawl about on the lower leaf surface.
**CAUSE** Sap-feeding insects, *Edwardsiana rosae*, which have two or three generations during the summer.
**CONTROL** Most insecticides used against aphids also control leafhoppers. These include heptenophos with permethrin, dimethoate, pirimiphos-methyl, permethrin, malathion, fenitrothion and pyrethrum.

## ROSE POWDERY MILDEW

**SYMPTOMS** Powdery, white fungal growth develops on the upper leaf surface (*see* **p.20**). Leaves may be attacked when very young and extensive distortion then results. On young foliage the fungal growth frequently develops on both leaf surfaces. The fungal growth is superficial and can be rubbed off but the leaf tissue beneath is discoloured. Powdery mildew growth may also develop on the stems, flower buds and thorns. When flower buds are severely infected, the buds may fail to open fully. Affected leaves fall early.
**CAUSE** The fungus *Sphaerotheca pannosa*, which overwinters as stem infections and also in dormant buds. The spores are air-borne. Powdery

mildew is encouraged by dry conditions around the roots and by moist air around the foliage. Roses growing up against walls are often particularly badly damaged because of the drying effect of the bricks and also the rain-shadow effect.

**CONTROL** Prune out badly infected stems and spray with a suitable fungicide such as triforine with bupirimate, mancozeb, carbendazim, penconazole or sulphur. Keep roses adequately watered, particularly during dry weather, but avoid wetting the foliage. Apply a good bulky mulch to encourage soil moisture retention. Check current catalogues and choose roses described as showing good disease resistance.

## ROSE ROOT APHID

**SYMPTOMS** This dark brown aphid, 3–4mm in length, feeds by sucking sap from rose roots. It would probably go unnoticed if it did not climb up the stems to deposit eggs there in the autumn. Recently-laid eggs are brown, but they soon become shiny black (*see p.35*). The eggs are about 1mm long, shaped like rugby balls and may be thickly clustered on the lower part of the stems. They will hatch in early spring and soon afterwards the young aphids go down into the soil.

**CAUSE** Sap-feeding aphids, *Maculolachnus submacula.* The root-feeding activities of this aphid seem to have little impact on roses and infestations can vary considerably in numbers from year to year.

**CONTROL** Many of the eggs can be removed when roses are being pruned. Where this

is not possible, the aphids can be sprayed like other rose aphids (*see p.170*) when they have hatched.

## ROSE RUST

**SYMPTOMS** Bright orange spots develop on the upper leaf surface, with corresponding bright orange spore masses beneath (*see p.16*). Severely infected leaves may fall prematurely. The leaf infections develop in early summer and are the most common and most obvious. Later in the summer, or in very early autumn, dark brown spore masses develop beneath the leaf as the winter spores are produced to replace the summer ones. Defoliation at this stage is almost inevitable. Occasionally, spring infections may develop, but these are usually only seen on species roses as a rupturing of the stem and the production of large quantities of bright orange spores which burst out from within the stem. The stem above may then wither as the crack allows secondary organisms to invade.

**CAUSE** The fungus *Phragmidium*, in particular *P. tuberculatum* and, less frequently, *P. mucronatum.* Moist air is essential for the development and spread of the infection, as the spores are carried on air currents. The fungus may persist within the stems of plants which have shown spring infections. The spores can also overwinter on the soil surface and on fallen debris, stems, fences and stakes.

**CONTROL** Prune out stems showing spring infections. This should be done promptly and thoroughly, so that no fungus is left within

adjacent tissue. Improve air circulation by pruning and by avoiding planting too closely. Spray with a suitable fungicide such as myclobutanil, penconazole, triforine with bupirimate or mancozeb.

## ROSE SICKNESS/ REPLANT PROBLEMS

**SYMPTOMS** Newly planted roses fail to thrive and may show signs of dieback. There is no obvious cultural problem, such as poor planting, very dry or very wet soil, or any signs of pest attack. The roots may appear poorly developed, compact and dark in colour and there may be rotting of the finer roots. Roses are most often affected by this, but other plants, in particular apple, cherry, peach, pear, quince and certain plums may be affected too. When planted on a site previously occupied by the same species, certain plants fail to thrive. However, if they are removed promptly and replanted on a fresh site they may recover.

**CAUSE** The precise cause is uncertain and may well vary from plant to plant and also with location. The main causes are soil-living nematodes (eelworms) which may transmit viruses. Soil-borne fungi such as *Thielaviopsis basicola* and *Pythium* may also play a part, as do nutritional factors such as nutrient depletion.

**CONTROL** Wherever possible do not allow the problem to arise, as there is no reliable control for plants other than clearing the soil from the roots, followed by replanting on a fresh site. If necessary, change the soil in the bed or area before planting. Soil must be changed to a depth of at least 45cm, preferably

deeper. If replacing individual roses, change a cube of soil measuring at least 45 x 45 x 45cm. Feeding with a high nitrogen fertilizer and incorporating materials high in nitrogen into the planting hole may also help minimize damage.

## ROSE SLUGWORM

**SYMPTOMS** Yellowish-green, caterpillar-like larvae, up to 14mm long with pale brown heads, feed on the foliage during summer. They graze away the lower leaf surface, causing the remaining damaged tissues to dry up and become whitish-brown (*see p.18*).

**CAUSE** Larvae of a sawfly, *Endelomyia aethiops,* which has two generations during the summer. Larvae of the second generation in mid- to late summer are often more numerous and therefore capable of more extensive damage.

**CONTROL** Slugworm larvae are relatively easy to control with most contact insecticides, including pyrethrum and derris. Others which can be used are permethrin, bifenthrin, pirimiphos-methyl and fenitrothion.

## ROSE VIRUSES

**SYMPTOMS** Symptoms show up most clearly on leaves, usually as vein clearing, yellow flecking or mottling (*see p.21*). The markings tend to be quite mild, and not as obvious as on most other virus-infected plants, but the plant may show some distortion and stunting. Symptoms may be confused with those caused by weedkiller damage, the latter being far more common than virus symptoms.

**CAUSE** Various viruses, including strawberry latent ringspot and rose mosaic.
**CONTROL** None is available. Some viruses may be spread by soil-living nematodes, but other methods may be involved as well. Infected plants showing severe symptoms should be removed and destroyed.

## ROSY APPLE APHID

**SYMPTOMS** Pinkish-grey insects, up to 2mm long, cluster on the young foliage of apples in the spring. Their feeding causes leaves at the shoot tips to become curled and yellowish. The aphids also feed on the young fruitlets and this can result in them staying small, with a pinched appearance around the eye end of the fruits (*see* **p.48**). The distribution of damaged fruits can be very variable, with some branches producing normal fruits, while others have stunted ones at harvest time.
**CAUSE** A sap-feeding insect, *Dysaphis plantaginea*, which overwinters on the tree as eggs. These hatch at bud burst and the aphids are active on the tree until early or mid-summer, when they migrate to their summer host plants, which are the wild flowers known as plantains.
**CONTROL** The egg stage can be controlled with a tar oil wash during winter. Otherwise spray against the hatched aphids as the foliage emerges from the buds, using pirimicarb, heptenophos with permethrin, dimethoate, pirimiphos-methyl or fenitrothion.

## RUSTS

**SYMPTOMS** The symptoms may develop on the foliage and stems, depending both on the host plant and the rust involved. Spores, either as spore masses or pustules, are usually bright orange or dark brown, and their colour may vary with the time of year. Rusts may produce several distinct spore stages, commonly known as spring, summer and winter spores. Those produced early in the year are generally orange or bright yellow and those later in the year usually brown. Occasionally the rust produces gelatinous masses containing the spores, or "cluster cups". Infected areas usually discolour and may wither and die off. Certain rusts have alternate hosts: although unrelated, both plants are needed for the fungus to complete its life cycle. On others, the life cycle needs only one host.
**CAUSE** Various fungi, the most common being species of *Puccinia*, *Uromyces*, *Phragmidium*, *Melampsora* and *Gymnosporangium*. The spores need a moist environment to germinate and infect, and so rust infections are generally most severe in damp conditions.
**CONTROL** Remove infected leaves. Spray plants with a suitable fungicide, such as penconazole, bupirimate with triforine, mancozeb, myclobutanil (check the labels for the suitability of each chemical for the host plant that you have in mind). Improve air circulation. Where available, grow resistant cultivars.
• *see also* BEAN RUST *p.105;* BLUEBELL RUST *p.109;* FUCHSIA RUST *p.129;* HOLLYHOCK RUST *p.136;* IRIS RUST *p.139;* LEEK RUST *p.144;* PERIWINKLE RUST *p.158;* PLUM RUST *p.161;* RHODODENDRON RUST *p.169;* ROSE RUST *p.172.*

## SALT DAMAGE

**SYMPTOMS** Leaf browning and curling, often associated with the leaf veins. Leaves may be undersized and fall prematurely. Dieback may also occur. Most tree species could be affected, but symptoms are particularly common on sycamore, maple, spruce, alder, lime and plane.
**CAUSE** Sodium chloride, usually as a result of a coastal situation, excessive use of salt on roads in winter, or to the presence of a salt dump for road salting being placed too close to trees. Occasionally, stormy weather may produce salt-laden winds, which cause salt damage a considerable distance inland, as well as to trees growing by the sea.
**CONTROL** Avoid contamination and where feasible, alert the local council to potential problems.

## SAWFLIES

Sawflies have caterpillar-like larvae, but they belong to the insect order called the Hymenoptera. They are therefore more closely related to ants, bees and wasps than to butterflies or moths, which belong to the Lepidoptera order. Sawfly larvae can be distinguished from moth and butterfly larvae by counting the clasping prolegs on the abdomen. Both types of larvae have three pairs of jointed legs on the thorax, but sawfly larvae have at least seven pairs of prolegs, while Lepidoptera caterpillars have five or less. Sawfly larvae are mainly leaf-feeders and many of the pest species are capable of causing extensive defoliation. Other feeding habits include:
**boring into developing fruits** – *see* APPLE SAWFLY *p.101;* PLUM SAWFLY *p.162;*
**causing galls or leaf distortion** – *see* ROSE LEAF-ROLLING SAWFLY *p.171;* WILLOW BEAN GALL SAWFLIES *p.190;*
**leaf-mining** – *see* GEUM SAWFLIES *p.131.*
• *see also* AQUILEGIA SAWFLY *p.102;* ARUNCUS SAWFLY *p.102;* GERANIUM SAWFLY *p.131;* GOOSEBERRY SAWFLY *p.133;* PEAR AND CHERRY SLUGWORM *p.157;* ROSE SLUGWORM *p.172;* SOLOMON'S SEAL SAWFLY *p.177.*

## SCALE INSECTS

Scale insects are sap-feeding pests that attack the foliage and stems of many plants in greenhouses and gardens. The insects are hidden underneath waxy shells or scales that they secrete over their bodies. Newly-hatched scales crawl around looking for suitable places to feed, but they soon settle down and remain immobile for most of their lives. The colour of the protective scale is usually brown or greyish-white and it may be flat or domed. The size of the scale varies according to the species and is in the range of 1–6mm. Some scale insects excrete a sugary liquid called honeydew, which makes the foliage sticky and permits the growth of sooty moulds. Most scale insects lay their eggs under the protection of their own bodies, but the cushion scales deposit eggs among white, waxy fibres that are secreted by the female scales. The stage in the life cycle most vulnerable to an insecticide, such as malathion, is that of the newly-hatched nymphs. With most outdoor species of scales, these are present in mid-summer, but on indoor plants breeding may be continuous throughout the year.

ACTUAL LENGTH: 1–5MM

**SCALE INSECTS**

• *see also* BEECH BARK SCALE *p.106;* BROWN SCALE *p.111;* CUSHION SCALE *p.121;* EUONYMUS SCALE *p.126;* HEMISPHERICAL SCALE *p.135;* HORSE CHESTNUT SCALE *p.137;* HYDRANGEA SCALE *p.138;* JUNIPER SCALE *p.140;* MUSSEL SCALE *p.149;* OLEANDER SCALE *p.152;* SOFT SCALE *p.177;* VIBURNUM SCALE *p.185;* WOOLLY VINE SCALE *p.192.*

## SCIARID FLIES

*See* FUNGUS GNATS *p.129.*

## SCLEROTINIA

**SYMPTOMS** Affected tissues become discoloured, brown and wet as they rot. Large quantities of cotton wool-like, fluffy fungal growth develop and are scattered with large, black sclerotia (fungal resting bodies) which may be several millimetres long. The fungal growth and sclerotia may develop within the stems, too. This disease may affect most above ground plant parts, but fruit and stem base infections are most common (*see* **p.36**). In addition, it may develop on corms and tubers in store. **CAUSE** The fungus *Sclerotinia sclerotiorum,* which is encouraged by cool, damp conditions. The sclerotia fall into the soil where they remain dormant until spring.

They then produce spore-producing, cup-shaped fungal growths (called apothecia) which are responsible for the following year's infections. **CONTROL** Remove and burn affected plant parts promptly. Dispose of all debris at the end of the season. Do not grow susceptible plants on that site for at least four years following an outbreak of the disease.

## SCORCH

**SYMPTOMS** Pale, scorched-looking patches develop which are usually pale brown or bleached. Damaged areas usually dry out, becoming crispy and brown. Leaves or petals are most commonly affected but occasionally other above-ground plant parts may be damaged, too. Stems or trunks of trees and shrubs may show scorching only on the exposed side. Other growth remains normal. **CAUSE** Hot or bright sun is the most common cause. The problem is exacerbated by droplets of moisture on the leaf surface – the droplets help to further magnify the sun's rays and increase the damage. Soft- or hairy-leaved plants are often most severely affected. Similar damage may occur with chemical contamination; either an unsuitable pesticide is used, or as a result of injury by a contact weed-killer. Less frequently, damage is due to the presence of moisture of any sort on the foliage, particularly if the leaves remain wet overnight as temperatures drop. On bark, the injury is usually the result of the scorching effect of bright sun on thin or young bark.

**CONTROL** Take care when watering, and water in early evening when leaves are not exposed to direct sunlight, yet have time to dry off before nightfall. Ensure that chemicals are selected properly and used with great care. In greenhouses and conservatories, provide adequate shading during the summer months.
• *see also* HORSE CHESTNUT MARGINAL LEAF SCORCH *p.137;* IRIS SCORCH *p.139;* NARCISSUS LEAF SCORCH *p.150.*

## SEIRIDIUM CANKER

**SYMPTOMS** Foliage of cypresses (*Cupressus*), junipers (*Juniperus*), Leyland cypresses (× *Cupressocyparis leylandii*) and thuyas (*Thuja*) loses colour, becoming slightly yellowed and losing its lustre. It then turns brown and dies (*see* **p.35**). This deterioration is due to the development of a canker on the affected stem. The bark becomes slightly roughened and resin is produced from around the cankers. Close inspection reveals black, raised, pinprick-sized fungal bodies around the cankered area. These may be rounded or may have burst, leaving minute craters. The infection is often first noticed as many small areas of dying foliage, but it may attack larger branches and cause extensive dieback. The tree may be rendered very unsightly and can even be killed. **CAUSE** The fungus *Seiridium cardinale* (syn. *Coryneum cardinale*), which enters branches through twig crotches or fine cracks in the bark, and attacks most readily when the tree is growing slowly. The spores are produced from within the fungal bodies on the

cankers and are carried on the wind. The fungus kills by girdling the affected stems as the canker enlarges. **CONTROL** Prune out infected areas promptly, as this may prevent further damage occurring. There is no chemical control available. Check other susceptible trees in the area, such as *Cupressus macrocarpa, C. sempervirens* var. *horizontalis, C. glabra* and *Thuja plicata,* and remove any infected areas.

## SHANKING OF GRAPES

**SYMPTOMS** Individual berries within the bunch fail to colour up normally – those of black cultivars turn red and those of white cultivars remain translucent. They then become wrinkled and, if left on the bunch, may start to resemble raisins. They have a watery or unpleasantly sour taste. **CAUSE** A disorder which may result from many different cultural problems. Under- or over-watering are the most common causes; stagnant soil conditions and under-feeding may also be factors. **CONTROL** Cut out the affected berries and spray the foliage regularly with a foliar feed to boost the vine's vigour. Ensure adequate but not excessive watering and feeding. Check that drainage is not impeded.

## SHARKA

*See* PLUM POX *p.161.*

## SHIELDBUGS

When viewed from above, these insects have a shield-like shape and a broad, flattened beetle-like appearance, with sucking

mouthparts. Some prey on other insects, but most feed on plant sap, although their impact on cultivated plants is negligible. Several species occur in gardens; the more common types include green shieldbug (*Palomena prasina*), hawthorn shieldbug (*Acanthosoma haemorrhoidale*), birch shieldbug (*Elasmostethus interstinctus*) and dock shieldbug (*Coreus marginatus*). They are up to 12mm long and are green, brown or a combination of yellowish-green and reddish-brown. They are most often seen between late summer and early autumn, when they tend to sun themselves on plant leaves before seeking sheltered places in which to overwinter. Eggs are laid between mid-spring and early summer after the adult insects have become active again. The immature nymphs have a more rounded body shape, but increasingly come to resemble the adults as they mature.
**CONTROL** None required.

## SHOTHOLE

**SYMPTOMS** Holes develop on the leaves (*see p.30*), usually with an inconspicuous brown ring around the edge of each hole. Brown spots of dead tissue (*see p.18*) are present on the leaf.
**CAUSE** Various fungal and bacterial infections, including bacterial canker and occasionally powdery mildews, may be responsible. The pathogens kill off areas of leaf, producing leaf spots. The damaged areas then fall away, leaving holes.
**CONTROL** Determine the primary cause of the leaf damage and take the appropriate action.

## SHOT-HOLE BORERS

**SYMPTOMS** Branches and sometimes the trunk of fruiting and ornamental forms of plums, cherry, almond, and less frequently, apple and pear, have many round holes, 1–3mm in diameter, in the bark. Underneath the bark is a maze of tunnels (*see p.33*) where white, legless grubs have been feeding. Some species also bore into the heart wood. Sawdust may be seen coming from the shot-holes in the bark, where the dark brown, cylindrical adult beetles, which are 2–4mm long, have emerged in early summer.
**CAUSE** Several species are collectively known as shot-hole borers on fruit trees. These include *Scoltyus rugulosus*, *S. mali* and *Xyleborus dispar*. All are likely to attack branches or trees that are already in poor condition, and they are not a problem on vigorous trees.
**CONTROL** Chemical control gives poor results since much of the life cycle is protected under the bark. Prune out and burn infested branches and improve the tree's health by feeding in the spring and keeping it watered in dry summers. Unsuitable growing conditions, such as poorly drained soil or a weak root system arising from bad planting, will make trees prone to dieback and shot-hole borers.

## SILVER FIR ADELGID

**SYMPTOMS** The smaller stems on silver firs such as *Abies procera*, *A. grandis* and *A. balsamea* develop lumpy growths where small black insects covered with white, waxy fibres are feeding.
**CAUSE** A sap-feeding insect,

*Adelges piceae*, which is active between late spring and autumn.
**CONTROL** Little can be done on large trees, but those small enough to be sprayed can be treated with malathion or pirimiphos-methyl in the spring.

## SILVER LEAF

**SYMPTOMS** Single or small numbers of branches develop foliage which has a silver sheen. On some hosts, such as rhododendron, the silvering is almost imperceptible, or non-existent. If affected stems more than 2.5cm in diameter are cut across, they have a central brownish stain. Other limbs appear perfectly normal, but may show symptoms at a later date. Affected limbs die off or fail to leaf up the following spring. The symptoms may be confused with those caused by other problems, such as adverse weather conditions, drought, malnutrition or insect attack. Fruiting bodies, often grouped together and adhering fairly closely to the bark, may develop on dead wood. Their exposed surface is a dark purple-grey colour.
**CAUSE** The fungus *Chondrostereum purpureum*, a fresh-wound parasite that invades freshly created wounds on deciduous trees. Pruning cuts or pest attack are common points of entry. The spores are produced from the fruiting bodies found on infected limbs of living trees, or on fallen timber. The infection is spread on air currents or on pruning tools. The fungus produces a toxin which causes the upper leaf surface to separate off from the main leaf blade. The air which

accumulates between the two layers alters the light-reflecting qualities of the leaf, causing it to appear silvered. The leaves themselves do not carry the infection.
**CONTROL** There is no control available, but occasionally mildly affected trees may recover. Generally, however, the infection spreads throughout the tree unless infected branches are removed to a point at least 15cm past the point at which the staining in the wood ends. Infection is least likely to occur during the summer months, so all pruning of susceptible trees should be carried out at this time. Although no longer recommended in most cases, trees which are particularly prone to silver leaf infection should have all pruning wounds painted with a wound treatment. This should be carried out at the same time as the pruning. Locate any possible sources of infection and dispose of them. Plums and cherries are the trees showing greatest susceptibility to this disease, 'Victoria' plum being the most susceptible cultivar. The rootstock may also influence the likelihood of infection occurring: some, such as 'Brompton', are very prone, whereas others, such as 'Pixie', show a good degree of resistance.

## SLIME FLUX AND WETWOOD

**SYMPTOMS** Watery substances, sometimes slightly viscous and orange, pink or white in colour, ooze from the stems of affected plants. The slime flux usually first appears a few inches above soil level and may have an unpleasant odour. The bark beneath may be killed and the upper parts of the stem wilt and die. Not all stems on the plant are affected at the same time.

**CAUSE** Injury to the stem early in the season when the sap pressure is high causes sap to leak out. The high sugar content of the sap means that it rapidly becomes colonised by various yeasts, fungi and bacteria which are responsible for the thickening and discoloration.

**CONTROL** Cut out affected stems, pruning back to perfectly sound wood, even if this means going below soil level. Attempt to prevent stem injury.

## SLIME MOULD

**SYMPTOMS** Grey, off-white, yellow or orange growth develops, most commonly on stems, or on other plant parts. It is usually in the form of numerous tiny spheres which may appear to dissolve when wet, as the spores are released. The plant's growth is not affected and if the slime mould is washed or scraped off, the plant beneath is generally unharmed. Slime moulds are particularly common on grass (*see p.58*).

**CAUSE** Various species of slime mould – non-parasitic organisms, believed to be most closely related to fungi.

**CONTROL** No chemical control is available, but none is justified, as the slime moulds are unlikely to cause much damage. Wash off slime moulds with a strong jet of water.

## SLUGS

Several species of slugs occur in gardens. These include the grey field slug (*Deroceras reticulatus*), the large black slug (*Arion ater*), garden slug (*A. hortensis*) and keeled slugs (*Milax* species). They can damage a wide range of plants and are present throughout the year, continuing to feed in the winter if temperatures are above 5°C (40°F). Irregular holes are eaten in foliage (*see p.29*), flowers (*see p.43*) and stems (*see p.36*), but examination of plants by torchlight on a mild evening may be necessary to find slugs feeding. Slugs secrete a slimy mucilage from their bodies, which may leave a distinctive silvery deposit where they have been feeding. Some slugs live mainly underground and damage potato tubers (*see p.57*) and tulip bulbs.

ACTUAL LENGTH: 20–200MM

SLUG

**CONTROL** Slugs can never be eliminated from gardens, so control measures should be concentrated on protecting vulnerable plants, such as seedlings and soft new growth on herbaceous plants. Poisoned baits in the form of slug pellets containing metaldehyde or methiocarb can be scattered thinly amongst the plants. Poor results will be achieved during cold or dry weather, when slugs are relatively inactive. Metaldehyde causes slugs to produce an excessive amount of slime and die of dehydration, but they may recover under wet conditions. Methiocarb has a more direct toxic action and is better where damp conditions prevail. Slug pellets can kill cats and dogs, so take care in their use. Less toxic to pets are slug killers based on aluminium sulphate. This is mainly active against young slugs and is best applied in mild, damp weather in the spring and autumn. Non-chemical controls include hand-removal after dark on mild evenings, and of beer traps. Jam jars or similar containers half filled with beer are sunk in the ground almost to ground level. Slugs, attracted by the odour, fall in and drown. Biological control in the form of a nematode, *Phasmarhabditis hermaphrodita*, can be used. It needs to be applied to soils which are moist, well-drained and above 5°C (40°F); the best times for use are spring and autumn. The microscopic nematodes penetrate the bodies of slugs when they enter the soil to seek shelter during the day, and release bacteria which cause a fatal disease in infected slugs. Under suitable conditions it gives a significant reduction in slug numbers for at least six weeks. This is a useful method of tackling under-ground slugs which damage bulbs and potato tubers, since they come to the soil surface infrequently and can therefore not be controlled by slug pellets. Damage to potato tubers can be reduced by lifting the crop as soon as the tubers have matured. There is some variation in the susceptibility of potato cultivars to slugs. The cultivars 'Maris Piper', 'Cara', 'Golden Wonder', 'Kondor' and 'Maris Bard' can be heavily attacked, while 'Pentland Ivory', 'Pentland Dell', 'Wilja', 'Stemster', 'Charlotte' and 'Estima' are less susceptible.

## SMALL BULB FLY

*See* NARCISSUS BULB FLY *p.150.*

## SMALL ERMINE MOTHS

**SYMPTOMS** The stems and foliage of various plants, including hawthorn, blackthorn, bird cherry, willow, apple, euonymus and sedum, become covered with dense, whitish, silk webbing in summer (*see p.38*). Whitish-yellow caterpillars with a row of black dots down each side of their bodies, up to 20mm long, feed on leaves. Infestations can be heavy enough to cause severe defoliation. The adult moths are silvery-white with many black spots on their wings; they emerge from pupae in the webbing in late summer.

**CAUSE** Caterpillars of several species of small ermine moths, *Yponomeuta* species. Most have one generation a year but *Y. vigintipunctata*, which attacks sedum, has two.

**CONTROL** Light infestations can be dealt with by pruning out the webbed shoots. Otherwise spray forcefully to penetrate the webbing with permethrin, bifenthrin, pyrethrum or pirimiphos-methyl, or use the bacterial spray *Bacillus thuringiensis*.

## SMUTS

**SYMPTOMS** Affected areas of the plant are discoloured and may be distorted (*see p.35*). In a few cases, such as dahlia smut (*see p.120*), slight leaf spotting is all that is seen. In other plants, swellings develop and erupt to reveal large masses of soot-black spores (*see p.44*). Stems, leaves, flowers and seeds may be affected. Some smut infections (for example, anther smuts) are systemic within the plant, but most are restricted only to discrete areas of the plant.

**CAUSE** Various species of the fungi *Entyloma*, *Ustilago* and *Urocystis*, which are spread by rain or water splash from nearby infected plant matter, or from the soil surface.
**CONTROL** Remove affected areas from plants as soon as seen, preferably before the spores are released. Avoid splashing water around plants and clear up all debris at the end of the season. Grow new plants on a fresh site.
• *see also* DAHLIA SMUT *p.120*; DIANTHUS ANTHER SMUT *p.123*; WINTER ACONITE SMUT *p.191*.

## SNAILS

The most common snail pest is the garden snail, *Helix aspersa*, but banded snails, *Cepaea* species, can also be troublesome. Many plants are eaten by snails between spring and autumn. Irregular holes are rasped in the foliage (*see **p.29***), and the stems and flowers (*see **p.43***), particularly of annuals and herbaceous plants. Snails are mostly active after dark or in wet weather. They secrete a slimy mucilage from their bodies and this often dries to leave a silvery deposit. Snails are less numerous in areas with acid soils, which lack the calcium salts necessary to form a snail's shell.

ACTUAL LENGTH: 20–200MM

**SNAIL**

**CONTROL** Snails can be controlled by the methods used against slugs (*see facing page*). Biological control is less effective, however; snails live mainly above soil level and are less likely to be infected by the bacterium-carrying nematodes.

## SNOWDROP GREY MOULD

**SYMPTOMS** Fuzzy, grey fungal growth, typical of a grey mould type of infection, develops around the foliage and/or flower stems of infected snowdrops. The plant starts to rot and as the infection spreads back, the bulb may be killed. Infection often first appears as the foliage comes above ground. Small black sclerotia (fungal resting bodies) develop around the bulb and foliage. The sclerotia drop off and overwinter in the soil.
**CAUSE** The fungus *Botrytis galanthina*, which is encouraged by mild winters and wet spring weather. The spores are spread by rain or watering splash and on air currents.
**CONTROL** Remove infected plants promptly, together with the soil in their immediate vicinity. Spraying with carbendazim may help to control the disease. Do not grow snowdrops on that site for at least five years.

## SOFT ROT

*See* BACTERIAL SOFT ROT *p.104*.

## SOFT SCALE

**SYMPTOMS** Yellowish-brown, flat, oval scales, up to 3 or 4mm long, occur on the underside of leaves and are often clustered along the larger leaf veins. Bay, citrus, ivy, ferns, *Ficus* species and *Schefflera* are frequently attacked, but this pest can infest a wide range of indoor plants. They excrete a sugary substance called honeydew that makes the upper leaf surface sticky and allows the growth of sooty mould.
**CAUSE** A sap-feeding insect, *Coccus hesperidum*, which breeds continuously throughout the year on house plants and in heated greenhouses.
**CONTROL** Spray the underside of the leaves thoroughly with malathion. Heavily infested plants will need three applications at two-week intervals in order to control nymphs as they hatch from eggs that are concealed beneath the adult scales' bodies. Alternatively, in greenhouses in the summer, a parasitic wasp, *Metaphycus helvolus*, can be introduced for biological control.

## SOLOMON'S SEAL SAWFLY

**SYMPTOMS** Greyish-white caterpillar-like larvae with black heads and up to 20mm long eat the foliage of Solomon's seal in early to mid-summer, often reducing plants to bare stalks (*see **p.31***). The adult insects have black bodies and two pairs of smoky-grey wings. They emerge when the plant is coming into flower, and females lay rows of eggs in the stems, causing purplish scars up to 25mm in length.
**CAUSE** Larvae of a sawfly, *Phymatocera aterrima*.
**CONTROL** In spite of severe leaf loss, plants affected by this sawfly normally make adequate growth and flower the following year, so control may not be necessary. If required the larvae can be removed by hand or sprayed with pyrethrum, derris, permethrin or bifenthrin when damage begins in early summer.

## SOOTY MOULD

Sooty mould is a black or sometimes greyish-green non-parasitic fungus that grows on foliage and other surfaces that have an accumulation of honeydew on them. Various sap-feeding insects, including aphids, whiteflies, mealybugs and some scale insects and suckers, excrete honeydew. This falls down onto the upper surface of leaves below where the pests are feeding. Sooty mould can therefore be found on plants that are themselves free of pests, but which are growing beneath trees that are infested with aphids. Heavy coatings of sooty mould spoil a plant's appearance and reduce the amount of light and air reaching the foliage. There is no direct control for sooty mould, but dealing with the pest that is causing the problem will allow new growth to remain clean.

## SPARROWS

The house sparrow, *Passer domesticus*, pecks flowers of crocus (*see **p.46***), primrose and polyanthus to shreds. Sparrows also eat the flower buds of other plants, including runner bean and wisteria, although loss of flower buds on these plants is more usually due to dry soil conditions than bird damage. Freshly sown grass seed is liable to be eaten and sparrows can cause an uneven soil surface by dust-bathing in seed beds. Seedling lettuce and beetroot plants can be defoliated.
**CONTROL** Netting is the only certain means of protecting flowers, seeds and seedlings. An alternative is to place sticks in the soil and criss-cross the area to be protected with black cotton. Sparrows have difficulty in seeing the cotton and are deterred from feeding. Encourage rapid germination and growth through the

vulnerable seedling stage by sowing in warm soil and keeping the young plants watered in dry weather.

## SPLITTING

**SYMPTOMS** Fruits and occasionally stems may show cracking. This usually runs longitudinally and initially the rest of the plant appears normal. The splitting allows other potential pathogens to enter, and so dieback or rotting may occur later. The cracks may not necessarily become infected and may dry or heal over.
**CAUSE** An erratic supply of water, or great fluctuations in temperature are the most common causes. Erratic availability of nutrients may be involved.
**CONTROL** Concentrate on improving growing conditions by regular watering and feeding, and use a mulch to help prevent soil drying out completely. Keep a watch on damaged areas, and in the case of affected fruits, remove them so they are not infected by secondary organisms.

## SPRINGTAILS

**SYMPTOMS** Active white insects, up to 2mm long, live in the compost of pot plants (*see p.54*), especially those growing in peat-based types. They are most frequently seen when plants are being watered, as this washes them out of the bottom of the pot or flushes them up onto the soil surface.
**CAUSE** There are many species of springtails, but those associated with pot plants are generally *Onychiurus* species. These are wingless insects with a prominent pair of antennae on their heads.

**CONTROL** Springtails feed on decaying plant material and fungi. No control measures are required, as they will not harm the roots of established plants.

## SQUIRRELS

The grey squirrel, *Sciurus carolinensis*, can be delightful to watch as it scampers around a garden, but squirrels can be very destructive. They dig up and eat crocus corms and tulip bulbs, eat flower buds and shoot tips, take nuts (*see p.52*), strawberry and tree fruits, and strip the bark from trees such as sycamore, beech and ash. If bark is lost from all or most of the circumference of the trunk, the growth above that point will dry up and die. Squirrels can also cause annoyance by taking food put out for birds.
**CONTROL** Shooting, traps and poisoned baits are used against squirrels; however, these methods are not only impractical or undesirable in gardens but are also unlikely to be effective unless used over a wider area than an individual garden. Squirrels are very mobile animals and others will soon move in to occupy the territory of those which have been removed. Netting can give protection to fruits and flowering shrubs such as magnolia and camellia during periods when squirrels are showing an interest in them. Wire netting is better used for permanent structures such as fruit cages, as squirrels can easily bite through plastic. Netting can also be placed over areas where bulbs and corms have been planted, to deter squirrels from digging them up. Squirrel-proof bird feeders are available from most garden centres.

## STACHYS CASE-BEARING CATERPILLAR

**SYMPTOMS** In early summer, the foliage of *Stachys* species develops a number of brown blotches where the internal tissues have been eaten out. The small caterpillars responsible for this damage spin silk cases around their bodies. Hairs and other fragments of the host plant's leaves are incorporated into the cases. The caterpillars are almost entirely hidden but, when feeding, the larvae will partly emerge in order to bore into the leaf. The feeding area is limited to how far the caterpillar can reach inside the leaf without leaving its protective case.
**CAUSE** Caterpillars of a tiny moth, *Coleophora lineola*, which feed on various types of *Stachys* and related plants. It is most frequently seen on lambs' ears, *Stachys byzantina* (formerly *S. lanata*), where the caterpillars incorporate leaf hairs into the case to give them a silvery-white, furry covering up to 4mm long and 2–3mm wide.
**CONTROL** The caterpillars can be removed by hand or controlled by spraying with permethrin, bifenthrin, pirimiphos-methyl or malathion when damage is detected in late spring.

## STEM AND BULB EELWORM

This microscopic nematode, *Ditylenchus dipsaci*, occurs as several biological races which have different ranges of host plants. The types that are most likely to be encountered in gardens are narcissus eelworm (*see p.150*), onion eelworm (*see p.153*) and phlox eelworm (*see p.159*). These pests

cause distinctive forms of distortion on the host plants indicated, and often cause their death.

## STORAGE ROTS

**SYMPTOMS** Fruits, roots and tubers in store may start to rot. In some cases the infection may have taken place while the plant was growing, but the storage conditions are such that rotting is encouraged. In other instances, previously perfectly healthy plants or plant parts may actually become infected in store, often from neighbouring diseased material. The rotting may be accompanied by the development of fungal growth on the surface or spreading within the plant.
**CAUSE** A very wide range of fungi and bacteria, some of which are described as secondary, enter only through injured areas; others, which are primary pathogens, cause damage to previously sound and uninjured tissue.
**CONTROL** Store only perfect fruits, roots and tubers. Inspect stores frequently and remove anything showing signs of deterioration, however slight. Provide storage conditions as close as possible to those advised.

## STRAWBERRY BLACK EYE

**SYMPTOMS** Strawberry flowers develop a distinct dark brown or black centre. The flower petals and the growth of the rest of the plant usually appear quite healthy and normal.
**CAUSE** Frost damage to the young flower which injures the centre. Flowers damaged in this way do not later form fruit.

**CONTROL** Use cloches, fleece or similar material to protect crops, particularly early cultivars, from frost. Pinch out affected flowers.

## STRAWBERRY GREEN PETAL

**SYMPTOMS** The flowers are reduced in size and bear green petals. Plants produce few if any fruits and those which are formed are very small and often distorted. The plant itself is somewhat stunted and distorted and may show yellow markings on the leaves. Once flowering is over, the older leaves may turn distinctly red.

**CAUSE** A mycoplasma known as strawberry green petal, which is transmitted by leafhoppers.

**CONTROL** Remove and burn infected plants promptly and do not propagate from them. Spray against leafhoppers.
• *see also* LEAFHOPPERS *p.142*.

## STRAWBERRY GREY MOULD

**SYMPTOMS** Fruits develop normally, but as they start to ripen and colour, pale brown necrotic patches develop on their skin. The fruit then deteriorates rapidly, becoming soft and brown. Shortly afterwards a fuzzy, grey fungal growth develops over the surface (*see **p.50***). Numerous spores are given off if the fruit is disturbed.

**CAUSE** The fungus *Botrytis cinerea*, which is very common on living and dead plant material. The spores enter the open flowers but remain dormant until the fruits start to ripen. The spores are readily spread by rain or water splash and on air currents. Strawberries which are not infected at the flowering stage, and have

formed healthy fruits which have been injured in some way (by slugs, for example), may subsequently become infected by spores from these rotting fruits. The fungus may carry over from year to year on plant debris or as sclerotia in the soil.

**CONTROL** Control is difficult, as the fungus is so widespread. Remove all dead and injured plant parts and fruits before they become infected. Remove infected fruits promptly. Do not leave plant debris lying around the garden. Spray at dusk, at flowering time, with a suitable fungicide such as carbendazim to prevent most of the infection that may occur at flowering.

## STRAWBERRY LEAF SPOT

**SYMPTOMS** Reddish-purple spots develop on the leaves (*see **p.12***), each with a grey centre. Off-white fungal growth and, occasionally, numerous tiny, raised, black fungal fruiting bodies may develop on the spots. The spots may join together and cover much of the leaf surface (*see **p.17***). They may also spread to the flower and leaf stems. Although spoiling the appearance of the foliage, this infection may occur in spring but is usually fairly late in the season and so does little real harm to the vigour of the plant.

**CAUSE** The fungus *Mycosphaerella fragariae*, which forms sclerotia (black fungal resting bodies) which may overwinter in the soil and then cause the earlier infections in the spring.

**CONTROL** Remove badly infected leaves. Clear up all infected debris at the end of the season.

## STRAWBERRY MITE

**SYMPTOMS** Strawberry plants make poor growth in the latter half of the summer, with new leaves appearing puckered and failing to expand to their full size. Similar symptoms can be caused by some virus diseases. The strawberry mites, which are whitish-brown and less than 1mm long, live in the crown of the plant and feed on the developing leaves. A x10 hand lens is necessary to see them.

**CAUSE** A tarsonemid mite, *Phytonemus pallidus*, which feeds by sucking sap. It tends to be more troublesome in hot, dry summers.

**CONTROL** None of the pesticides available to amateur gardeners will control strawberry mite. Destroy affected plants and start a new bed with certified stock strawberry plants which will be free of problems such as strawberry mite and virus diseases.

## STRAWBERRY RED CORE

**SYMPTOMS** Groups of stunted plants appear in late spring, bearing reddish-brown, strangely stiff leaves. The roots are discoloured and have a red inner core.

**CAUSE** The fungus *Phytophthora fragariae*, which is encouraged by heavy, wet soils and produces resistant spore stages from the deteriorating root material, releasing these into the soil where they may persist for more than ten years. The fungus is easily introduced on contaminated plants or on soil adhering to plants, footwear, tools etc.

**CONTROL** None available.

Remove infected plants immediately, together with the soil in the vicinity of their roots, and burn them. Do not grow strawberries on that site again and avoid moving soil from that area to other parts of the garden.

## STRAWBERRY SEED BEETLES

**SYMPTOMS** Seeds are removed from the surface of ripening strawberry fruits, causing small, dried-up brown patches. The culprits are fast-moving black beetles, about 15mm long, with slender reddish-brown or black legs. They are mainly active after dark. In addition to eating the seeds, they sometimes eat into the fruits, but it is difficult to distinguish this form of damage from that caused by slugs.

**CAUSE** Carabid or ground beetles, *Pterostichus* species and *Harpalus rufipes*. At other times of the year these beetles will feed on weed seeds.

ACTUAL LENGTH: 15MM

**STRAWBERRY SEED BEETLE**

**CONTROL** Keep the strawberry bed free of weeds to discourage a build-up of seed beetles. Jam jars sunk into the soil will act as pit-fall traps, but they will also capture beneficial ground beetles that are predatory in their feeding habits. Because ripe strawberries have such thin skins, chemical controls are undesirable while ripe fruit is present but, if necessary, plants with unripe fruits can be sprayed with bifenthrin at dusk.

## STRAWBERRY VIRUSES

**SYMPTOMS** Stunting and distortion of the whole plant, usually combined with stunting, distortion and yellowing of the leaves. The discoloration is seen as distinct patterns – mosaic, ring spot, streaking or yellowing around the leaf edge. Infected plants fail to thrive and may not flower or fruit properly.
**CAUSE** Various viruses, most commonly strawberry yellow edge or little leaf virus, crinkle (spread by aphids), arabis mosaic, strawberry ring spot and tomato black ring (spread by eelworms).
**CONTROL** None available. Remove and burn infected plants as soon as the symptoms are noticed. Control aphids and leaf-hoppers with appropriate insecticides. Do not grow strawberries on that site again. Do not propagate from infected plants, as their runners will also be infected, even if without symptoms. Always buy certified stocks of strawberries, guaranteed to be virus-free when purchased. Certain cultivars are known to be particularly susceptible to particular viral infections and these include 'Royal Sovereign' and 'Redgauntlet' (yellow edge).
• see also APHIDS p.99; LEAFHOPPERS p.142.

## STRAWBERRY YELLOWS

**SYMPTOMS** Leaves turn yellow. The entire plant may be affected or it may appear blotched yellow with part of the leaf remaining green. Symptoms usually appear most severe in mid-summer.
**CAUSE** The precise cause is not known, but it is thought to be largely the result of unsuitable soil conditions. A soil which is excessively wet or which is lacking adequate nutrients will generally encourage these symptoms.
**CONTROL** Determine which adverse soil conditions are involved and attempt to improve problems such as poor drainage. Apply a fertilizer containing trace elements.

## SUCKERS

These are small sap-feeding insects that cause damage on various garden plants during the spring and/or summer months. They are also known as psyllids and have very distinctive immature nymphs that appear squashed flat when viewed from the side. The adult insects have a more rounded body shape and resemble winged aphids. They are 2–3mm long when adult. An immature nymph has wing pads on its upper surface, which become larger each time the nymph sheds its skin, until the adult stage is reached with fully formed wings. Some suckers cause gall formation on their host plants while others live exposed on the leaf surface.
• see also ALDER SUCKER p.98; APPLE SUCKER p.102; ASH SUCKER p.103; BAY SUCKER p.104; BOX SUCKER p.109; EUCALYPTUS SUCKER p.126; PEAR SUCKER p.157.

## SWEETCORN SMUT

**SYMPTOMS** Individual kernels on the cob become hugely swollen and deformed (see **p.51**). Each is pale grey and ruptures to release large quantities of powdery black spores. These may become mixed with rain to produce a black liquid which runs down the plant. Occasionally, other parts of the plant may be affected. The infection is not systemic and healthy and infected cobs may develop on the same plant.
**CAUSE** The fungus *Ustilago maydis*, which is quite common if summer temperatures are high. The spores are carried on air currents and by water splash, and either infect other sweetcorn plants directly or persist in the soil.
**CONTROL** None available. Remove infected cobs before the swellings rupture. Remove and burn all infected plant debris at the end of the season. Do not grow sweetcorn on that site for at least five years.

## SWIFT MOTHS

**SYMPTOMS** Long, slender, white caterpillars, up to 35mm long, with brown heads are found in the soil (see **p.53**). They feed on plant roots, especially herbaceous perennials, and can also damage bulbs, corms and rhizomes. Heavy infestations cause plants to lack vigour; they are some-times killed. Some species take two years to complete their larval development, so the caterpillars may be unearthed at any time of year. The adult moths emerge and lay eggs from late spring to late summer.
**CAUSE** There are several species of swift moth, but the most troublesome in gardens are the ghost swift moth, *Hepialus humuli*, and the garden swift moth, *H. lupulinus*.
**CONTROL** Good weed control can help avoid infestations, as the moths often deposit eggs where there is a dense cover of vegetation. Any caterpillars found during digging or division of perennials should be crushed and thrown to the robins. Chemical control is difficult, as the older and more damaging larvae are more tolerant of pesticides. Insecticides such as lindane or pirimiphos-methyl can be added to the soil if valuable plants are under attack.

## SYCAMORE SOOTY BARK DISEASE

**SYMPTOMS** Leaves wilt and die back on infected branches, but remain attached to the tree. If the bark is pared off, green or dark yellow staining may be visible (see **p.36**). Although these are the first symptoms, the problem may only become apparent at a later stage, when black spore masses or blister-like outgrowths develop on the bark of affected stems and when the tree does not come into leaf in the spring. These enlarge and elongate and may be associated with the extensive death and breaking up of the bark. Death of the tree usually occurs within a couple of years.
**CAUSE** The fungus *Cryptostroma corticale*, which appears to be encouraged by warm temperatures and, in particular, by hot summers.
**CONTROL** Remove infected trees promptly in order to limit spread of the disease.

## TAR SPOT OF ACER

**SYMPTOMS** Large, slightly raised, shiny black blotches develop on the leaves (see **p.13**); each can measure up to 15mm in diameter.

**CAUSE** The fungus *Rhytisma*, in particular *R. acerinum*, which can overwinter on fallen infected leaves. In spring it produces spores in a gelatinous coating which adhere to young leaves as they are produced. Leaves may fall prematurely, but this is not always the case and the vigour of the tree is rarely significantly affected.

**CONTROL** None available, but it is not necessary. Rake up and dispose of affected leaves as they fall.

## TARSONEMID MITES

**SYMPTOMS** Mainly greenhouse and house plants are attacked by minute, whitish-brown mites that infest the developing leaves at the shoot tips and the flower buds. Their feeding activities cause distorted flowers with white or brown flecks on the petals. The new foliage fails to expand to its full size and becomes progressively more stunted and distorted as the infestation develops (*see* **p.27**). In severe attacks, growth comes to a complete halt and the stems have a brownish scarring. Tarsonemid mites can be a particular problem on cyclamens, begonias, pelargoniums, busy Lizzies and many other pot plants.

**CAUSE** Microscopic mites that feed by sucking sap. These include broad mite, *Polyphagotarsonemus latus*, and cyclamen mite, *Phytonemus pallidus*. The latter also occurs as races that attack some garden plants – *see* MICHAELMAS DAISY MITE *p.147* and STRAWBERRY MITE *p.179*.

**CONTROL** None of the pesticides available to amateur gardeners will control tarsonemid mites. Infested plants should be

destroyed. This pest usually comes into a greenhouse when an infested plant is acquired. Care should be taken to avoid this by purchasing plants from reputable suppliers.

## THRIPS

Thrips are narrow-bodied, elongate insects up to 2mm long. They feed by sucking sap and generally cause a fine silvery-white mottled discoloration of the upper leaf surface. Thrips will readily feed from the upper leaf surface, unlike most sap-feeding pests that remain on the underside of leaves. Adult thrips are usually black with two pairs of wings that are heavily fringed with hairs. The wings are folded back along the body when not in use and so are not readily visible. Some thrips have black and white areas on their wings, which give the insects a banded appearance. The immature nymphs are wingless and creamy-yellow in colour. Some species, like western flower thrips, can spread certain plant virus diseases on their mouthparts. To control specific thrips commonly found in gardens and greenhouses, *see* BANDED PALM THRIPS *p.104*; GLADIOLUS THRIPS *p.132*; ONION THRIPS *p.153*; PEA THRIPS *p.155*; PRIVET THRIPS *p.165*; WESTERN FLOWER THRIPS *p.189*. Others may be controlled by spraying thoroughly with pyrethrum, malathion, permethrin, pirimiphos-methyl or dimethoate when signs of activity are seen.

## THUNDERWORMS

Thunderworms derive their common name from the fact that they are often seen after

heavy rain, particularly between late spring and mid-summer. Under damp conditions the nematodes, *Mermis* species, emerge from the soil and climb onto plants where they can be seen writhing about. They are less than 1mm in diameter, but can be up to 60mm long. They are creamy-white, although blackish body contents can often be seen through the semi-transparent skin. Eggs are deposited on foliage and if the eggs and leaf are subsequently eaten by a suitable animal, such as a caterpillar, grasshopper or earwig, the eggs will hatch and develop as parasites in the insect's gut. Eventually the fully-formed nematode emerges and goes into the soil. Thunderworms can only develop inside insects and so pose no threat to humans or pets.

## TOADSTOOLS

*See* TURF TOADSTOOLS *p.184*.

## TOMATO BLIGHT

**SYMPTOMS** Brown patches develop on the leaves, causing them to dry and curl. The stems may also show blackening. Infected fruits develop a brown discoloration and may appear to shrink inwards. They may rot, either on the plant or a few days after being picked. Outdoor tomatoes are affected more frequently than those grown under glass. Potatoes may also be attacked (*see* POTATO BLIGHT *p.163*). Tomatoes usually succumb slightly later than potatoes.

**CAUSE** The fungus *Phytophthora infestans*, which is encouraged by moist or wet weather,

particularly if it occurs in late summer. The spores are carried on air currents or on water splash, and need a minute film of moisture on the foliage if they are to cause damage.

**CONTROL** Preventive spraying is worthwhile, particularly if the tomatoes are growing in an area where blight is common and if there are other potato or tomato crops in the vicinity. In a wet season this should be carried out as soon as the first tomatoes have set and then repeated according to the manufacturer's instructions. A copper-based fungicide or one based on mancozeb should be used.

## TOMATO BLOTCHY RIPENING

**SYMPTOMS** Patches of hard, yellow or green flesh remain unripened (*see* **p.50**). These areas may be anywhere on the fruit and, unlike tomato greenback (*see p.182*), this disorder does not follow a regular pattern. The patches are not clearly demarcated. The damage is only visible on mature fruits, and those on the lower trusses are generally the most frequently affected.

**CAUSE** This disorder is believed to be caused by some form of malnutrition, and a deficiency of potassium is probably the single most important factor. Dry soil or compost or a poorly functioning root system may also play a part. High temperatures in the greenhouse also increase the damage.

**CONTROL** Keep the plants well fed, in particular ensuring that a high potash feed is used. Keep the greenhouse well ventilated and the plants well watered.

## TOMATO CYST EELWORM

See POTATO CYST EELWORM p.163.

## TOMATO FRUIT SPLITTING

**SYMPTOMS** Tomatoes develop normally but start to split, usually just before picking (*see* **p.50**). The splits may become infected by secondary organisms such as grey mould (*see p.134*), and may then rot. Occasionally the split surface temporarily dries up and the fruit does not rot.
**CAUSE** Erratic growth and fruit swelling. This is usually caused by an irregular supply of water or wildly fluctuating temperatures.
**CONTROL** Ensure that plants are adequately watered at all times and are not subjected to temperature extremes. In greenhouses, provide shading and ventilation. Remove affected fruits before secondary organisms invade.

## TOMATO GHOST SPOT

**SYMPTOMS** On unripe fruits, pale green or yellow rings of discoloration appear. As the fruits develop and colour, the rings turn yellow or pale orange. The fruit's development is not affected and it is perfectly edible.
**CAUSE** The fungus *Botrytis cinerea*. Spores are readily spread by rain or water splash and on air currents. The fungus is widespread on living and dead plant matter. The rings of discoloration result from a hypersensitive reaction to the fungal spores.
**CONTROL** As affected fruits rarely deteriorate, there is no need to remove them, but eliminate sources of infection in the area.

## TOMATO GREENBACK

**SYMPTOMS** As greenhouse tomatoes begin to ripen, a partial or complete ring of clearly demarcated leathery tissue remains around the stalk end. It does not ripen properly, remaining either green or yellow (*see* **p.50**). The rest of the plant develops normally.
**CAUSE** A deficiency which is under genetic control, and is brought about by several different factors either alone or, more usually, in combination. Low levels of potassium and phosphorus can be involved but hot, bright sunlight which causes heat injury is probably the most important single factor.
**CONTROL** Provide adequate shading for the plants, particularly once the fruits start to develop. Do not remove too much of the plant's foliage, as this acts as a good shading material. Ventilate the greenhouse to prevent excessively high temperature levels. Keep the plants well fed, ensuring adequate supplies of potash and phosphate in particular. Choose tomatoes which show resistance to this disorder. The following are suitable, but in some cases it appears that the resistance may be breaking down, and even these may show some symptoms in very hot weather: 'Cherry Wonder', 'Craigella', 'Estrella', 'Euro-cross BB', 'Golden Sunrise', 'Grenadier', 'Matador', 'Moneymaker', 'Shirley', 'Tigercross', 'Tigrella', 'Totem' and most modern F1 hybrids.

## TOMATO LEAF MOULD

**SYMPTOMS** Yellow blotches develop on the upper leaf surface. Beneath each spot is a grey or faintly purple velvety fungal growth. The lower leaves are usually attacked first, but all leaves may eventually be affected. The leaves turn brown, wither and dry but remain hanging on the plant. A secondary infection of grey mould may mask the symptoms of leaf mould. Greenhouse tomatoes are frequently affected, those grown out of doors rarely developing this disease.
**CAUSE** The fungus *Fulvia fulva*, which spreads rapidly in warm, humid conditions. The spores are carried on air currents and may also be spread by insects, and on clothing. The fungus overwinters on infected plant debris and on greenhouse structures.
**CONTROL** Remove and burn all infected plants at the end of the season and thoroughly clean out the greenhouse. Spray infected plants with a suitable fungicide, such as mancozeb or carbendazim. Ensure adequate ventilation and try to minimize excessively humid conditions by spacing plants out well and by careful removal of some of the foliage. Grow tomatoes which show resistance to this disease, such as 'Cyclon', 'Dombello', 'Estrella', 'Eurocross BB', 'Grenadier', 'Seville Cross', 'Shirley' and 'Sonatine'.

## TOMATO MOTH

**SYMPTOMS** The foliage and fruits of tomatoes are eaten by caterpillars. Newly-hatched larvae feed together by grazing away the lower leaf surface, causing the remaining upper surface to become papery-white. In greenhouses, such damage can often be seen first on leaves which are growing close to the glass. As they grow larger, the caterpillars disperse and eat holes right through the foliage, and bore into green and ripening fruits. When fully grown, the caterpillar is up to 35mm long and either brown or pale green in colour, with a yellow line – marked by many tiny white dots – along each side of the body.
**CAUSE** Caterpillars of a moth, *Lacanobia oleracea*, which is also sometimes known as the bright-line brown-eye moth.

ACTUAL LENGTH: 35MM

**CATERPILLAR OF A TOMATO MOTH**
**CONTROL** Look for leaves bearing newly-hatched caterpillars and remove them. Otherwise spray with pyrethrum, permethrin or bifenthrin, or use biological control with *Bacillus thuringiensis* when caterpillars are seen.

## TOMATO PITH NECROSIS

**SYMPTOMS** The upper leaves wilt and turn yellow as the fruits start to ripen. Dry, dark brown to black patches develop on the stems. If cut longitudinally, the internal tissues of the stem can be seen to be discoloured, usually black in the pith cavity or in the pith itself.
**CAUSE** The bacterium *Pseudomonas corrugata*, probably spread by water splash and by handling, and possibly on any tools used when removing side shoots.
**CONTROL** Remove and burn infected plants promptly to reduce the risk of the infection spreading.

## TOMATO VIRUSES

**SYMPTOMS** Several different viruses may affect tomatoes and the symptoms they produce vary, not only with the virus involved, but also depending on the tomato cultivar and the prevalent growing conditions. Common leaf symptoms include yellow mottling or mosaicing, distortion and the production of unusually small leaves. With tomato mosaic, in addition to these symptoms, the youngest leaves may curl downwards. With fern leaf, the leaf blades are extensively narrowed and malformed and may be mistakenly assumed to be damaged by weedkiller. Other symptoms of virus infection may include yellow streaking, stunting and the bronzing of the young fruits, yellow streaking of the fruits, or failure of the plant to set fruit.

**CAUSE** The viruses most commonly involved, either alone or in combination, are tomato spotted wilt, streak, tomato mosaic, potato virus X and cucumber mosaic. All these viruses are readily spread by handling and in some cases by aphids or other sap-feeding pests. Many of these viruses have very wide host ranges, so they may be introduced from unrelated plants in the garden, or may be spread to these from infected tomatoes.

**CONTROL** Remove and dispose of infected plants promptly. Avoid handling plants unnecessarily and always wash hands thoroughly after handling. Control aphids and other sap-feeding pests. Remove and burn all crop debris at the end of the season. Grow tomato cultivars described as showing resistance. At present the following are listed as showing resistance to tomato mosaic virus: 'Cumulus', 'Cyclon', 'Dombello', 'Estrella' and 'Shirley'.

## TORTRIX MOTHS

*See* CARNATION TORTRIX MOTH, *p.114.*

## TULIP FIRE

**SYMPTOMS** As the foliage emerges above ground it may be withered, distorted and covered in numerous buff-coloured flecks (*see **p.22***), or scorched. It soon becomes covered in fuzzy, grey fungal growth. Elliptical grey-brown marks develop on the stalks. The plants may fail to mature, though some may flower, or flower buds may develop but fail to open. Those which open normally may develop numerous small, bleached spots on the petals, each one being caused by the plant's hypersensitive reaction to the spores. The flower stems may topple over and in wet weather the whole plant may become covered in the fungal growth and then rot rapidly. Bulbs may rot or may appear firm but bear black sclerotia (fungal resting bodies) clustered around the neck (*see **p.55***).

**CAUSE** The fungus *Botrytis tulipae*, which overwinters as pinhead-sized sclerotia in the soil or on infected bulbs. The spores produced on all the above-ground parts of the plant are readily spread on air currents and by water splash.

**CONTROL** Remove infected foliage promptly. Do not plant tulips on the same ground for more than two years in succession, as this may encourage build-up of the sclerotia in the soil. Remove and dispose of severely infected plants, together with the soil in the immediate vicinity of the roots. Check bulbs carefully before planting and only use those which are apparently healthy and free from signs of sclerotia. Late planting may also decrease the chances of the disease developing. On sites where the disease has been identified, do not replant with tulips for at least three years. At the end of the season lift tulip bulbs and dust them with sulphur, or dip them in a solution of carbendazim according to the manufacturer's instructions.

## TULIP GREY BULB ROT

**SYMPTOMS** Bulbs fail to emerge above ground, leaving gaps in the bed. If shoots are produced, they are extensively distorted and soon wither and die off. There are no signs of fungal growth on the affected foliage. Infected bulbs turn grey and dry as they start to rot. The roots and the basal plate may be all that remain. Soil tends to adhere to infected bulbs, especially at the noses. A dense, white or grey felty fungal growth develops between the scales and on the outside of the bulb. Black sclerotia (fungal resting bodies) develop in amongst the fungal growth and may reach up to 7mm in size, remaining viable in the soil for up to five years.

**CAUSE** The fungus *Rhizoctonia tuliparum*, which may attack many bulbs, including alliums, amaryllis, brodiaeas, chionodoxas, colchicums, crocus, eranthis, fritillaries, gladioli, hyacinths, iris, ixias, lilies, daffodils and narcissi, snowdrops, scillas and tulips.

**CONTROL** Remove and burn affected plants, together with the soil around them. Do not grow susceptible bulbs on that soil for at least five years.

## TURF DOLLAR SPOT

**SYMPTOMS** Grass dies off in patches, becoming straw-yellow (*see **p.58***). Patches usually measure between 7.5–10cm in diameter.

**CAUSE** The fungus *Sclerotinia homeocarpa*, which is most prevalent during warm, damp spells in early autumn, and on lawns which have become compacted.

**CONTROL** Improve drainage by spiking the lawn in autumn. Treat infected areas with a suitable fungicide, such as carbendazim. Avoid using grass seed mixtures containing a high proportion of *Festuca* species (the fescues) as these are particularly prone to this infection.

## TURF FAIRY RINGS

**SYMPTOMS** Rings develop on the turf, increasing in size as the fungus spreads. Around the periphery of the ring one or more zones of very lush, green grass develop. There is also an inner ring of similarly luxuriant grass. The area between these two rings is dead and brown, or the soil may be completely bare. Occasionally, incomplete rings or other formations develop. The fungus grows just beneath the soil surface and around the bases of the grasses, forming a dense, off-white fungal growth or mycelium. Pale brown, slender toadstools, each 4–10cm tall, may develop

at the outer part of the middle zone in summer or autumn (see **p.59**).
**CAUSE** The fungus *Marasmius oreades*. The death of the middle zone of grass is caused by drought – the dense fungal mycelium is so water-repellent that water cannot penetrate to the grass roots in this area.
**CONTROL** There is no chemical control available on the gardening market. Rake up and dispose of all toadstools, preferably before their caps open to release spores. Mow the infected area of grass separately and collect up and burn mowings from this area in order to limit the risk of the fungus spreading. Dig out the affected grass and topsoil to a depth of at least 30cm. In addition, dig out the grass in the area, extending about 30cm beyond the edges of the ring to the same depth. Dispose of all affected grass and soil carefully, taking care not to contaminate other areas of lawn or grass. Fill the trench with topsoil and re-seed.

## TURF OPHIOBOLUS PATCH

**SYMPTOMS** Bleached patches develop and spread. Fungal growth may be visible in among dead, brown roots and stolons of the grasses.
**CAUSE** The fungus *Gaeumanomyces graminis*, which is very damaging on lawns composed largely of fine grasses. It is also very prevalent on chalky or limey soils or on lawns which have been limed.
**CONTROL** Remove affected areas and several inches of soil beneath the remains of the grasses. Use fresh topsoil to fill in, and then re-seed. Avoid liming the soil.

## TURF RED THREAD

**SYMPTOMS** Patches of reddish or bleached grass develop, 7.5–8cm in diameter. In amongst the grasses and attached to them, dark pink, horn-like, gelatinous, branching strands of fungus develop (see **p.59**). These then turn pale pink and slightly fluffy. Grass is rarely killed, but it is weakened and its appearance spoiled.
**CAUSE** The fungus *Corticium fuciforme*, which is most common on lawns composed mainly of fine lawn grasses such as the fescues. Perennial rye grass and annual meadow grass are rarely affected. Inadequate nitrogen levels and poor aeration encourage the disease. It is usually most prevalent in the year following a drought, because the dry weather will have made the usual fertilizing routine very difficult.
**CONTROL** Improve drainage by aeration and ensure the grass is adequately fertilized. Apply nitrogen as sulphate of ammonia and this may cure the problem. If a fungicide application proves necessary, carbendazim can be used, but additional nitrogen must also be applied or else the problem invariably recurs.

## TURF SNOW MOULD

**SYMPTOMS** Small patches of dying grass develop and gradually turn brown, increase in size and even join together. In damp weather the dying patches may become covered in a pale pink, fluffy fungal growth (see **p.59**). Snow mould is most commonly seen in the autumn and during mild spells in winter.
**CAUSE** The fungus *Monographella nivalis* and

occasionally other fungi. Snow mould is often encouraged by poor aeration and by damp conditions. If the grass is walked on in winter after a fall of snow, the disease frequently develops on these areas.
**CONTROL** Carry out regular routine spiking, and use any other methods for improving aeration. Avoid walking on the grass during wet or cold weather. Treat the lawn with a suitable fungicide such as carbendazim.

## TURF THATCH FUNGAL MYCELIUM

**SYMPTOMS** Bleached, yellowed or occasionally reddish patches of grass develop on the lawn: the grasses are severely stressed or may be killed (see **p.58**). A dense mat of fungal growth grows around the bases of the grasses.
**CAUSE** Various fungi, often described as thatch fungi, which are not directly pathogenic but live on organic debris, including thatch (remains of old grasses, mowings and other debris) which is found at the base of the grasses. These fungi form a strongly water-repellent layer, which causes the grass to die from drought stress. Lawns which have not been properly maintained, and newly-laid turf, are the most susceptible.
**CONTROL** Carry out all necessary lawn maintenance: scarification, spiking and weed and moss control. Do not allow mowings to be redistributed over the lawn. Buy turf only from a reliable, reputable supplier and check it carefully on delivery. Where necessary, the only solution may be the removal of badly infested areas followed by re-seeding or re-turfing.

## TURF TOADSTOOLS

**SYMPTOMS** The grass grows normally, but toadstools develop in random arrangements on the lawn. Toadstools may also develop in straight lines, following the path of underground tree roots or other buried woody debris.
**CAUSE** Various fungi – including ink caps, field mushrooms and puff balls – which do not attack the grasses but feed on organic debris in the soil.
**CONTROL** Brush off toadstools as soon as they appear, preferably before they release their spores. If the problem is recurrent, the only long-term solution may be to excavate beneath the grass and remove the tree root or other material on which the fungus is feeding.

## TURNIP GALL WEEVIL

**SYMPTOMS** The adult beetles lay eggs on the roots of turnip, swede and leafy brassicas such as cabbage, cauliflower, Brussels sprouts and broccoli. The feeding activities of the white, legless grubs cause the plant to produce rounded swellings that enclose the grubs. This can be confused with roots swollen by infection with clubroot disease (see *p.118*). They can be distinguished by cutting the swellings in half. Turnip weevil galls are hollow and contain grubs, or have circular exit holes if the grubs have left to pupate in the soil, whereas clubroot galls are solid and less regularly rounded.
**CAUSE** Larvae of a beetle, *Ceutorhynchus pleurostigma*. Infestations on leafy brassica roots have little impact on their growth, but this pest is

more troublesome when it is developing on edible roots such as turnip and swede.

ACTUAL LENGTH: UP TO 4MM

**TURNIP GALL WEEVIL LARVA**

**CONTROL** Where turnip gall weevil has been a problem, apply lindane or pirimiphos-methyl dust to the seed row, or planting hole when seedlings are transplanted.

## TWO-SPOTTED MITE

*See* GLASSHOUSE RED SPIDER MITE *p.132*.

## VAPOURER MOTH

**SYMPTOMS** Hairy caterpillars, up to 30mm long, occur during late spring to late summer on a wide range of trees and shrubs (*see p.31*). They are greyish-black with numerous orange-red spots on the body; there are four clumps of pale yellow hairs, resembling shaving brushes, on the upper body surface and long tufts of black hairs pointing forward from the head end, and backwards from the rear end. The caterpillars generally occur in small numbers, but are sometimes abundant enough to cause significant damage.
**CAUSE** Caterpillars of a moth, *Orgyia antiqua*, emerging in mid- to late summer. The female is wingless and after mating she lays her eggs on the silk cocoon from which she has emerged. The eggs are the overwintering stage.
**CONTROL** Hand picking will

deal with light infestations. Otherwise use pyrethrum, permethrin, bifenthrin, or use *Bacillus thuringiensis* for biological control.

## VERTICILLIUM WILTS

**SYMPTOMS** Plants show wilting of the foliage (*see p.29*). Most or all of the leaves on an affected branch or stem show the symptoms, but the whole plant is unlikely to be affected simultaneously. The leaves may show interveinal discoloration (usually as yellowing or browning between the veins) and they then die. Stem death follows shortly, but it may be several years before a large shrub is killed completely. Smaller shrubs or herbaceous perennials and annuals may be killed within a single season. If the bark is removed from an affected stem, the vascular tissue (conducting tissue) is seen to be discoloured, usually by brown or purple-brown streaks which run the length of the stem (*see p.33*), but are usually most apparent towards the base. Roots have a central core of discoloured conducting tissue.
**CAUSE** *Verticillium dahliae* and *V. albo-atrum*. These are both found in infested soil, on plant debris and within the plant sap. *V. dahliae* forms fungal resting bodies which may persist in the soil. The host ranges are wide, and as well as many cultivated plants, several common garden weeds may harbour the infection too.
**CONTROL** None available. Remove affected plants promptly, together with the soil in the immediate vicinity of their roots. As the infection can be spread on

pruning tools, always thoroughly clean any tools after they have been used on an infected plant. Avoid replanting any of the following on the same site: maples, Indian bean tree (*Catalpa bignonioides*), Judas tree, smoke bush (*Cotinus*), golden rain tree (*Koelreuteria paniculata*), stag's horn sumach, lime, berberis, daphnes, quince, apples, pears or *Prunus*.

## VIBURNUM BEETLE

**SYMPTOMS** The foliage of viburnums, particularly *Viburnum tinus*, *V. opulus* and *V. lantana*, is reduced to a lacework of damaged leaves during the summer. Most of the damage is caused by creamy-yellow, black-spotted beetle larvae, up to 7mm long, during late spring to early summer. The greyish-brown adult beetles, 5–6mm long, emerge from pupae in the soil in late summer and cause additional, but less extensive, damage.
**CAUSE** Adults and larvae of a beetle, *Pyrrhalta viburni*. This pest overwinters on its host plant as eggs, which are placed in small pits in the bark of the stems.
**CONTROL** Look for signs of damage on the new foliage in late spring, and spray with permethrin, bifenthrin, pirimiphos-methyl, fenitrothion or pyrethrum while the larvae are still small and at their most damaging.

## VIBURNUM SCALE

**SYMPTOMS** The upper leaf surface on *Viburnum tinus* and ivies (*Hedera* species) becomes sticky with honeydew excreted by this pest. Black sooty moulds often develop on the

honeydew. The scales live on the underside of leaves, usually next to a leaf vein. They are pale yellow, flat, oval creatures up to 4mm long. When fully mature in late spring, the females are pale brown and they cover themselves with a fluffy, white waxy material in which they deposit their eggs.
**CAUSE** A sap-feeding insect, *Lichtensia viburni*. It is more troublesome on plants growing against a wall or in other sheltered places.
**CONTROL** Spray thoroughly with malathion, especially on the underside of the leaves, in mid-summer to control the more vulnerable newly-hatched scale nymphs.

## VIBURNUM WHITEFLY

**SYMPTOMS** The only viburnum affected is laurustinus (*Viburnum tinus*), but this pest also attacks strawberry trees (*Arbutus*). In mid-summer, white-winged insects, 1–2mm long, occur on the underside of the younger leaves. The immature nymphs are flat, oval, yellowish-green, scale-like objects found on the lower leaf surface in late summer. More easily seen are the overwintering pupal stages; these are black, oval objects 1mm long, which are encrusted with a white, waxy substance (*see p.15*). Heavy infestations may cause soiling of the foliage with honeydew and sooty moulds.
**CAUSE** Adults and nymphs of a sap-feeding insect, *Aleurotrachelus jelinekii*.
**CONTROL** Light infestations can be ignored, since the pest has little direct impact on its host plant's growth. If necessary, spray the underside of the leaves with permethrin, bifenthrin,

pirimiphos-methyl, pyrethrum or an insecticidal soap, when adult whitefly are present.

## VINE ERINOSE MITE

**SYMPTOMS** During mid- to late spring, parts of grape vine leaves bulge upwards (*see p.15*); the underside of these deformed areas is covered in a dense mat of fine, whitish hairs. As the summer progresses, the hairs darken and become yellowish or brown. The upper surface of the blistered areas may also become discoloured.
**CAUSE** Microscopic gall mites, *Colomerus vitis*, which suck sap from the foliage and secrete chemicals into the leaf to induce the excessive growth of hairs. Apart from the leaf distortion, the mites seem to have little impact on the plant's vigour and ability to produce grapes.
**CONTROL** No effective chemicals are available to amateur gardeners. Removing affected leaves may check infestations, but since the mites are of little consequence, heavy defoliation would be more harmful than tolerating their presence.

## VINE WEEVIL

**SYMPTOMS** Both the adult and larval stages cause damage. The larvae are plump, creamy-white, legless grubs with brown heads. They are up to 10mm long and have slightly curved bodies. They live in the soil (*see p.53*) and feed on roots, which are either severed or, on woody plants, have the outer bark removed. This results in plants making slow growth, followed by wilting and death when insufficient roots remain to sustain the plant. The grubs also bore into the

tubers and corms of plants such as begonia and cyclamen. A wide range of garden and indoor plants is attacked and plants grown in pots or other containers are particularly at risk. Larval damage is mostly seen between early autumn and mid-spring. The adult weevils are 9mm long, with dull black, pear-shaped bodies and antennae that are bent at an angle about half-way along their length. They are slow-moving beetles that emerge at night and eat irregular notches from the leaf margins of many plants (*see p.29*), but especially rhododendrons, *Euonymus japonica*, and hydrangea. The adults can be found at any time from spring to autumn.
**CAUSE** Larvae and adults of a beetle, *Otiorhynchus sulcatus*. All the adults are female and capable of laying many hundreds of eggs over a period of several months.

ACTUAL LENGTH: 9MM

**ADULT VINE WEEVIL**

**CONTROL** The insecticides available to amateur gardeners give disappointing results against vine weevil. They are not powerful enough to kill the older grubs and have insufficient persistence to give any lasting protection. Biological control with one of the pathogenic nematodes, *Heterorhabditis megidis* or *Steinernema carpocapsae*, is the best means of protecting plants. Nematodes should be watered into the potting compost in late summer, while the soil is warm and moist, but before the grubs

have grown large enough to cause serious damage. Nematodes are ineffective in heavy soils, or those which are dry or colder than 14°C (57°F). Adult weevils can be searched for by torchlight and removed from plants. The leaf damage caused by the adults is rarely sufficient to affect plant growth, but if it is causing concern, plants can be sprayed at dusk on mild evenings with pirimiphos-methyl or bifenthrin, or lindane dust can be used.

## VIOLET GALL MIDGE

**SYMPTOMS** Leaves become greatly thickened and fail to unfurl (*see p.27*). If these are broken open, orange-white maggots or white silk cocoons, 2mm long, will be found inside the galls. Galled leaves are present throughout the year, but are often found in the winter after normal foliage has died down.
**CAUSE** A small fly, *Dasineura affinis*, which has three or four generations during the summer when females deposit their eggs on the developing leaves.
**CONTROL** Heavy infestations may kill plants, but usually plants survive despite being galled. Affected leaves can be removed by hand, but it is difficult to prevent reinfestation. If necessary, spray the plants with dimethoate or pirimiphos-methyl in late spring and early summer when the current year's infestation begins.

## VIOLET ROOT ROT

**SYMPTOMS** Plants are stunted and chlorotic, but the most characteristic symptoms are

seen on the affected roots, tubers and rhizomes. Their surface is covered in a dense network of dark purple fungal strands and often have considerable quantities of soil adhering to them (*see p.57*). Closely packed areas of fungal threads form into sclerotia which may fall into the soil as the tissue beneath starts to rot and turns brown, but this may be largely the effect of secondary organisms.
**CAUSE** The fungus *Helicobasidion purpureum*, which is most prevalent in wet, acid soils and persists in the soil as sclerotia. It may also remain viable on plant tissue for a considerable time.
**CONTROL** Lift and burn infected plants to minimize disease carry-over. Where possible, remove infected plants before they start to disintegrate and release sclerotia into the soil. Improve soil drainage. Do not grow susceptible plants on that soil.

## VIRUS VECTORS

Some virus diseases affecting plants can be spread by certain pests and these are described as virus vectors. It is mainly sap-feeding insects such as some aphids, leafhoppers and thrips that are involved. When they insert their needle-like mouthparts into virus-infected plants, they will pick up some virus particles which can be transmitted to other plants when they move on to new host plants. The champion virus vector seems to be the peach potato aphid, *Myzus persicae*, which feeds on a wide range of plants and is capable of transmitting more than a hundred viruses. Some soil-dwelling eelworms, such as some *Xiphinema*, *Trichodorus* and *Longidorus* species, can transmit virus diseases when they feed on roots. The spread of viruses can be limited by controlling insect pests where feasible, and by removing plants affected by damaging virus diseases. Raspberries, strawberries and daffodils are susceptible to eelworm-vectored viruses. New plants should be put into fresh ground, since if they are planted where virus-infected plants were previously growing, eelworms may infect them with viruses when they start feeding on the roots.

## VIRUSES

**SYMPTOMS** The most common and widely spread symptoms of viral infection are stunting of the whole plant, and distortion (*see* **pp.25, 27**). The leaves and occasionally other above-ground parts of the plant develop markings (*see* **p.23**). These are usually yellow, though on some plants such as orchids they may be dark brown or black. The markings are commonly seen as streaks, mosaic patterns, flecks, mottles or ring spots. Flowers may show distortion or fail completely, or they may form but show "flower breaking" where pale streaks of colour develop on dark petals (*see* **p.42**), or coloured streaks develop on pale petals. Infected plants generally crop poorly or fail to crop at all and in many cases death is premature. Some viruses may infect certain plants without causing visible symptoms, and this is called latent infection. For example, cucumber mosaic virus (*see*

*p.120*) can exist in a number of common weeds which, although looking perfectly normal, can then act as a source of viral infection for other plants.
**CAUSE** A wide range of viruses, most of which may be found alone or in combination with other viruses. There are various methods of transmission, the most common being by sap-feeding insects, in particular aphids. Leafhopper, whitefly, plant hoppers, thrips, scale insects, mites and even beetles may also be involved. In most cases the infection is non-persistent, that is to say the insect can acquire the virus after feeding on an infected plant for only a few seconds, but can then transfer it to other plants only within a short period of time. Cucumber mosaic virus (*see p.120*) is an example of this. Some insect-borne viruses are persistent, that is, the insect cannot transmit the virus for several hours after feeding on an infected plant, but is then capable of transmitting it for a long time afterwards, and sometimes for the duration of its life. Tomato spotted wilt, which is transmitted by western flower thrips, is an example of this. Some viruses are transmitted by soil-inhabiting nematodes and fungi. Viruses may also be transmitted by handling, by grafting, by tubers, bulbs and corms, and during routine operations such as disbudding or removal of side shoots.
**CONTROL** None available. Remove and dispose of affected plants promptly. Wash hands and tools thoroughly after handling infected plants. Do not put replacement plants of a similar type on the same

site, and control potential vectors. Good weed control is essential too. Do not propagate from infected plants. Where possible, choose virus-resistant cultivars and buy plants certified as virus-free.

## VOLES

Voles are rodents with generally shorter tails and more rounded heads than the mice and rats that they resemble. They feed on a wide range of plant material, including seeds, flower buds, bulbs, corms and plant roots. They also gnaw bark from the trunks and branches of young trees and shrubs (*see* **p.36**). They may eat bark away around the entire stem (girdling), which prevents the uptake of water and nutrients, leading to the death of branches or the whole plant. Several species can occur in gardens, including the mouse-sized bank vole, *Clethrionys glareolus*, and the short-tailed field vole, *Microtus agrestis*. The water vole, *Arvicola amphibius*, is larger, and can be confused with a rat. Vole numbers can fluctuate widely, especially of the short-tailed field vole, and it is usually only in years when they are abundant that they cause serious damage to garden plants.
**CONTROL** Where voles are causing problems in a garden, some protection can be given by setting mouse or rat traps, as appropriate, in places where voles are active. Pieces of dessert apple seem to be a particularly attractive bait for the traps. These need to be placed under covers made of logs or bricks to keep birds away and prevent pets from harming themselves.

## WALNUT BLISTER MITE

**SYMPTOMS** During early summer leaves develop a oval, upwardly bulging areas up to 20mm long and about 10mm wide. The underside of these blistered areas is covered with a dense mass of creamy-white hairs which darken and become brown later in the summer. The upper surface of galled areas may become purplish-yellow.
**CAUSE** Microscopic gall mites, *Eriophyes erineus*, which suck sap from the underside of the foliage and cause the distorted growth. Most walnuts may be affected and sometimes a large proportion of leaves will have galls by late summer.
**CONTROL** No effective insecticides are available to gardeners. Fortunately, the mite has no impact on the tree's vigour or cropping.

## WALNUT BLOTCH

**SYMPTOMS** Leaves develop brown, necrotic blotches, and premature leaf fall occurs. Similar lesions may develop on the fruits, turning them from green to black.
**CAUSE** The fungus *Marssonina juglandis*, which overwinters on fallen leaves.
**CONTROL** Rake up and dispose of affected leaves. If the tree is not too tall, consider spraying with a copper-based fungicide such as copper oxychloride.

## WALNUT SOFT SHELL

**SYMPTOMS** The walnuts appear to develop normally, but the shell fails to form properly at one end. It may be thin or totally absent, so the nut within may become damaged or rotted by organisms which can gain entry through the faulty area. **CAUSE** The precise cause is unclear, but soft shell is believed to be largely the result of poor feeding and an erratic supply of water at that stage when the nuts are developing. **CONTROL** Keep trees fed, particularly if they have been cropping well. Mulch to help soil moisture retention.

## WASPS

Social wasps, such as *Vespula vulgaris* and *V. germanica*, and tree nesting species, such as *Dolichovespula media* and *D. sylvestris*, are important as plant pests because of their interest in ripe fruits. On relatively soft-skinned fruits, such as grapes and plums, they are capable of initiating damage (*see* **p.49**), but on apple and pears they generally enlarge damage started by birds. Wasps eat the soft inner tissues and gradually hollow out large cavities (*see* **p.47**). Some species of wasp make their papery nests above ground level in the branches of trees and shrubs. This can make hedge-cutting hazardous. **CONTROL** If wasp nests can be located, they can be controlled by placing some insecticide dust, such as bendiocarb or pirimiphos-methyl, in the nest entrance at dusk when wasps have stopped flying. There is no point in spraying fruit trees against wasps, since any killed will soon be replaced by others. Ripening fruits can be protected by enclosing some of the better fruit trusses in bags made from old nylon tights or muslin.

## WATER CORE

*See* APPLE WATER CORE *p.102.*

## WATER LILY APHID

**SYMPTOMS** Olive-green or brownish insects, about 2mm long, cluster on the upper surface of water lily leaves and on the flowers during the summer. Other floating and emergent pond plants are also liable to be attacked. Heavy attacks result in poor growth and flowering, with water lily leaves appearing dirty due to the accumulation of aphids and their whitish cast skins. **CAUSE** A sap-feeding aphid, *Rhopalosiphum nymphaeae*. It overwinters as an egg on wild and cultivated plums and cherries, feeding on the foliage of these plants in the spring before migrating to water lilies in early summer. **CONTROL** Although many insecticides will control aphids, none of them can be used on pond plants, as they will kill fish, frogs and other pond wildlife. Forceful spraying with plain water will reduce infestations by knocking aphids off the plants.

## WATER LILY BEETLE

**SYMPTOMS** Damage starts in late spring and continues throughout the summer. Yellowish-brown beetles, 7mm long, eat irregularly shaped slots in the lily pads (*see* **p.30**). In mid-summer the larvae, which have elongate bodies, up to 8mm long, that are black on the upper surface but pale yellow underneath, continue this damage. Infested leaves are unsightly and deteriorate rapidly as secondary rots develop on the damaged tissues. The adult beetles also feed on the flowers.

**CAUSE** Adults and larvae of a beetle, *Galerucella nymphaeae*. The entire life cycle of eggs, larvae, pupae and adults takes place on the upper surface of water lily leaves, and there are usually two generations during the summer.

ACTUAL LENGTH: 7MM

**WATER LILY BEETLE**

**CONTROL** Insecticides cannot be used, as they will be harmful to fish and other animals in the pond. With small ponds, adult beetles, larvae and pupae can be removed by hand. Forceful spraying with water will dislodge the larvae, some of which will fail to find their way back onto the leaves.

## WATER LILY LEAF SPOT

**SYMPTOMS** Red-brown spots, sometimes with a distinct yellow band around the edge, develop on the leaves. The upper leaf surface is generally most affected, but spotting may also occur beneath the leaves. Concentric zoning and numerous tiny, black pinprick-sized fungal bodies may also be visible. As the spots mature, they darken, turning a brownish-black colour. The centre of the spot may disintegrate, leaving a hole. **CAUSE** The fungus *Ramularia nymphaerum*, which is carried by water splash. **CONTROL** Remove affected leaves promptly. No chemical control is available and none should be attempted, as it may pose a threat to pond life.

## WATERLOGGING

**SYMPTOMS** The plant appears to be suffering from drought, with wilting foliage (*see* **p.34**) and, in extreme cases, flowers (*see* **p.71**). The leaves may be yellowed and may fall early. Inspection of the roots may reveal that they have disintegrated, often peeling readily, to leave just the dead vascular strands. **CAUSE** Overwatering, excessive rain or poor drainage may all be involved. **CONTROL** Attempt to improve soil conditions and avoid over-application of water. On heavy soils ensure that a large area of soil is lightened by the incorporation of plenty of grit, bulky organic matter and other materials before planting. The application of a foliar feed to any remaining green leaves may help to counteract the effects of slight waterlogging; it stimulates root production to replace some of those roots which have been lost.

## WEEDKILLER DAMAGE – BROADLEAVED

**SYMPTOMS** The leaf blades are distinctly narrowed (*see* **p.27**), often to such an extent that the leaf veins are so closely drawn together that they appear parallel. Petioles and stems may show spiral twisting, and the foliage may become cupped (*see* **p.27**). Fruits may not form or if they do, they may be distorted. Tomatoes produce plum-shaped fruits which are hollow with few, if any, seeds and little flesh. The stems of brassicas develop roughened gall-like growths at or just above soil level and these may appear to be adventitious roots (*see* **p.39**). Affected plants may live, but they can fail to crop normally.

**CAUSE** The plant has taken up broadleaved weedkiller, probably applied to control weeds nearby, into its tissues. Contamination may occur in many ways, such as inadvertent spraying, spray drift, use of watering cans or spraying equipment which have been insufficiently washed out, and mulching with lawn mowings which contain traces of weedkillers.

**CONTROL** Take great care when spraying and never do it on hot, windy or gusty days. Protect nearby plants with polythene sheeting or bags. Where possible, use a watering can and dribble bar rather than a sprayer, as these produce larger droplets that are less likely to cause drift problems. Keep equipment used for applying weedkillers clearly labelled and use it for this purpose only. Always wash it out thoroughly after use. Prune out severely damaged areas on trees or shrubs and feed and water to promote new replacement growth.

## WEEDKILLER DAMAGE – CONTACT

**SYMPTOMS** Brown scorched spots develop on the foliage (*see **pp.23, 34***) and any other part of the plant on which the chemical lands. If the weedkiller is applied on bulbs before the leaves have died down properly, bleached yellow or white foliage is produced the following season, which soon withers and dies. It may take several years for the bulb to recover fully.

**CAUSE** Contamination by the the contact weedkillers paraquat and diquat.

**CONTROL** As for broadleaved weedkiller damage.

## WEEDKILLER DAMAGE – GLYPHOSATE

**SYMPTOMS** Stunted rosettes of short shoots and malformed, strap-like leaves develop in place of normal side shoots. The symptoms resemble those caused by a witches' broom infection (*see p.191*).

**CAUSE** The widely-used weedkiller glyphosate, which may cause problems on a wide range of plants, in particular those which form suckers.

**CONTROL** As for broadleaved weedkiller damage (*see p.188*).

## WEEVILS

Weevils are beetles that are characterized by having clubbed antennae that can be bent at right angles about halfway along their length. The front of the head projects forward to form a snout or rostrum which may be long, as in the nut weevil (*see p.151*), or short, as in vine weevil (*see p.186*).

• *see also* APPLE BLOSSOM WEEVIL *p.100;* BEECH LEAF-MINING WEEVIL *p.106;* FIGWORT WEEVILS *p.126;* LEAF WEEVILS *p.143;* PEA AND BEAN WEEVIL *p.155;* TURNIP GALL WEEVIL *p.184.*

## WESTERN FLOWER THRIPS

**SYMPTOMS** Mainly house plants and greenhouse plants are affected, but this insect can also attack outdoor plants during the summer. On some host plants such as tomato and cucumber, the thrips feed mainly on the foliage, causing a silvery mottling of the upper leaf surface. On ornamentals such as African violets, streptocarpus, gloxinias, chrysanthemums, verbenas, pelargoniums, fuchsia, cyclamen and busy

Lizzies, it is mainly flowers that are attacked (*see **p.42***). They cause a pale flecking on the petals, and heavy infestations cause rapid deterioration in the blooms or prevent the buds from opening. Adult western flower thrips are 2mm long and have narrow, elongate, brown bodies. The immature nymphs have a similar shape and are pale yellow. This pest can transmit tomato spotted wilt virus, which causes dieback and distorted growth on many plants.

**CAUSE** A sap-feeding insect, *Frankliniella occidentalis*, which originates from North America, but became widely established in Europe after 1986. It breeds continuously in heated greenhouses and has many generations a year.

**CONTROL** Many insecticides used against other thrips have little impact on this species. Biological control is available with a predatory mite, *Amblyseius degenerans*, but this has to be introduced before heavy infestations arise. Alternatively, malathion-based insecticides will give some control, but it is difficult to treat thrips hidden within flower buds or in folded leaves. When buying house plants, check flowers carefully and reject any with thrips or signs of feeding damage on petals.

## WHIPTAIL OF BRASSICAS

**SYMPTOMS** Leaves become mottled yellow and growth is reduced.

**CAUSE** An inadequate supply or availability of molybdenum (a trace element) in the soil or compost. This is particularly common on acid soils and is most likely to affect seedlings, especially brassicas.

**CONTROL** Lime the soil. Water the soil around seedlings with a solution of sodium or ammonium molybdate.

## WHITE BLISTER

**SYMPTOMS** Raised, shiny white fungal pustules develop, primarily on the lower leaf surface of brassicas (*see **p.14***). Occasionally these may develop on other above-ground plant parts. The pustules are arranged either singly or in concentric rings. Affected leaves may become puckered and distorted, and the upper leaf surface develops yellow, slightly sunken pits.

**CAUSE** The fungus *Albugo candida*, which is encouraged by humid or moist air. The spores are spread on air currents or by water splash, or occasionally by insects. A resilient resting spore stage may also be formed and this may persist in the soil.

**CONTROL** Remove infected leaves. Remove and dispose of severely infected plants. Where feasible, grow resistant varieties of vegetable such as the Brussels sprout 'Saxon'.

## WHITEFLIES

Whiteflies are sap-feeding insects, about 2mm long, and, when adult, have white wings. They live on the underside of leaves and readily fly up when the plant is disturbed. They lay eggs and these hatch into flat, oval, scale-like nymphs. Like the adults, they excrete a sugary excrement called honeydew. This drops down onto the upper surfaces of leaves below, which become sticky and covered with black sooty moulds. The final nymphal stage, sometimes called a pupa, is thicker, often with hairs or white, waxy secretions on the upper surface. Some garden plants, such as rhododendron, honeysuckle and *Phillyrea*, are occasionally attacked by whiteflies specific to these plants. If necessary, control them as for azalea whitefly.

• see also AZALEA WHITEFLY *p.103*; CABBAGE WHITEFLY *p.113*; GLASSHOUSE WHITEFLY *p.132*; VIBURNUM WHITEFLY *p.185*.

## WILLOW ANTHRACNOSE

**SYMPTOMS** Leaves curl, turn yellow and fall prematurely. Extensive defoliation may occur by mid-summer. Pinhead-sized, dark brown spots may develop on infected leaves and these produce spores. Raised, rough lens-shaped cankers develop on affected stems and these may cause dieback if they enlarge sufficiently to girdle the stem. Although symptoms may appear alarming, and recurrent attacks will weaken the tree, this disease is unlikely to cause the death of an otherwise healthy tree.
**CAUSE** The fungus *Marssonina salicicola*, which is encouraged by moist conditions. The spores from the leaves and the cankers are carried on water splash. The fungus overwinters in the stem cankers, on bud scales and on fallen leaves.
**CONTROL** Rake up and dispose of fallen infected leaves. Prune out severely cankered stems, if possible, without spoiling the weeping habit of the tree. Consider growing the resistant Peking willow *Salix babylonica* var. *Pekinensis* 'Pendula' or the hybrid *S.* × *sepulcralis*.

## WILLOW BARK APHID

**SYMPTOMS** Dense colonies of brownish-black aphids, which are up to 5mm long, form on the bark of willow trunks and branches, mainly in late summer (*see p.36*). The upper surface of the aphids is marked with small black dots, with two larger tubercles towards the rear of the abdomen. Large quantities of sugary honey-dew are excreted by the aphids and this makes the stems and ground below sticky. Sooty moulds often grow on the honeydew, which also attracts wasps and flies.
**CAUSE** A sap-feeding insect, *Tuberolachnus salignus*.
**CONTROL** On willow trees small enough to be sprayed, this aphid can be controlled by using pirimicarb, dimethoate, heptenophos with permethrin, pirimiphos-methyl or malathion.

## WILLOW BEAN GALL SAWFLIES

**SYMPTOMS** Hard swellings form in the leaf blades during the summer months (*see p.17*). They are yellowish-green or red, and their rounded oval shape resembles that of certain bean seeds. The galls are initiated when the female sawfly lays her eggs in the leaf. She secretes chemicals that induce the leaf to produce a gall around each egg. The egg later hatches into a caterpillar-like larva that gradually hollows out the gall.
**CAUSE** Several sawflies, *Pontania* species, cause this type of gall on willows, and there are often two generations occurring between late spring and early autumn.

**WILLOW BEAN GALLS**

**CONTROL** Apart from causing galls on the foliage, these sawflies have no harmful effect on willows. Affected leaves can be removed from small trees if they are considered unsightly, but spraying would not be justified.

## WILLOW BLACK CANKER

**SYMPTOMS** In late spring or early summer angular, red-brown spots develop on the leaves; they spread, causing the leaves to shrivel and turn black, but remain hanging on the stem. The fungus then spreads into the stem and may produce lens-shaped, flattened cankers. It may be confused with willow scab (*see below*) or willow anthracnose (*see above*).
**CAUSE** The fungus *Glomerella miyabeana*, which is encouraged by warm temperatures and high humidity.
**CONTROL** Prune out affected shoots.

## WILLOW LEAF BEETLES

**SYMPTOMS** Both the adult and larval stages damage the leaves of willow, poplar and aspen. The adult beetles are about 4mm long and metallic blue-black or brassy green in colour. The larvae have elongate bodies up to 6mm long, and are mainly black on the upper surface, but creamy-yellow below. Both the larvae and adult beetles

graze away part of the leaf surface, causing the remaining damaged tissues to dry up and turn brown or white (*see p.18*). The larvae are gregarious and often feed side by side in groups of up to 30 or more.
**CAUSE** Several beetles and their larvae, including *Phyllodecta vitellinae*, *P. vulgatissima* and *Plagiodera versicolora*, cause damage to willows and poplars from late spring to early autumn. There are usually two generations.
**CONTROL** Only young trees suffer significant damage, although the foliage on older trees is often disfigured. Small trees can be sprayed with bifenthrin, pirimiphos-methyl or fenitrothion, particularly if heavily infested in early summer.

## WILLOW SCAB

**SYMPTOMS** Similar to willow anthracnose and willow black canker (*see above*). In early spring irregular spots develop on the young foliage, spreading and causing it to turn black and die. Olive-brown pustules develop on the lower leaf surface and on affected stems. Black lesions develop on the stems and may cause girdling (*see p.41*).
**CAUSE** The fungus *Fusicladium saliciperdum*, which overwinters as pustules on the stems.
**CONTROL** Prune out affected stems.

## WILLOW WATERMARK DISEASE

**SYMPTOMS** In mid- to late spring, young leaves turn red and wither. Ooze, which contains the bacteria, is produced from the stems.

Leaves deteriorate further and affected branches die back. The tree loses much of its growth, but is rarely killed. The wood develops a watermark-like stain internally and infected cut surfaces show complete or partial red or black rings. The disease is not common, occurring only in certain parts of the country. *Salix alba* var. *coerulea* and other white willows may be affected, but it is uncertain whether other species are also susceptible.
**CAUSE** The bacterium *Erwinia salicis*, which persists in the wood and produces toxins which cause the leaf discoloration and dieback. The bacteria enter the tree via wounds, and are probably spread by birds, insects and on air currents.
**CONTROL** Remove and burn affected trees.

## WILTS

**SYMPTOMS** Wilting of the foliage followed by dieback. The whole plant is rarely affected simultaneously and may take several seasons to die. Smaller or herbaceous or annual plants may be killed in one season. Staining may develop, associated with the vascular or conducting tissues in stems and roots.
**CAUSE** Various fungi, including *Verticillium*, *Fusarium* and *Ceratocystis* species. Much damage is caused by the blocking of vascular elements and the subsequent deterioration of stems and leaves. The fungi may be spread by pruning, root contact or grafting, or by contaminated sap coming into contact with that of a healthy plant.
**CONTROL** No chemical controls are available. Where feasible, grow resistant

plants. Remove and dispose of affected plants with the soil in the vicinity of the roots.
• *see also* ASTER WILT *p.103;* CLEMATIS WILT *p.118;* DROUGHT *p.124;* DUTCH ELM DISEASE *p.124;* FUSARIUM WILT *p.130;* PEONY WILT *p.158;* VERTICILLIUM WILTS *p.185;* WATERLOGGING *p.188.*

## WIND DAMAGE

**SYMPTOMS** Leaves appear scorched. One side or face of the plant is usually more severely affected. The most damage occurs on the side facing the prevailing wind. Flowers may be similarly affected (*see **p.45***). If winds are salt-laden, the effect is usually much more severe and the whole plant may suffer extreme dieback or even be killed. Trees on exposed sites may develop a lop-sided growth habit, known as "krumholtz", where they grow away from the prevailing wind.
**CAUSE** Wind. Often made worse if wind tunnels are created or if a windbreak is removed.
**CONTROL** Where possible, plant or erect windbreaks to prevent wind damage. If stems are killed, prune these out to prevent secondary organisms invading.

## WINTER ACONITE SMUT

**SYMPTOMS** Raised, blister-like grey outgrowths develop on stems and leaves, erupting to release large quantities of black, powdery spores. Severely affected parts die back and general growth and vigour is reduced.
**CAUSE** The fungus *Urocystis eranthidis*, which appears to be systemic (that is, found throughout all parts of the plant and not just on the

area showing the symptoms) within the plant.
**CONTROL** Remove and dispose of affected plants.

## WINTER MOTH

**SYMPTOMS** Yellowish-green caterpillars with paler lines along their bodies feed on the foliage (*see **p.31***) of many fruit trees and other deciduous trees and shrubs between bud burst and early spring. The caterpillars, up to 25mm long, walk with a looping action, as they have only two pairs of clasping legs on their abdomens. They bind leaves together loosely with silk threads and are often hidden from view. The caterpillars also eat fruit blossom and make holes in apple fruitlets, causing them to develop as badly misshapen fruits.
**CAUSE** Caterpillars of a moth, *Operophtera brumata*. The adult moths emerge between late autumn and mid-winter. Only males have normal wings; the females have swollen bodies with tiny wings and are incapable of flight. When they emerge from pupae in the soil, the females have to crawl up the tree trunk in order to lay eggs on the branches.
**CONTROL** Place sticky grease bands around tree trunks, especially of fruit trees, in mid-autumn to stop female moths reaching the branches. Control newly-hatched caterpillars by spraying with permethrin, bifenthrin, pirimiphos-methyl or fenitrothion as the leaves emerge from the buds.

## WIREWORMS

**SYMPTOMS** Wireworms are slender, orange-brown beetle larvae that have three pairs of short legs near the head

end (*see **p.54***). They are up to 25mm long and have a small protuberance on the underside of the rear end of the abdomen. Wireworms live in the soil and can kill seedlings by biting into the stems just below soil level. In late summer, vegetables such as potato tubers (*see **p.56***) and carrots can be extensively tunnelled by the larvae.
**CAUSE** The larval stage of click beetles (*see p.118*), including *Agriotes lineatus*, *A. obscurus*, *A. sputator* and *Athous haemorrhoidalis*. They are found in the greatest numbers in newly-dug grassland; their numbers will gradually decline over the next three or four years if the ground is kept in cultivation.
**CONTROL** Where wireworms are troublesome, dig up potato tubers as soon as they have matured to limit damage. Protect seedlings by dusting seed rows with lindane or pirimiphos-methyl.

## WITCHES' BROOMS

**SYMPTOMS** Extensively branched twigs develop, clustered together to form "witches' brooms" in the crown of the tree (*see **p.41***). In winter they may at first be mistaken for birds' nests and can spoil the outline of the tree or hedge, but in summer they appear as dense, leafy areas and so are less obtrusive.
**CAUSE** Various fungi, in particular species of *Taphrina*, spread by water splash and on air currents. Occasionally other factors, in particular mite infestations, may be responsible.
**CONTROL** Prune out the growths, if desired. Spread seems to be very slow, but removal will help to limit the problem. No chemical

control is available.
• see also BIRCH WITCHES'
BROOMS p.107; HORNBEAM
WITCHES' BROOMS p.137.

## WOODLICE

**SYMPTOMS** Grey or pinkish-brown animals (see **p.53**), with a hard segmented covering over their bodies, are seen. During the day, woodlice hide in dark places and are often found under pots, logs and stones, or in the surface layers of compost heaps. They sometimes damage seedlings, but established plants are not harmed.
**CAUSE** Several species of woodlice are common in gardens, including *Oniscus asellus*, *Porcellio scaber* and *Armadillidium vulgare*. The last-mentioned rolls itself up into a ball when disturbed and is commonly known as a pillbug.

ACTUAL LENGTH: 4–15MM

**WOODLOUSE**

**CONTROL** Woodlice are such abundant animals in most gardens that there is little prospect of achieving any lasting reduction in their numbers. Fortunately, they feed mainly on dead plant material rather than growing plants. Where susceptible plants such as seedlings are being damaged, they can be given some protection by applying lindane or bendiocarb dust.

## WOOLLY APHID

**SYMPTOMS** A fluffy, white, waxy substance develops on the bark of apple trees and ornamental *Malus*, pyracantha and cotoneaster

in summer (see **p.35**). The wax is secreted from the bodies of pinkish-brown aphids. During mid- to late spring they are mainly seen around old pruning cuts and in splits in older bark. Later in the summer the aphids spread to young branches where their feeding causes soft, knobbly swellings to develop. These galls often split during frosty weather creating wounds that apple canker (see p.100) can infect.
**CAUSE** A sap-feeding insect, *Eriosoma lanigerum*. Most aphids overwinter as eggs but this species overwinters as immature nymphs under loose bark.
**CONTROL** Watch for signs of woolly aphid during the spring and, if present, spray with pirimicarb, dimethoate, heptenophos with permethrin or pirimiphos-methyl. Heavy infestations in late summer are difficult to control.
• see also BEECH WOOLLY APHID p.106.

## WOOLLY VINE SCALE

**SYMPTOMS** This pest attacks many woody plants, including grape vine, alder, currants, gooseberry and pyracantha. When mature in late spring, females deposit eggs in a thick pad of white, waxy fibres (see **p.39**). The scales are dark brown and up to 6mm long. If they are removed by hand, the egg masses can be drawn out in long waxy threads. Heavily infested plants show a lack of vigour and the foliage is soiled by sooty moulds.
**CAUSE** A sap-feeding scale insect, *Pulvinaria vitis*. Nymphs hatch in early summer and feed initially on the underside of leaves. Before leaf fall they move

back to the bark where they complete their development and overwinter as adults.
**CONTROL** On fruit trees, apply a tar oil wash in early to mid-winter when the plants are fully dormant. Otherwise spray with malathion in mid-summer against the newly-hatched nymphs.

## WORM CASTS

See EARTHWORMS p.125.

## WORMS

See CUTWORMS p.121;
EARTHWORMS p.125; EELWORMS p.125; ENCHYTRAEID WORMS p.126; FLATWORMS p.127; THUNDERWORMS p.181; WIREWORMS p.191.

## YELLOW-TAIL MOTH

**SYMPTOMS** An adult moth with white wings and body, except for the end of the abdomen, which is covered with yellow hairs, lays eggs on a range of trees and shrubs in late summer. It overwinters as young caterpillars which feed during spring and early summer. The hairy caterpillars, up to 30mm long, are black with an orange-red double stripe running down the upper body and white markings along the sides. They eat both the foliage and flowers of their hosts.
**CAUSE** Caterpillars of a moth, *Euproctis similis*.

ACTUAL LENGTH: UP TO 30MM

**CATERPILLAR OF YELLOW-TAIL MOTH**

**CONTROL** Infestations are generally light and can be tolerated on established plants. If heavier attacks occur it may be necessary to spray with permethrin, bifenthrin, pirimiphos-methyl, fenitrothion or pyrethrum. Biological control with *Bacillus thuringiensis* can also be used.

## YUCCA LEAF SPOT

**SYMPTOMS** Circular brown spots develop on affected leaves. On variegated yuccas, the spots are usually most prevalent on the variegated parts of the leaf (see **p.18**). The spots show concentric ringing and numerous, pin-prick-sized, raised, black fruiting bodies develop on them. The spots enlarge and may join up.
**CAUSE** The fungi *Cercospora concentrica* and *Coniothyrium concentricum*, which produce similar symptoms. The spores are spread by water splash.
**CONTROL** Remove severely affected leaves. Spray badly affected plants with a fungicide such as mancozeb.

## ZUCCHINI YELLOW MOSAIC VIRUS

**SYMPTOMS** Bright yellow mosaic patterns on leaves of marrows, courgettes and squashes. Foliage is often distorted and the whole plant stunted. If plants reach their flowering and fruiting stage, fruits are very knobbly and distorted and the skin is blotched dark green.
**CAUSE** A mosaic virus, spread by aphids.
**CONTROL** Remove infected plants immediately.

# INDIVIDUAL PLANT PROBLEMS

COMMON PROBLEMS TO which different garden plants are subject are listed here grouped by plant type, then under the common name of the plant within its type. Where common names are not known, the index (*pp.218–222*) lists these under plants' botanical names, for ease of access. For each individual plant problem, cross-references are given both to its detailed entry in the A–Z of Pests, Diseases and Disorders, and to any illustrations of the problem's effects in the Gallery of Symptoms.

# GARDEN TREES

ONCE ESTABLISHED, trees are usually sufficiently vigorous for many problems to seem insignificant. However, if seriously attacked by a pest or a disease, an established tree's size can make it difficult – sometimes impossible – to treat. Many problems can be avoided by careful maintenance. With deciduous trees, the autumn leaf-fall also helps, removing many pathogens on foliage.

A number of common problems may affect a wide range of unrelated trees (*see right*). Some, however, are more host-specific; trees that are particularly prone to certain problems are listed below. Resistant species or cultivars may be suggested in the entry for the problem in question in the *A–Z of Pests, Diseases and Disorders*.

## COMMON PROBLEMS

The most common non-host-specific tree ailments are:

ALGAE 98
APHIDS 99
BRACKET FUNGI 110
CATERPILLARS 115
CORAL SPOT 119
DROUGHT 124
FROST DAMAGE 129
GANODERMA 130
HONEY FUNGUS 136
IRON DEFICIENCY AND LIME-INDUCED
   CHLOROSIS 139
LEAF SPOTS (FUNGAL) 143
LIGHTNING DAMAGE 70
MAGNESIUM DEFICIENCY 146

PHYTOPHTHORA 159
POWDERY MILDEWS 165
SCORCH 174
VERTICILLIUM WILTS 185
VIRUSES 187
WATERLOGGING 188
WEEDKILLER DAMAGE 188–9
WIND DAMAGE 191

# A

**Alder (*Alnus*)**
ALDER SUCKER **32**, 98
GALL MITES 130
PHYTOPHTHORA 159

**Almond (*Prunus dulcis*)**
APHIDS 99
BACTERIAL CANKER **35**, 104
BLOSSOM WILT **41**, 108
BROWN ROT **47**, **49**, 111
BULLFINCHES **46**, 112
FASCIATION **43**, 126
HONEY FUNGUS **34**, 136
IRON DEFICIENCY AND
   LIME-INDUCED CHLOROSIS
   **24**, 139
PEACH LEAF CURL **26**, 156
PHYTOPHTHORA 159
POWDERY MILDEWS **25**, 165
SHOTHOLE **18**, **30**, 175
SHOT-HOLE BORERS **33**, 175
SILVER LEAF 175

**Amelanchier** *see* SHRUBS
AND CLIMBERS 196–7

**Ash (*Fraxinus*)**
ASH CATKIN GALL MITE 102
ASH SUCKER 103
BACTERIAL CANKER **35**, 104
DIEBACK 123
GANODERMA 130
INONOTUS HISPIDUS 138
NECTRIA CANKER 151

# B

**Bay (*Laurus nobilis*)**
BAY SUCKER **26**, 104
CARNATION TORTRIX MOTH
   **20**, 114
FROST DAMAGE (FOLIAGE
   AND STEMS) **28**, 129

HORSE CHESTNUT SCALE **40**,
   137
POWDERY MILDEWS **25**, 165
SOFT SCALE 177

**Beech (*Fagus*)**
BEECH BARK DISEASE **35**, 106
BEECH BARK SCALE **35**, 106
CANKER 114
CORAL SPOT **41**, 119
FELT GALL MITE **19**, 126
FOMES ROOT AND BUTT ROT
   **37**, 128
GALL MIDGES 130
GANODERMA 130
HONEY FUNGUS **34**, 136
LEAF MINERS 143
MERIPILUS GIGANTEUS **39**, 147
PHYTOPHTHORA 159
POWDERY MILDEWS **25**, 165
SQUIRRELS **52**, 178
WOOLLY BEECH APHID **35**, 192

**Birch (*Betula*)**
APHIDS 99
BRACKET FUNGI **37**, **38**, 110
FELT GALL MITE **19**, 126
FOMES ROOT AND BUTT ROT
   **37**, 128
HONEY FUNGUS **34**, 136
LEAF SPOT (FUNGAL) **12**, **13**,
   143
LEAF WEEVILS 143
PHYTOPHTHORA 159
POWDERY MILDEWS **25**, 165
RUSTS 173
SAWFLIES 173
WITCHES' BROOMS **41**, 191

# C

**Catalpa**
MAGNESIUM DEFICIENCY **22**,
   146
PHYTOPHTHORA 159
VERTICILLIUM WILTS **33**, 185

**Cedar (*Cedrus*)**
BRACKET FUNGI **37**, **38**, 110
CEDAR APHID **32**, 115
HONEY FUNGUS **34**, 136

**Cercidiphyllum japonicum**
FROST DAMAGE (FOLIAGE
   AND STEMS) **28**, 129
VERTICILLIUM WILTS **33**, 185

**Cherry, flowering** *see*
**Prunus, ornamental**

**Cherry laurel** *see* **Laurel,
cherry**

**Crab apple (*Malus*)**
APPLE CANKER **36**, 100
APPLE SCAB **16**, 101
BLOSSOM WILT **41**, 108
BROWN ROT **49**, 111
CORAL SPOT **41**, 119
FIREBLIGHT **35**, 127
GANODERMA 130
HONEY FUNGUS **34**, 136
LEAF MINERS 142
LEAF WEEVILS 143
PAPERY BARK **40**, 154
POWDERY MILDEWS **25**, 165
SILVER LEAF 175
WINTER MOTH **31**, 191
WOOLLY APHID **35**, 192

**Cypress, Leyland
cypress (*Cupressus,
x Cupressocyparis
leylandii*)**
CROWN GALL **38**, 119
CYPRESS APHID 121
FOMES ROOT AND BUTT ROT
   **37**, 128
HONEY FUNGUS **34**, 136
JUNIPER SCALE **37**, 140
KABATINA SHOOT BLIGHT
   140
PHOMOPSIS CANKER 159
PHYTOPHTHORA 159
SEIRIDIUM CANKER 174

# D

**Dogwood (*Cornus*)**
DOGWOOD ANTHRACNOSE 123
HORSE CHESTNUT SCALE **40**,
   137
PHYTOPHTHORA 159
POWDERY MILDEWS **25**, 165

**Douglas fir** *see* **Fir,
Douglas**

# E

**Elder** *see* SHRUBS AND
CLIMBERS 196–7

**Elm (*Ulmus*)**
BARK BEETLES **33**, 104
CORAL SPOT **41**, 119
DUTCH ELM DISEASE 124
GALL MITE 130
GANODERMA 130
HONEY FUNGUS **34**, 136
HORSE CHESTNUT SCALE **40**,
   137
LEAF SPOT (FUNGAL) **12**, **13**,
   143
PHYTOPHTHORA 159
WITCHES' BROOMS **41**, 191

**Eucalyptus**
EUCALYPTUS SUCKER 126
OEDEMA **19**, 152
PHYTOPHTHORA 159
SILVER LEAF 175

# F

**False acacia (*Robinia
pseudoacacia*)**
LAETIPORUS SULPHUREUS **38**,
   141
PHYTOPHTHORA 159

SAWFLIES 173
VERTICILLIUM WILTS **33**, 185

**Fir, Douglas
(*Pseudotsuga menziesii*)**
DOUGLAS FIR ADELGID 124
FOMES ROOT AND BUTT ROT
   **37**, 128
PHYTOPHTHORA 159

**Fir, silver (*Abies alba*)**
HONEY FUNGUS **34**, 136
SILVER FIR ADELGID 175

# H

**Hawthorn (*Crataegus*)**
APHIDS 99
APPLE CANKER **36**, 100
COTONEASTER WEBBER MOTH
   **37**, 119
CROWN GALL **38**, 119
FIREBLIGHT **35**, 127
FOMES ROOT AND BUTT ROT
   **37**, 128
HAWTHORN GALL MIDGE 135
HONEY FUNGUS **34**, 136
LEAF SPOT (FUNGAL) **12**, **13**,
   143
PHYTOPHTHORA 159
POWDERY MILDEWS **25**, 165
RUSTS 173
SILVER LEAF 175

**Hazel (*Corylus*)** *see*
FRUITS AND NUTS 204–5

**Holly (*Ilex*)**
CUSHION SCALE **15**, 121
HONEY FUNGUS **34**, 136
LEAF MINERS 142
LEAF SPOT (FUNGAL) **12**, **13**,
   143
PHYTOPHTHORA 159

**Honey locust (*Gleditsia*)**
CORAL SPOT **41**, 119

GLEDITSIA GALL MIDGE **27**, 133
MERIPILUS GIGANTEUS **39**, 147

**Hornbeam (*Carpinus*)**
APHIDS 99
CORAL SPOT **41**, 119
LEAF SPOT (FUNGAL) **12**, **13**, 143
WINTER MOTH **31**, 191
WITCHES' BROOMS **41**, 191

**Horse chestnut (*Aesculus hippocastanum*)**
BLEEDING CANKER 108
GANODERMA 130
HONEY FUNGUS **34**, 136
HORSE CHESTNUT LEAF BLOTCH **20**, 137
HORSE CHESTNUT SCALE **40**, 137
LEAF SCORCH 143
PHYTOPHTHORA 159
SLIME FLUX AND WETWOOD 175

# J

**Judas tree (*Cercis siliquastrum*)**
CORAL SPOT **41**, 119
VERTICILLIUM WILTS **33**, 185

**Juniper (*Juniperus*)**
CONIFER RED SPIDER MITE 119
JUNIPER APHIDS 140
JUNIPER SCALE **37**, 140
JUNIPER WEBBER MOTH **32**, 140
KABATINA SHOOT BLIGHT 140
PESTALOTIOPSIS 158
PHOMOPSIS CANKER 159
PHYTOPHTHORA 159
RUSTS 173

# L

**Laburnum**
BLACKFLY 108
DOWNY MILDEW 124
HONEY FUNGUS **34**, 136
LABURNUM LEAF MINER **18**, 141
LEAF SPOT (FUNGAL) **12**, **13**, 143
PHYTOPHTHORA 159
POWDERY MILDEWS **25**, 165
SILVER LEAF 175
VIRUSES 187

**Larch (*Larix*)**
BRACKET FUNGI **37**, 110
FOMES ROOT AND BUTT ROT **37**, 128
HONEY FUNGUS **34**, 136
LARCH ADELGID **32**, 142
NEEDLE BLIGHTS AND CASTS OF PINES **32**, 151

**Laurel, cherry (*Prunus laurocerasus*)**
BACTERIAL CANKER **35**, 104
HONEY FUNGUS **34**, 136
LEAF SPOT (FUNGAL) **12**, **13**, 143
MAGNESIUM DEFICIENCY **22**, 146
PHYTOPHTHORA 159
POWDERY MILDEWS **25**, 165
SHOTHOLE **18**, **30**, 175
SILVER LEAF 175

**Lime (*Tilia*)**
APHIDS 99
FELT GALL MITE **19**, 126
GANODERMA 130
HORSE CHESTNUT SCALE **40**, 137
LEAF SPOT (FUNGAL) **12**, **13**, 143
LIME NAIL GALL MITE **14**, 146
PHYTOPHTHORA 159
VERTICILLIUM WILTS **33**, 185

**Liquidambar**
HONEY FUNGUS **34**, 136
IRON DEFICIENCY AND LIME-INDUCED CHLOROSIS **24**, 139

# M

**Magnolia**
CAPSID BUGS **29**, 114
CORAL SPOT **41**, 119
FROST DAMAGE **28**, **44**, 129
GREY MOULD **45**, 134
HONEY FUNGUS **34**, 136
HORSE CHESTNUT SCALE **40**, 137
LEAF SPOT (FUNGAL) **12**, **13**, 143
IRON DEFICIENCY AND LIME-INDUCED CHLOROSIS **24**, 139
SQUIRRELS **52**, 178
VIRUSES 187

**Maple (*Acer*)**
ACER GALL MITE **17**, 98
APHIDS 99
CORAL SPOT **41**, 119
FELT GALL MITE **19**, 126
GANODERMA 130
HONEY FUNGUS **34**, 136
HORSE CHESTNUT SCALE **40**, 137
POWDERY MILDEWS **25**, 165
SQUIRRELS **52**, 178
SYCAMORE SOOTY BARK DISEASE **36**, 180
VERTICILLIUM WILTS **33**, 185
*See also* **Sycamore**

**Monkey puzzle (*Araucaria araucana*)**
HONEY FUNGUS **34**, 136
PHYTOPHTHORA 159

**Mountain ash/rowan (*Sorbus aucuparia*)**
APPLE LEAF MINER **19**, 101

CORAL SPOT **41**, 119
FIREBLIGHT **35**, 127
HONEY FUNGUS **34**, 136
LEAF WEEVILS 143
LEOPARD MOTH 144
MOUNTAIN ASH BLISTER MITE 149
NECTRIA CANKER 151
SILVER LEAF 175

**Mulberry** *see* FRUITS AND NUTS 204–5

# O

**Oak (*Quercus*)**
ACORN GALL WASP **52**, 98
APHIDS 99
FISTULINA HEPATICA 127
GALL WASPS 130
GANODERMA 130
GRIFOLA FRONDOSA 134
HONEY FUNGUS **34**, 136
LAETIPORUS SULPHUREUS 141
LEAF MINERS 142
LEAF SPOT (FUNGAL) **12**, **13**, 143
OAK PHYLLOXERA 152
POWDERY MILDEWS **25**, 165

# P

**Pear, ornamental (*Pyrus*)**
APPLE AND PEAR CANKER **36**, 100
HONEY FUNGUS **34**, 136
POWDERY MILDEWS **25**, 165

**Pine (*Pinus*)**
BRACKET FUNGI **37**, **38**, 110
CONIFER APHIDS 119
FOMES ROOT AND BUTT ROT **37**, 128
HONEY FUNGUS **34**, 136
PINE ADELGID **39**, 160
RUSTS 173

**Plane (*Platanus*)**
GANODERMA 130
INONOTUS HISPIDUS 138
POWDERY MILDEWS **25**, 165
SALT DAMAGE 173

**Poplar (*Populus*)**
APPLE SCAB **16**, 101
BACTERIAL CANKER **35**, 104
HONEY FUNGUS **34**, 136
LEAF SPOT (FUNGAL) **12**, **13**, 143
PEACH LEAF CURL **26**, 156
POPLAR CANKER 162
RUSTS 173
SILVER LEAF 175
WILLOW LEAF BEETLES **18**, 190

**Prunus, ornamental**
APPLE LEAF MINER **19**, 101

BACTERIAL CANKER **35**, 104
BLOSSOM WILT **41**, 108
BULLFINCHES **46**, 112
CHERRY BLACKFLY **24**, 116
CROWN GALL **38**, 119
FASCIATION **43**, 126
HONEY FUNGUS **34**, 136
MAGNESIUM DEFICIENCY **22**, 146
PEACH LEAF CURL **26**, 156
PEAR AND CHERRY SLUGWORM **19**, 157
PHYTOPHTHORA 159
POWDERY MILDEWS **25**, 165
ROSE SICKNESS/REPLANT PROBLEMS 172
SHOTHOLE **18**, **30**, 175
SHOT-HOLE BORERS **33**, 175
SILVER LEAF 175
WINTER MOTH **31**, 191
WITCHES' BROOMS **41**, 191

# R

**Rowan** *see* **Mountain ash**

# S

**Silver fir** *see* **Fir, silver**

**Spruce (*Picea*)**
CONIFER APHIDS 119
CONIFER RED SPIDER MITE 119
FOMES ROOT AND BUTT ROT **37**, 128
GREEN SPRUCE APHID 134
HONEY FUNGUS **34**, 136
PINEAPPLE GALL ADELGID **38**, 160
RUSTS 173

**Strawberry tree (*Arbutus unedo*)**
ARBUTUS LEAF SPOT **14**, 102
PHYTOPHTHORA 159
VIBURNUM WHITEFLY **15**, 185

**Sumach** *see* SHRUBS AND CLIMBERS 196–7

**Sweet chestnut (*Castanea sativa*)**
HONEY FUNGUS **34**, 136
LAETIPORUS SULPHUREUS **38**, 141
PHYTOPHTHORA 159

**Sycamore (*Acer pseudoplatanus*)**
ACER GALL MITE **17**, 98
APHIDS 99
CORAL SPOT **41**, 119
FELT GALL MITE **19**, 126
GANODERMA 130
HONEY FUNGUS **34**, 136
HORSE CHESTNUT SCALE **40**, 137
LEAF SCORCH 143

POWDERY MILDEWS **25**, 165
SQUIRRELS **52**, 178
SYCAMORE SOOTY BARK DISEASE **36**, 180
TAR SPOT OF ACER **13**, 180
VERTICILLIUM WILTS **33**, 185
WINTER MOTH **31**, 191

# T

**Thuja**
CONIFER APHIDS 119
HONEY FUNGUS **34**, 136
JUNIPER SCALE **37**, 140
KABATINA SHOOT BLIGHT 140
SEIRIDIUM CANKER 174

**Tulip tree (*Liriodendron*)**
HONEY FUNGUS **34**, 136
LEAF SPOT (FUNGAL) **12**, **13**, 143
MAGNESIUM DEFICIENCY **22**, 146
SLIME FLUX AND WETWOOD 175

# W

**Walnut (*Juglans*)** *see* FRUITS AND NUTS 204–5

**Willow (*Salix*)**
ANTHRACNOSE 99
APHIDS 99
DAEDALEOPSIS CONFRAGROSA **37**, 122
GALL MITES 130
HONEY FUNGUS **34**, 136
PHELLINUS IGNARIUS 159
RUSTS 173
WILLOW BARK APHID **36**, 190
WILLOW BEAN GALL SAWFLIES **17**, 190
WILLOW BLACK CANKER 190
WILLOW LEAF BEETLES **18**, 190
WILLOW SCAB **41**, 190
WILLOW WATERMARK DISEASE 190

# YZ

**Yew (*Taxus*)**
BIG BUD MITE **41**, 107
CARNATION TORTRIX MOTH **20**, 114
LAETIPORUS SULPHUREUS **38**, 141
PHYTOPHTHORA 159
VINE WEEVIL **29**, **53**, 186

**Zelkova**
DUTCH ELM DISEASE 124

# SHRUBS AND CLIMBERS

CORRECT PRUNING is one of the keys to maintaining healthy shrubs and climbers, thinning overcrowded and old growth to prevent a stagnant environment. Prompt recognition of problems is also vital; the removal of much badly affected growth can spoil an attractively shaped shrub or carefully trained climber.

Some common problems may affect many unrelated plants (*see right*). Others are more host-specific; plants particularly prone to certain problems are listed below. Resistant species or cultivars may be suggested in the entry for the problem in question in the *A–Z of Pests, Diseases and Disorders*.

## COMMON PROBLEMS

The most common non-host-specific ailments affecting shrubs and climbing plants are:

| | |
|---|---|
| ALGAE 98 | IRON DEFICIENCY AND LIME-INDUCED CHLOROSIS 139 |
| APHIDS 99 | MAGNESIUM DEFICIENCY 146 |
| CORAL SPOT 119 | PHYTOPHTHORA 159 |
| DEER 122 | POWDERY MILDEWS 165 |
| FROST DAMAGE 129 | RABBITS 165 |
| HONEY FUNGUS 136 | VERTICILLIUM WILTS 185 |
| LEAF SCORCH 143 | VIRUSES 187 |
| LEAF SPOT (FUNGAL) 143 | WEEDKILLER DAMAGE 188–9 |

## A

**Actinidia**
GLASSHOUSE RED SPIDER MITE **22**, **23**, 132
HONEY FUNGUS **34**, 136

**Amelanchier**
FIREBLIGHT **35**, 127

**Aucuba**
FROST DAMAGE (FOLIAGE AND STEMS) **28**, 129
PHYTOPHTHORA 159

## B

**Bamboo**
APHIDS 99

**Barberry (*Berberis*)**
APHIDS 99
HONEY FUNGUS **34**, 136
LEAF SPOT (FUNGAL) **12**, **13**, 143
POWDERY MILDEWS **49**, 165
RUSTS 173

**Box (*Buxus*)**
BOX SUCKER **26**, 109
LEAF SPOT (FUNGAL) **12**, **13**, 143
MUSSEL SCALE **36**, 149
PHYTOPHTHORA 159
RUSTS 173

**Broom (*Cytisus*)**
BROOM GALL MITE **39**, 110
FROST DAMAGE (FOLIAGE AND STEMS) **28**, 129
HONEY FUNGUS **34**, 136
SILVER LEAF 175
WATERLOGGING **34**, 188

**Buddleja**
APHIDS 99
CAPSID BUGS **29**, **46**, 114
EARWIGS **46**, 125
FIGWORT WEEVILS 126

GLASSHOUSE RED SPIDER MITE **22**, **23**, 132
HONEY FUNGUS **34**, 136
LEAF AND BUD EELWORM **25**, 142
LEAF SPOT (FUNGAL) **12**, **13**, 143
MULLEIN MOTH 149
VIRUSES **23**, **27**, 187

## C

**Camellia**
APHIDS 99
CAMELLIA GALL **26**, 113
CAMELLIA YELLOW MOTTLE **22**, 114
CARNATION TORTRIX MOTH **20**, 114
CUSHION SCALE **15**, 121
HONEY FUNGUS **34**, 136
LEAF SPOT (FUNGAL) **12**, **13**, 143
PESTALOTIOPSIS 158
PHYTOPHTHORA 159
SQUIRRELS **52**, 178
VINE WEEVIL **29**, 186

**Carpenteria**
LEAF SPOT (FUNGAL) **12**, **13**, 143

**Caryopteris**
CAPSID BUGS **29**, **46**, 114
HONEY FUNGUS **34**, 136

**Ceanothus**
DROUGHT **44**, 124
FROST DAMAGE **28**, **44**, 129
HONEY FUNGUS **34**, 136
IRON DEFICIENCY AND LIME-INDUCED CHLOROSIS **24**, 139
MUSSEL SCALE **36**, 149

**Clematis**
APHIDS 99
CAPSID BUGS **29**, **46**, 114
CATERPILLARS 115
CLEMATIS GREEN FLOWER 117
CLEMATIS WILT 118

EARWIGS **46**, 125
GLASSHOUSE RED SPIDER MITE **22**, **23**, 132
LEAF SPOT (FUNGAL) **12**, **13**, 143
POWDERY MILDEWS **49**, 165
SLIME FLUX AND WETWOOD 175
VIRUSES **23**, **27**, 187

**Cotinus**
POWDERY MILDEWS **49**, 165
VERTICILLIUM WILTS **29**, 185

**Cotoneaster**
APHIDS 99
BROWN SCALE **36**, 111
COTONEASTER WEBBER MOTH **37**, 119
FIREBLIGHT **35**, 127
HONEY FUNGUS **34**, 136
LEAF SPOT (FUNGAL) **12**, **13**, 143
MUSSEL SCALE **36**, 149
SILVER LEAF 175
VIRUSES **23**, **27**, 187
WOOLLY APHID **35**, 192

## D

**Daphne**
CUCUMBER MOSAIC VIRUS **21**, **25**, 120
FASCIATION **43**, 126
GREY MOULD **34**, 134
HONEY FUNGUS **34**, 136
LEAF SPOT (FUNGAL) **12**, **13**, 143
VERTICILLIUM WILTS **29**, 185
VIRUSES **22**, **23**, 187

**Dogwood (*Cornus*)**
DOGWOOD ANTHRACNOSE 123
HORSE CHESTNUT SCALE **40**, 137
MUSSEL SCALE **36**, 149
PHYTOPHTHORA 159
POWDERY MILDEWS **49**, 165

## E

**Elaeagnus**
CORAL SPOT **41**, 119
LEAF SPOT (FUNGAL) **12**, **13**, 143

**Elder (*Sambucus*)**
BLACKFLY 108
GLASSHOUSE RED SPIDER MITE **22**, **23**, 132
HONEY FUNGUS **34**, 136
LEAF SPOT (FUNGAL) **12**, **13**, 143
VIRUSES **23**, **27**, 187

**Escallonia**
BROWN SCALE **36**, 111
SILVER LEAF 175
VIRUSES **23**, **27**, 187

**Euonymus**
BLACKFLY 108
CROWN GALL **38**, 119
CUSHION SCALE **15**, 121
EUONYMUS SCALE* **21**, 126
POWDERY MILDEWS **49**, 165
SMALL ERMINE MOTHS **38**, 176
VINE WEEVIL* **29**, 186
VIRUSES **22**, **23**, 187
(*E. japonica* only)

## F

**Fatsia**
CAPSID BUGS **29**, **46**, 114

**Firethorn (*Pyracantha*)**
BROWN SCALE **36**, 111
CORAL SPOT **41**, 119
FIREBLIGHT **35**, 127
PYRACANTHA LEAF MINER 166
PYRACANTHA SCAB 166
WOOLLY APHID **35**, 192
WOOLLY VINE SCALE **39**, 192

**Flowering currant (*Ribes*)**
AMERICAN GOOSEBERRY MILDEW **49**, 98

BROWN SCALE **36**, 111
CORAL SPOT **41**, 119
CROWN GALL **38**, 119
CURRANT AND GOOSEBERRY LEAF SPOT 120
CURRANT BLISTER APHID **27**, 121
DIEBACK 123
GREY MOULD **34**, 134
HONEY FUNGUS **34**, 136
IRON DEFICIENCY AND LIME-INDUCED CHLOROSIS **24**, 139
RUSTS 173
WOOLLY VINE SCALE **39**, 192

**Forsythia**
BULLFINCHES **46**, 112
CAPSID BUGS **29**, **46**, 114
FORSYTHIA GALL **40**, 128
HONEY FUNGUS **34**, 136
LEAF SPOT (FUNGAL) **12**, **13**, 143
POWDERY MILDEWS **49**, 165

**Fuchsia, hardy**
CAPSID BUGS **29**, **46**, 114
ELEPHANT HAWK MOTH **30**, 125
FLEA BEETLES **29**, 127
FUCHSIA RUST **18**, 129
GLASSHOUSE RED SPIDER MITE **22**, **23**, 132
LEAF SPOT (FUNGAL) **12**, **13**, 143

## H

**Hawthorn (*Crataegus*)**
see GARDEN TREES 194–5

**Heathers (*Erica, Calluna*)**
HONEY FUNGUS **34**, 136
LEAF SPOT (FUNGAL) **12**, **13**, 143
PHYTOPHTHORA 159
POWDERY MILDEWS **49**, 165

**Hebe**
APHIDS 99
CARNATION TORTRIX MOTH **20**, 114

DOWNY MILDEW 124
LEAF SPOT (FUNGAL) **12**, **13**, 143

**Honeysuckle (*Lonicera*)**
HONEYSUCKLE APHID 137
LEAF SPOT (FUNGAL) **12**, **13**, 143
POWDERY MILDEWS **49**, 165
SILVER LEAF 175
THRIPS 181
WHITEFLIES 189

**Hydrangea**
APHIDS 99
CAPSID BUGS **29**, **46**, 114
FOOT AND ROOT ROT **55**, 128
FROST DAMAGE (FOLIAGE AND STEMS) **28**, 129
GLASSHOUSE RED SPIDER MITE **22**, **23**, 132
GREY MOULD **34**, 134
HONEY FUNGUS **34**, 136
HYDRANGEA SCALE **15**, 138
IRON DEFICIENCY AND LIME-INDUCED CHLOROSIS **24**, 139
LEAF SPOT (FUNGAL) **12**, **13**, 143
MAGNESIUM DEFICIENCY **22**, 146
POWDERY MILDEWS **49**, 165
VINE WEEVIL **29**, 186
VIRUSES **23**, **27**, 187

**Hypericum**
RUSTS 173

**I**

**Ivy (*Hedera*)**
BLACKFLY 108
BRYOBIA MITES 111
LEAF SPOT (FUNGAL) **12**, **13**, 143
SOFT SCALE 177
VIBURNUM SCALE **15**, 185
VINE WEEVIL **29**, 186
VIRUSES **22**, **23**, 187

**J**

**Japanese quince (*Chaenomeles*)**
APHIDS 99
BROWN ROT **47**, 111
BROWN SCALE **36**, 111
GREY MOULD **34**, 134
IRON DEFICIENCY AND LIME-INDUCED CHLOROSIS **24**, 139

**Jasmine (*Jasminum*)**
GLASSHOUSE RED SPIDER MITE **22**, **23**, 132
GREY MOULD **34**, 134
VIRUSES **23**, **27**, 187

**Jerusalem sage (*Phlomis*)**
LEAFHOPPERS 142

**L**

**Laurel (*Laurus*)**
BACTERIAL CANKER **35**, 104
HONEY FUNGUS **34**, 136
LEAF SPOT (FUNGAL) **12**, **13**, 143
POWDERY MILDEWS **49**, 165

**Lavatera**
LAVATERA LEAF AND STEM ROT 142
LEAF SPOT (FUNGAL) **12**, **13**, 143
RUSTS 173

**Lavender (*Lavandula*)**
CUCKOO SPIT **39**, **43**, 120
GREY MOULD **34**, 134
HONEY FUNGUS **34**, 136
LAVENDER SHAB 142
LEAF SPOT (FUNGAL) **12**, **13**, 143

**Lilac (*Syringa*)**
HONEY FUNGUS **34**, 136
LEAF SPOT (FUNGAL) **12**, **13**, 143
LILAC BLIGHT **33**, 145
LILAC LEAF MINER **19**, 145
POWDERY MILDEWS **49**, 165
PRIVET THRIPS **22**, 165
SILVER LEAF 175

**M**

**Magnolia** *see* GARDEN TREES 194–5

**Mahonia**
MAHONIA RUST **12**, **24**, 146
POWDERY MILDEWS **49**, 165

**Mexican orange (*Choisya*)**
CARNATION TORTRIX MOTH **20**, 114
FROST DAMAGE **28**, **44**, 129
GLASSHOUSE RED SPIDER MITE **22**, **23**, 132
SCORCH 174
SNAILS **29**, 177

**Mock orange (*Philadelphus*)**
BLACKFLY 108
LEAF SPOT (FUNGAL) **12**, **13**, 143
VIRUSES **23**, **27**, 187

**Morning glory (*Ipomoea*)**
FROST DAMAGE (FOLIAGE AND STEMS) **28**, 129
GLASSHOUSE RED SPIDER MITE **22**, **23**, 132
MAGNESIUM DEFICIENCY **22**, 146
VIRUSES **23**, **27**, 187

**N**

**New Zealand flax (*Phormium*)**
LEAF SPOT (FUNGAL) **12**, **13**, 143
PHORMIUM MEALYBUG 159

**P**

**Photinia**
APPLE SCAB **16**, 101
FIREBLIGHT **35**, 127
LEAF SPOT (FUNGAL) **12**, **13**, 143
PEAR SCAB **48**, 157

**Privet (*Ligustrum*)**
APHIDS 99
HONEY FUNGUS **34**, 136
LEAF SPOT (FUNGAL) **12**, **13**, 143
LILAC LEAF MINER **19**, 145
PRIVET THRIPS **22**, 165
VIRUSES **23**, **27**, 187

**R**

**Rhododendron**
APHIDS 99
AZALEA WHITEFLY 103
BUD, DRY 112
CATERPILLARS 115
CROWN GALL **38**, 119
FROST DAMAGE **28**, **44**, 129
GREY MOULD **34**, 134
HONEY FUNGUS **34**, 136
IRON DEFICIENCY AND LIME-INDUCED CHLOROSIS **24**, 139
LEAF SCORCH 143
PHYTOPHTHORA 159
RHODODENDRON BUD BLAST **41**, 168
RHODODENDRON LACEBUG **21**, 168
RHODODENDRON LEAFHOPPER 169
RHODODENDRON PETAL BLIGHT **42**, 169
RHODODENDRON POWDERY MILDEW **15**, 169
RUSTS 173
SILVER LEAF 175
VINE WEEVIL **29**, 186

**Rock rose (*Helianthemum*)**
APHIDS 99
LEAF SPOT (FUNGAL) **12**, **13**, 143
POWDERY MILDEWS **49**, 165

**Rose (*Rosa*)**
CAPSID BUGS **29**, **46**, 114
CROWN GALL **38**, 119
DEER **41**, 122

DIEBACK 123
DOWNY MILDEW 124
GLASSHOUSE RED SPIDER MITE **22**, **23**, 132
GREY MOULD **34**, 134
HONEY FUNGUS **34**, 136
LEAF-CUTTING BEES **30**, 142
MAGNESIUM DEFICIENCY **22**, 146
NITROGEN DEFICIENCY 151
POTASSIUM DEFICIENCY **24**, 162
RABBITS 166
ROBIN'S PIN CUSHION **39**, 170
ROSE APHIDS **43**, 170
ROSE BALLING **44**, 170
ROSE BLACK SPOT **13**, 171
ROSE LEAFHOPPER **21**, 171
ROSE LEAF-ROLLING SAWFLY **27**, 171
ROSE POWDERY MILDEW **20**, 171
ROSE ROOT APHID **35**, 172
ROSE RUST **16**, 172
ROSE SICKNESS/REPLANT PROBLEMS 172
ROSE SLUGWORM **18**, 172
VIRUSES **23**, **27**, 187
WEEDKILLER DAMAGE – BROADLEAVED **27**, 188
WEEDKILLER DAMAGE – GLYPHOSATE 189
WINTER MOTH **31**, 191

**Rosemary (*Rosmarinus*)**
BROWN SCALE **36**, 111
CUCKOO SPIT **39**, **43**, 120
FROST DAMAGE (FOLIAGE AND STEMS) **28**, 129
HONEY FUNGUS **34**, 136
LEAFHOPPERS 142

**Russian vine (*Polygonum baldschuanicum*)**
LEAF MINERS 142

**S**

**Skimmia**
HORSE CHESTNUT SCALE **40**, 137
IRON DEFICIENCY AND LIME-INDUCED CHLOROSIS **24**, 139
MAGNESIUM DEFICIENCY **22**, 146
PHYTOPHTHORA 159
VINE WEEVIL **29**, 186
WATERLOGGING **34**, 188

**Solanum**
FLEA BEETLES **29**, 127
GREY MOULD **34**, 134
LEAF SPOT (FUNGAL) **12**, **13**, 143
VERTICILLIUM WILTS **29**, 185

**Stranvaesia**
FIREBLIGHT **35**, 127
HONEY FUNGUS **34**, 136

**Sumach (*Rhus*)**
FROST DAMAGE **28**, **44**, 129
LEAF SPOT (FUNGAL) **12**, **13**, 143
VERTICILLIUM WILTS **29**, 185

**Sun rose (*Cistus*)**
APHIDS 99

**Sweet bay (*Laurus*)**
BAY SUCKER **26**, 104
CARNATION TORTRIX MOTH **20**, 114
HORSE CHESTNUT SCALE **40**, 137
SOFT SCALE 177

**T**

**Trumpet vine (*Campsis*)**
BROWN SCALE **36**, 111
GLASSHOUSE RED SPIDER MITE **22**, **23**, 132

**V**

**Viburnum**
BLACKFLY 108
CAPSID BUGS **29**, **46**, 114
CARNATION TORTRIX MOTH **20**, 114
GREY MOULD **34**, 134
HONEY FUNGUS **34**, 136
LEAF SPOT (FUNGAL) **12**, **13**, 143
VIBURNUM BEETLE 185
VIBURNUM SCALE 185
VIBURNUM WHITEFLY **15**, 185

**W**

**Weigela**
BROWN SCALE **36**, 111
CAPSID BUGS **29**, **46**, 114
HONEY FUNGUS **34**, 136
LEAF AND BUD EELWORM **25**, 142
LEAF SPOT (FUNGAL) **12**, **13**, 143

**Wisteria**
BROWN SCALE **36**, 111
FROST DAMAGE **28**, **44**, 129
GLASSHOUSE RED SPIDER MITE **22**, **23**, 132
HONEY FUNGUS **34**, 136
LEAF SPOT (FUNGAL) **12**, **13**, 143
MAGNESIUM DEFICIENCY **22**, 146

**Y**

**Yucca**
BLACKFLY 108
SNAILS **29**, 177
YUCCA LEAF SPOT **18**, 192

# HERBACEOUS PERENNIALS

PERENNIALS DO NOT suffer from many serious problems. The annual dying-back of foliage and stems gives an opportunity for the removal and disposal of plant tissue harbouring many diseases and pests. Regular division of established clumps also helps to keep plants healthy. A number of common problems may affect a wide range of unrelated plants (*see right*). Some, however, are more host-specific (*see below*). Resistant species or cultivars may be suggested in the entry for the problem in question in the *A–Z of Pests, Diseases and Disorders*.

## COMMON PROBLEMS

The most common non-host-specific ailments affecting herbaceous perennials are:

| | |
|---|---|
| APHIDS 99 | LEAF SPOT (FUNGAL) 143 |
| CATERPILLARS 115 | POWDERY MILDEWS 165 |
| CROWN ROT 120 | RABBITS 166 |
| DEER 122 | SLUGS 176 |
| DOWNY MILDEW 124 | SNAILS 177 |
| DROUGHT 124 | SWIFT MOTH 180 |
| FROST DAMAGE 129 | VIRUSES 187 |
| GREY MOULD 134 | WATERLOGGING 188 |

# A

**Agapanthus**
VIRUSES **42**, 187

**Alstroemeria**
VIRUSES **42**, 187

**Alyssum**
CLUBROOT **55**, 118
DOWNY MILDEW 124
FLEA BEETLES **29**, 127
FOOT AND ROOT ROT **55**, 128
FUSARIUM WILT 130
POWDERY MILDEWS **25**, 165
SCLEROTINIA **36**, 174
WHITE BLISTER **14**, 189

**Anchusa**
SCLEROTINIA **36**, 174

**Anemone**
DOWNY MILDEW 124
GREY MOULD **45**, 134
PLUM RUST **22**, 161
POWDERY MILDEWS **25**, 165
SMUTS **35**, 176
VIRUSES **42**, 187

**Astilbe**
POWDERY MILDEWS **25**, 165

**Aubrieta**
APHIDS 99
DOWNY MILDEW 124
FLEA BEETLES **29**, 127
POWDERY MILDEWS **25**, 165
WHITE BLISTER **14**, 189

**Auricula**
(*Primula auricula*)
CROWN ROT **20**, 120
GLASSHOUSE LEAFHOPPER **20**, 132
ROOT APHIDS **55**, 170
VINE WEEVIL **29**, 186

# C

**Californian poppy**
(*Romneya*)
CATERPILLARS 115

PHYTOPHTHORA 159
VERTICILLIUM WILTS **29**, 185

**Campanula**
DOWNY MILDEW 124
GREY MOULD **45**, 134
LEAF SPOT (FUNGAL) **12**, **13**, 143
PHYTOPHTHORA 159
POWDERY MILDEWS **25**, 165
RUSTS 173
VERTICILLIUM WILTS **29**, 185
VIRUSES **42**, 187

**Candytuft (*Iberis*)**
CLUBROOT **55**, 118
DOWNY MILDEW 124
LEAF SPOT (FUNGAL) **12**, **13**, 143
POWDERY MILDEWS **25**, 165

**Canna**
HONEY FUNGUS **34**, 136

**Carnation (*Dianthus*)**
APHIDS 99
FUSARIUM WILT 130
GREY MOULD **45**, 134
LEAF SPOT (FUNGAL) **12**, **13**, 143
LEAFY GALL **39**, 143
POWDERY MILDEWS **25**, 165
RUSTS 173
SLUGS **29**, 176
SMUTS **35**, 176
SNAILS **29**, 177
VERTICILLIUM WILTS **29**, 185
VIRUSES **42**, 187

**Centaurea**
LETTUCE DOWNY MILDEW **22**, 144
POWDERY MILDEWS **25**, 165
RUSTS 173
SCLEROTINIA **36**, 174

**Chrysanthemum**
(*Dendranthema*)
APHIDS 99
CAPSID BUGS **29**, 114
CATERPILLARS 115
CHRYSANTHEMUM EELWORM 117
CHRYSANTHEMUM LEAF MINER **25**, 117

CHRYSANTHEMUM WHITE RUST **14**, 117
CROWN GALL **38**, 119
EARWIGS **46**, 125
FUSARIUM WILT 130
LEAF SPOT (FUNGAL) **12**, **13**, 143
MELON COTTON APHID **43**, 147
POWDERY MILDEWS **25**, 165
RUSTS 173

**Columbine (*Aquilegia*)**
APHIDS 99
AQUILEGIA SAWFLY 102
LEAF MINERS 142
LEAF SPOT (FUNGAL) **12**, **13**, 143
POWDERY MILDEWS **25**, 165

**Convolvulus**
FROST DAMAGE (FOLIAGE AND STEMS) **28**, 129
WATERLOGGING **34**, 188

**Coral flower**
(*Heuchera*)
LEAF SPOT (FUNGAL) **12**, **13**, 143
LEAFY GALL **39**, 143
VINE WEEVIL **29**, 186

**Cosmos**
FROST DAMAGE **28**, **44**, 129
POWDERY MILDEWS **25**, 165

**Cranesbill (*Geranium*)**
CAPSID BUGS **29**, 114
DOWNY MILDEW 124
GERANIUM SAWFLY 131
LEAFY GALL **39**, 143
POWDERY MILDEWS **25**, 165
VINE WEEVIL **29**, 186

# D

**Daisy (*Bellis perennis*)**
FASCIATION **43**, 126
POWDERY MILDEWS **25**, 165

**Daylily (*Hemerocallis*)**
HEMEROCALLIS GALL MIDGE **45**, 135

LEAF SPOT (FUNGAL) **12**, **13**, 143

**Delphinium**
DELPHINIUM LEAF MINER **18**, 122
DELPHINIUM MOTH 123
FASCIATION **43**, 126
FOOT AND ROOT ROTS **55**, 128
GREY MOULD **45**, 134
LEAF SPOT (BACTERIAL) **13**, **29**, 143
LEAF SPOT (FUNGAL) **12**, **13**, 143
POWDERY MILDEWS **25**, 165
SCLEROTINIA **36**, 174
SLUGS **29**, 176

# E

**Elephants' ears**
(*Bergenia*)
LEAF AND BUD EELWORM **25**, 142
LEAF SPOT (FUNGAL) **12**, **13**, 143
VINE WEEVIL **29**, 186

**Epimedium**
LEAF-CUTTING BEES **30**, 142
LEAF SPOT (FUNGAL) **12**, **13**, 143
MAGNESIUM DEFICIENCY **22**, 146
VINE WEEVIL **29**, 186

**Euphorbia**
FASCIATION **43**, 126
GREY MOULD **45**, 134

# F

**Foxglove (*Digitalis*)**
CATERPILLARS 115
DOWNY MILDEW 124
GLASSHOUSE LEAFHOPPER **20**, 132
LEAF SPOT (FUNGAL) **12**, **13**, 143
POWDERY MILDEWS **25**, 165

# G

**Gentian**
RUSTS 173

**Geraniums, hardy** *see*
**Cranesbill**

**Geum**
APHIDS 99
DOWNY MILDEW 124
GEUM SAWFLIES 131

**Globe thistle (*Echinops*)**
APHIDS 99
LEAF MINERS 142

**Goatsbeard**
(*Aruncus dioicus*)
ARUNCUS SAWFLY **32**, 102
POWDERY MILDEWS **25**, 165

**Golden rod (*Solidago*)**
POWDERY MILDEWS **25**, 165

# H

**Hellebore (*Helleborus*)**
APHIDS 99
DOWNY MILDEW 124
HELLEBORE LEAF BLOTCH **20**, 135
SLUGS **29**, 176

**Himalayan poppy**
(*Meconopsis*)
DOWNY MILDEW 124
PHYTOPHTHORA 159
SLUGS **29**, 176

**Hollyhock (*Alcea*)**
CATERPILLARS 115
HOLLYHOCK RUST **17**, 136
SLUGS **29**, 176
SNAILS **29**, 177

**Honesty (*Lunaria*)**
CLUBROOT **55**, 118
POWDERY MILDEWS **25**, 165
VIRUSES **42**, 187
WHITE BLISTER **14**, 189

**Hosta**
GREY MOULD **45**, 134

LEAF SPOT (FUNGAL) **12**, **13**, 143
SLUGS **29**, 176
SNAILS **29**, 177
VINE WEEVIL **29**, 186
VIRUSES **42**, 187

**House leek (Sempervivum)**
RUSTS 173
VINE WEEVIL **29**, 186

## I

**Ice plant (Sedum spectabile)**
VINE WEEVIL **29**, 186

## J

**Japanese anemone (Anemone × hybrida, A. hupehensis)**
GREY MOULD **45**, 134
LEAF AND BUD EELWORM **25**, 142
POWDERY MILDEWS **25**, 165
VIRUSES **42**, 187

## L

**Lily-of-the-valley (Convallaria)**
GREY MOULD **45**, 134
LEAF SPOT (FUNGAL) **12**, **13**, 143

**Lobelia**
DAMPING OFF **34**, 122
FOOT AND ROOT ROTS **55**, 128

GLASSHOUSE RED SPIDER MITE **22**, **23**, 132
SLUGS **29**, 176
SNAILS **29**, 177
VIRUSES **42**, 187

**Lupin (Lupinus)**
FOOT AND ROOT ROTS **55**, 128
GLASSHOUSE RED SPIDER MITE **22**, **23**, 132
LEAF SPOT (FUNGAL) **12**, **13**, 143
LUPIN APHID **41**, 146
POWDERY MILDEWS **25**, 165
SCLEROTINIA **36**, 174
SLUGS **29**, 176
SNAILS **29**, 177
VIRUSES **42**, 187

**Lychnis**
GREY MOULD **45**, 134

## M

**Michaelmas daisy (Aster novi-belgii)**
FUSARIUM WILT 130
GREY MOULD **45**, 134
LEAF SPOT (FUNGAL) **12**, **13**, 143
MICHAELMAS DAISY MITE **45**, 147
POWDERY MILDEWS **25**, 165

## P

**Peony (Paeonia)**
CATERPILLARS 115
HONEY FUNGUS **34**, 136
LEAF SPOT (FUNGAL) **12**, **13**, 143
PEONY WILT **43**, 158
VIRUSES **42**, 187

**Penstemon**
FROST DAMAGE (FOLIAGE AND STEMS) **28**, 129
GREY MOULD **45**, 134
LEAF AND BUD EELWORM **25**, 142
LEAF SPOT (FUNGAL) **12**, **13**, 143

**Periwinkle (Vinca)**
PERIWINKLE RUST **15**, 158

**Phlox**
CROWN GALL **38**, 119
LEAF SPOT (FUNGAL) **12**, **13**, 143
LEAFY GALL **39**, 143
PHLOX EELWORM **33**, 159
POWDERY MILDEWS **25**, 165
VERTICILLIUM WILTS **29**, 185

**Phygelius**
CAPSID BUGS **29**, 114
FIGWORT WEEVILS 126

**Pinks** see **Carnation**

**Polyanthus (Primula)**
APHIDS 99
BROWN CORE OF PRIMULAS 111
BRYOBIA MITES 111
CATERPILLARS 115
GLASSHOUSE LEAFHOPPER **20**, 132
GLASSHOUSE RED SPIDER MITE **22**, **23**, 132
GREY MOULD **45**, 134
LEAF SPOT (FUNGAL) **12**, **13**, 143
SLUGS **29**, 176
SPARROWS **46**, 177
VINE WEEVIL **29**, 186
VIRUSES **42**, 187

**Primrose** see **Polyanthus**

**Pulmonaria**
POWDERY MILDEWS **25**, 165

**Pyrethrum (Tanacetum coccineum)**
APHIDS 99
CHRYSANTHEMUM EELWORM 117
CHRYSANTHEMUM LEAF MINER **25**, 117
GREY MOULD **45**, 134
VIRUSES **42**, 187

## R

**Red-hot poker (Kniphofia)**
APHIDS 99
SNAILS **29**, 177

**Rock cress (Arabis)**
FLEA BEETLES **29**, 127
VIRUSES **42**, 187
WHITE BLISTER **14**, 189

## S

**Sage (Salvia)**
CAPSID BUGS **29**, 114
FOOT AND ROOT ROTS **55**, 128
LEAFHOPPERS 142
POWDERY MILDEWS **25**, 165
SLUGS **29**, 176
VERTICILLIUM WILTS **29**, 185

**Saxifrage (Saxifraga)**
RUSTS 173
VINE WEEVIL **29**, 186

**Snapdragon (Antirrhinum)**
ANTIRRHINUM RUST **15**, 99
DAMPING OFF **34**, 122
DOWNY MILDEW 124

FUSARIUM WILT 130
LEAF SPOT (BACTERIAL) **13**, **29**, 143
LEAF SPOT (FUNGAL) **12**, **13**, 143
POWDERY MILDEWS **25**, 165
SCLEROTINIA **36**, 174
SHOTHOLE **18**, **30**, 175
VERTICILLIUM WILTS **29**, 185
WILTS 191

**Stachys**
POWDERY MILDEWS **25**, 165
STACHYS CASE-BEARING CATERPILLAR 178

## V

**Verbascum**
FIGWORT WEEVILS 126
MULLEIN MOTH 149
POWDERY MILDEWS **25**, 165
WATERLOGGING **34**, 188

**Violet (Viola)**
APHIDS 99
GLASSHOUSE RED SPIDER MITE **22**, **23**, 132
LEAF SPOT (FUNGAL) **12**, **13**, 143
POWDERY MILDEWS **25**, 165
VIOLET GALL MIDGE **27**, 186

## W

**Water lily (Nymphaea)**
CHINA MARK MOTH **30**, 116
CHIRONOMID MIDGES 116
WATER LILY APHID 188
WATER LILY BEETLE **30**, 188
WATER LILY LEAF SPOT 188

# LAWNS

MANY GARDENERS happily tolerate a certain quantity of weeds in the lawn, and indeed some would say that a crop of daisies adds to a lawn's appeal. A possible exception is moss, which can be pernicious, particularly on damp ground; there are several effective moss-killers available, often combined with a lawn-feed mixture. Eradication of severe moss infestation may also have to be combined with re-seeding.

Pests and diseases, however, are always disfiguring and controls are necessary. In most cases of disease, chemical control must be combined with cultural improvements if the lawn's health is to be improved in the long term.

## COMMON PROBLEMS

The most common ailments affecting lawns are:

ANTS **59**, 99
BURROWING BEES **58**, 112
CHAFER GRUBS **58**, 116
EARTHWORMS **58**, 125
LEATHERJACKETS **58**, 143
MOLES **59**, 148
SLIME MOULD **58**, 176
TURF DOLLAR SPOT **58**, 183
TURF FAIRY RINGS **59**, 183
TURF OPHIOBOLUS PATCH 184
TURF RED THREAD **59**, 184
TURF SNOW MOULD **59**, 184

TURF THATCH FUNGAL MYCELIUM **58**, 184
TURF TOADSTOOLS 184

# BULBOUS PLANTS

THE PLANTING OF bulbs, corms and tubers in a suitable site is crucial to their health. Many will not tolerate soils which become heavy or wet in winter, soon succumbing to disease. In most cases, too, it is essential for the foliage to be left in place for several weeks after flowering in order to allow the bulbs to build and store up energy reserves for the following year.

Division of established clumps allows for the removal of less vigorous or unhealthy-looking portions before problems can spread. Many bulbs are prone to virus infections, and regular inspection and removal of those showing symptoms is advised. Other problems are more host-specific; the particular problems affecting various plants are listed below. Resistant species or cultivars may be suggested in the entry for the problem in question in the *A–Z of Pests, Diseases and Disorders*.

---

### COMMON PROBLEMS

The most common non-host-specific ailments affecting bulbs are:

SLUGS 176                     VIRUSES 187

---

## A

**Acidanthera**
GLADIOLUS CORM ROT **55**, 131
GLADIOLUS DRY ROT 131
GREY MOULD **45**, 134
LEAF SPOT (FUNGAL) **12**, **13**, 143

**Autumn crocus (*Colchicum*)**
SLUGS **29**, **43**, 176
SMUTS **35**, 176
TULIP GREY BULB ROT 183

## B

**Bluebell (*Hyacinthoides*)**
BLUEBELL RUST 109
NARCISSUS BULB FLY **54**, 150
VIRUSES **42**, 187

## C

**Colchicum** *see* **Autumn crocus**

**Crinum**
NARCISSUS LEAF SCORCH **29**, 150

**Crocosmia** *see* **Montbretia**

**Crocus**
BLUE MOULD ON BULBS **54**, 109
GLADIOLUS CORE ROT 131
GLADIOLUS DRY ROT 131
GREY MOULD **45**, 134
MICE **51**, 147
NARCISSUS BASAL ROT 149
SPARROWS **46**, 177
SQUIRRELS **52**, 178
TULIP GREY BULB ROT 183
VIOLET ROOT ROT **57**, 186

**Cyclamen, hardy**
BACTERIAL SOFT ROT 104
FOOT AND ROOT ROTS **55**, 128
FUSARIUM WILT 130
GREY MOULD **45**, 134
LEAF SPOT (FUNGAL) **12**, **13**, 143
POWDERY MILDEWS **25**, 165
SLUGS **29**, **43**, 176

## D

**Daffodil (*Narcissus*)**
BLINDNESS OF BULBS **41**, 108
BULB SCALE MITE 112
NARCISSUS BASAL ROT 149
NARCISSUS BULB FLY **54**, 150
NARCISSUS EELWORM **54**, 150
NARCISSUS LEAF SCORCH **29**, 150
NARCISSUS SMOULDER 150
NARCISSUS VIRUSES **23**, 150
POLLEN BEETLES **43**, 162
SLUGS **29**, **43**, 176
TULIP GREY BULB ROT 183
WEEDKILLER DAMAGE –
CONTACT 189

**Dahlia**
BLACKFLY 108
CAPSID BUGS **29**, **46**, 114
CROWN GALL **38**, 119
DAHLIA SMUT 122
EARWIGS **46**, 125
FOOT AND ROOT ROTS **55**, 128
GLASSHOUSE RED SPIDER
MITE **22**, **23**, 132
GREY MOULD **45**, 134
LEAF SPOT (FUNGAL) **12**, **13**, 143
LEAFY GALL **39**, 143
ONION THRIPS 153
PHYTOPHTHORA 159
POWDERY MILDEWS **25**, 165
SCLEROTINIA **36**, 174
VERTICILLIUM WILTS **29**, **33**, 185
VIRUSES **42**, 187

## E

**Erythronium**
GREY MOULD **45**, 134
RUSTS 173

## F

**Freesia**
APHIDS 99
BLUE MOULD ON BULBS **54**, 109
GLADIOLUS CORE ROT 131
GLADIOLUS DRY ROT 131
GLADIOLUS HARD ROT 132
GLASSHOUSE RED SPIDER
MITE **22**, **23**, 132
GREY MOULD **45**, 134
LEAF SPOT (FUNGAL) **12**, **13**, 143
SLUGS **29**, **43**, 176
VIRUSES **42**, 187

**Fritillary (*Fritillaria*)**
LILY BEETLE 145
SLUGS **29**, **43**, 176
VIRUSES **42**, 187

## G

**Gladiolus**
APHIDS 99
FUSARIUM BULB ROT 130
GLADIOLUS CORE ROT 131
GLADIOLUS CORM ROT **55**, 131
GLADIOLUS DRY ROT 131
GLADIOLUS SCAB AND NECK
ROT 132
GLADIOLUS THRIPS **45**, 132
LEAF SPOT (FUNGAL) **12**, **13**, 143
SLUGS **29**, **43**, 176
SNAILS **29**, **43**, 177
VIRUSES **42**, 187

**Grape hyacinth (*Muscari*)**
BLUEBELL RUST 109
SMUTS **35**, 176

## H

**Hyacinth (*Hyacinthus*)**
FUSARIUM BULB ROT 130
GREY MOULD **45**, 134
SLUGS **29**, **43**, 176
TULIP GREY BULB ROT 183
VIRUSES **42**, 187

## I

**Iris**
APHIDS 99
BLUE MOULD ON BULBS **54**, 109
IRIS INK DISEASE 138
IRIS RHIZOME ROT **55**, 139
IRIS SAWFLY **29**, 139
IRIS SCORCH 139
LEAF MINERS 142
LEAF SPOT (FUNGAL) **12**, **13**, 143
NARCISSUS BASAL ROT 149
RUSTS 173
SLUGS **29**, **43**, 176
SNAILS **29**, **43**, 177
TULIP GREY BULB ROT 183
VIRUSES **42**, 187

## L

**Lily (*Lilium*)**
APHIDS 99
LILY BEETLE 145
LILY DISEASE 145
SLUGS **29**, **43**, 176
VINE WEEVIL **29**, **53**, 186
VIRUSES **42**, 187

## M

**Montbretia (*Crocosmia*)**
GLADIOLUS DRY ROT 131

GLASSHOUSE RED SPIDER
MITE **22**, **23**, 132
IRIS INK DISEASE 138
LEAF SPOT (FUNGAL) **12**, **13**, 143

## O

**Ornithogalum**
SMUTS **35**, 176

## S

**Snowdrop (*Galanthus*)**
GLADIOLUS DRY ROT 131
NARCISSUS BULB FLY **54**, 150
NARCISSUS EELWORM **54**, 150
NARCISSUS LEAF SCORCH **29**, 150
SNOWDROP GREY MOULD 177

**Solomon's seal (*Polygonatum*)**
SLUGS **29**, **43**, 176
SOLOMON'S SEAL SAWFLY **31**, 177

## T

**Tulip (*Tulipa*)**
APHIDS 99
BLUE MOULD ON BULBS **54**, 109
DEER **41**, 122
MICE **51**, 147
SQUIRRELS **52**, 178
TULIP FIRE **22**, **55**, 183
TULIP GREY BULB ROT 183
VIRUSES **42**, 187

## W

**Winter aconite (*Eranthis*)**
SMUTS **35**, 176

# ANNUALS AND BIENNIALS

MANY PROBLEMS affecting annuals and biennials can be avoided by careful initial choice of plants, good planting, and regular feeding and watering. To avoid the foot and root rots to which many are prone, do not grow the same, or closely related, plants on any site for more than one year. Some problems affect a wide range of unrelated plants (*see right*); others are more host-specific (*below*). Resistant cultivars may be suggested in the entry for the problem in question in the *A–Z of Pests, Diseases and Disorders*.

## COMMON PROBLEMS

The most common non-host-specific ailments affecting annual and biennial plants are:

APHIDS 99
CUTWORMS 121
FOOT AND ROOT ROTS 128
FROST DAMAGE 129
LEAF SPOT (FUNGAL) 143
POTASSIUM DEFICIENCY 162
POWDERY MILDEWS 165

RABBITS 166
SLUGS 176
SNAILS 177

## A

**Anchusa** *see* HERBACEOUS PERENNIALS 198–9

**African marigold (*Tagetes erecta*)**
CUTWORMS **53**, 121
FOOT AND ROOT ROTS **55**, 128
GLASSHOUSE LEAFHOPPER **20**, 132
GLASSHOUSE RED SPIDER MITE **22**, **23**, 132
SLUGS **29**, **43**, 176
SNAILS **29**, **43**, 177

**Ageratum**
POWDERY MILDEWS **25**, 165

## B

**Busy Lizzie (*Impatiens*)**
APHIDS 99
FOOT AND ROOT ROTS **55**, 128
GLASSHOUSE LEAFHOPPER **20**, 132
GLASSHOUSE RED SPIDER MITE **22**, **23**, 132
GLASSHOUSE WHITEFLY 132
GREY MOULD **45**, 134
LEAF SCORCH 142
LEAF SPOT (FUNGAL) **12**, **13**, 143
TARSONEMID MITES **27**, 181
VINE WEEVIL **29**, **53**, 186
WATERLOGGING **34**, 188
WESTERN FLOWER THRIPS **42**, 189

## C

**Cabbage, Ornamental** *see* VEGETABLES AND HERBS 205–7

**Campanula** *see* HERBACEOUS PERENNIALS 198–9

**Canna** *see* HERBACEOUS PERENNIALS 198–9

**Castor-oil plant (*Ricinus communis*)**
HONEY FUNGUS **34**, 136

**Centaurea** *see* HERBACEOUS PERENNIALS 198–9

**China aster (*Callistephus*)**
APHIDS 99
CUTWORMS **53**, 121
FUSARIUM WILT 130
GREY MOULD **45**, 134
POWDERY MILDEWS **25**, 165
SCLEROTINIA **36**, 174
VIRUSES **23**, **25**, 187

**Convolvulus** *see* HERBACEOUS PERENNIALS 198–9

## E

**Evening primrose (*Oenothera*)**
LEAF SPOT (FUNGAL) **12**, **13**, 143

## F

**French marigold (*Tagetes patula*)** As for **African marigold**

**Forget-me-not (*Myosotis*)**
APHIDS 99
DOWNY MILDEW 124
POWDERY MILDEWS **25**, 165

## L

**Lobelia**
DAMPING OFF **34**, 122
FOOT AND ROOT ROTS **55**, 128
GLASSHOUSE RED SPIDER MITE **22**, **23**, 132
SLUGS **29**, **43**, 176

## M

**Marigold (*Calendula*)**
CROWN GALL **38**, 119
GREY MOULD **45**, 134
LEAF SPOT (FUNGAL) **12**, **13**, 143
POWDERY MILDEWS **25**, 165
SMUTS **35**, 176

## N

**Nasturtium (*Tropaeolum*)**
BLACKFLY 108
CABBAGE WHITE BUTTERFLY 113
DOWNY MILDEW 124
FLEA BEETLES **29**, 127
LEAFY GALL **39**, 143
VIRUSES **23**, **25**, 187

**Nemesia**
PHYTOPHTHORA 159

**Nicotiana**
GLASSHOUSE LEAFHOPPER **20**, 132
POWDERY MILDEWS **25**, 165
VIRUSES **23**, **25**, 187

## P

**Pansy (*Viola*)**
APHIDS 99
DOWNY MILDEW 124
FOOT AND ROOT ROTS **55**, 128
GLASSHOUSE RED SPIDER MITE **22**, **23**, 132
LEAF SPOT (FUNGAL) **12**, **13**, 143
PANSY SICKNESS 154
POWDERY MILDEWS **25**, 165
RUSTS 173
SLUGS **29**, **43**, 176
SMUTS **35**, 176

## Petunia

**Petunia**
APHIDS 99
DAMPING OFF **34**, 122
PETUNIA FOOT ROT 159
POWDERY MILDEWS **25**, 165
SLUGS **29**, **43**, 176
VIRUSES **23**, **25**, 187

**Poppy (*Papaver*)**
DOWNY MILDEW 124
PEDICEL NECROSIS 158
POWDERY MILDEWS **25**, 165
VERTICILLIUM WILTS **29**, **33**, 185

## S

**Salvia**
APHIDS 99
CAPSID BUGS **29**, **46**, 114
SLUGS **29**, **43**, 176

**Statice**
GREY MOULD **45**, 134
POWDERY MILDEWS **25**, 165

**Stocks (*Matthiola*)**
CLUBROOT **55**, 118
DOWNY MILDEW 124
FLEA BEETLES **29**, 127
GREY MOULD **45**, 134
VIRUSES **23**, **25**, 187
WHITE BLISTER **14**, 189

**Sweet pea (*Lathyrus*)**
APHIDS 99
CROWN GALL **38**, 119
CUCUMBER MOSAIC VIRUS **21**, **25**, 120
DOWNY MILDEW 124
FOOT AND ROOT ROTS **55**, 128
FUSARIUM WILT 130
GREY MOULD **45**, 134
LEAF SPOT (FUNGAL) **12**, **13**, 143
LEAFY GALL **39**, 143
POLLEN BEETLES **43**, 162
POWDERY MILDEWS **25**, 165
SLUGS **29**, **43**, 176
SNAILS **29**, **43**, 177
VIRUSES **23**, **25**, 187

**Sweet William (*Dianthus barbatus*)**
LEAF SPOT (FUNGAL) **12**, **13**, 143
RUSTS 173

## T

**Tobacco plant** *see* **Nicotiana**

## V

**Verbena**
APHIDS 99
FOOT AND ROOT ROTS **55**, 128
GLASSHOUSE LEAFHOPPER **20**, 132
GLASSHOUSE RED SPIDER MITE **22**, **23**, 132
WESTERN FLOWER THRIPS **42**, 189

**Viola** *see* **Pansy**

## W

**Wallflower (*Cheiranthus*)**
CLUBROOT **55**, 118
CROWN GALL **38**, 119
DOWNY MILDEW 124
FLEA BEETLES **29**, 127
GREY MOULD **45**, 134
LEAFY GALL **39**, 143
RABBITS 166
VIRUSES **23**, **25**, 187
WHITE BLISTER **14**, 189

## Z

**Zinnia**
FOOT AND ROOT ROTS **55**, 128
GREY MOULD **45**, 134
LEAF SPOT (FUNGAL) **12**, **13**, 143

# GREENHOUSE AND HOUSE PLANTS

A SHELTERED environment not only protects plants, but also pests and diseases. They may therefore build up very rapidly, and exotic species may flourish that would have little chance to cause serious problems out of doors. However, an enclosed environment also lends itself to effective and thorough treatment, and to the use of biological controls (*see pp.90–91*) to eradicate pests.

Some common problems may affect a wide range of unrelated plants (*see right*). Others are more host-specific (*see below*). Resistant species or cultivars may be suggested in the entry for the problem in question in the *A–Z of Pests, Diseases and Disorders*.

## COMMON PROBLEMS

The most common non-host-specific ailments affecting plants grown in the greenhouse or conservatory, or as house plants, are:

APHIDS 99
CARNATION TORTRIX MOTH 114
FOOT AND ROOT ROTS 128
FUNGUS GNATS 129
GLASSHOUSE LEAFHOPPER 132
GLASSHOUSE RED SPIDER MITE 132
GLASSHOUSE WHITEFLY 132
LEAF SCORCH 143
MEALYBUGS 147

POWDERY MILDEWS 165
ROOT MEALYBUGS 170
SCALE INSECTS 173
SLUGS 176
SPRINGTAILS 178
VINE WEEVIL 186
VIRUSES 187
WATERLOGGING 188
WESTERN FLOWER THRIPS 189

## A

**Abutilon**
APHIDS 99
GLASSHOUSE WHITEFLY 132
SOFT SCALE 177
VIRUSES 23, 187

**African violet (*Saintpaulia*)**
APHIDS 99
CROWN ROT 120
GREY MOULD 34, 134
LEAF SCORCH 143
MEALYBUGS 32, 37, 147
PHYTOTOXICITY 160
POWDERY MILDEWS 49, 165
ROOT MEALYBUGS 55, 170
TARSONEMID MITES 27, 181
VINE WEEVIL 53, 186
WESTERN FLOWER THRIPS 42, 189

**Amaryllis (*Hippeastrum*)**
BULB SCALE MITE 112
NARCISSUS BULB FLY 54, 150
NARCISSUS LEAF SCORCH 29, 150
VIRUSES 23, 187

**Angels' trumpets (*Brugmansia/Datura*)**
GLASSHOUSE RED SPIDER MITE 22, 23, 132
GLASSHOUSE WHITEFLY 132
MEALYBUGS 32, 37, 147
POTATO BLIGHT 28, 56, 163
TOMATO BLIGHT 51, 181
VIRUSES 23, 187

**Arum lily (*Zantedeschia*)**
APHIDS 99
BACTERIAL SOFT ROT 104
LEAF SPOT (FUNGAL) 12, 13, 143
PHYTOPHTHORA 159
VIRUSES 23, 187

**Asparagus fern**
CARNATION TORTRIX MOTH 20, 114
MEALYBUGS 32, 37, 147
OLEANDER SCALE 152

**Aspidistra**
GLASSHOUSE RED SPIDER MITE 22, 23, 132
LEAF SPOT (FUNGAL) 12, 13, 143

## B

**Banana (*Musa*)**
GLASSHOUSE RED SPIDER MITE 22, 23, 132
LEAF SCORCH 143
LEAF SPOT (FUNGAL) 12, 13, 143
MEALYBUGS 32, 37, 147

**Begonia**
APHIDS 99
FOOT AND ROOT ROTS 55, 128
GREY MOULD 34, 134
LEAF AND BUD EELWORM 25, 142
LEAF SCORCH 143
LEAF SPOT (FUNGAL) 12, 13, 143
OEDEMA 19, 152
PHYTOPHTHORA 159
POWDERY MILDEWS 49, 165
TARSONEMID MITES 27, 181
VINE WEEVIL 53, 186
VIRUSES 23, 187

**Bougainvillea**
APHIDS 99
MEALYBUGS 32, 37, 147

## C

**Cacti**
GLASSHOUSE RED SPIDER MITE 22, 23, 132

MEALYBUGS 32, 37, 147
PHYTOPHTHORA 159
ROOT MEALYBUGS 55, 170
WATERLOGGING 34, 188

**Calceolaria**
APHIDS 99
FOOT AND ROOT ROTS 55, 128
PHYTOPHTHORA 159
VIRUSES 23, 187

**Canna** *see* HERBACEOUS PERENNIALS 198–9

**Carnation (*Dianthus*)**
APHIDS 99
CARNATION TORTRIX MOTH 20, 114
FUSARIUM WILT 130
GLASSHOUSE RED SPIDER MITE 22, 23, 132
LEAFY GALL 39, 143
POWDERY MILDEWS 49, 165

**Christmas cactus (*Schlumbergera*)**
MEALYBUGS 32, 37, 147

**Chrysanthemum (*Dendranthema*)**
CHRYSANTHEMUM EELWORM 117
CHRYSANTHEMUM LEAF MINER 25, 117
CHRYSANTHEMUM PETAL BLIGHT 45, 117
CHRYSANTHEMUM RAY BLIGHT 117
CHRYSANTHEMUM WHITE RUST 14, 117
CROWN GALL 38, 119
FUSARIUM WILT 130
GLASSHOUSE LEAFHOPPER 20, 132
GLASSHOUSE RED SPIDER MITE 22, 23, 132
GLASSHOUSE WHITEFLY 132
GREY MOULD 34, 134
LEAF SPOT (FUNGAL) 12, 13, 143

LEAFY GALL 39, 143
MELON COTTON APHID 43, 147
PHYTOPHTHORA 159
RUSTS 173
SCLEROTINIA 36, 174
VERTICILLIUM WILTS 29, 33, 185
VIRUSES 23, 187
WESTERN FLOWER THRIPS 42, 189

**Cineraria (*Senecio* × *hybridus*)**
APHIDS 99
CHRYSANTHEMUM LEAF MINER 25, 117
CROWN GALL 38, 119
DOWNY MILDEW 124
FOOT AND ROOT ROTS 55, 128
FUSARIUM WILT 130
GREY MOULD 34, 134
LEAF SPOT (FUNGAL) 12, 13, 143
LEAFY GALL 39, 143
LETTUCE DOWNY MILDEW 22, 144
PHYTOPHTHORA 159
RUSTS 173
SCLEROTINIA 36, 174
VERTICILLIUM WILTS 29, 33, 185
VIRUSES 23, 187

**Citrus fruits (*Citrus, Fortunella*)**
APHIDS 99
GLASSHOUSE RED SPIDER MITE 22, 23, 132
MEALYBUGS 32, 37, 147
PHYTOPHTHORA 159
SOFT SCALE 177

**Clivia**
LEAF SPOT (FUNGAL) 12, 13, 143
MEALYBUGS 32, 37, 147
VERTICILLIUM WILTS 29, 33, 185

**Coleus (*Solenostemon*)**
APHIDS 99
GREY MOULD 34, 134
VERTICILLIUM WILTS 29, 33, 185

**Croton (*Codiaeum*)**
FROST DAMAGE 28, 44, 129
GLASSHOUSE RED SPIDER MITE 22, 23, 132
MEALYBUGS 32, 37, 147
WATERLOGGING 34, 188

**Cyclamen**
APHIDS 99
BACTERIAL SOFT ROT 104
CARNATION TORTRIX MOTH 20, 114
FOOT AND ROOT ROT 55, 128
FUSARIUM WILT 130
GLASSHOUSE RED SPIDER MITE 22, 23, 132
GREY MOULD 34, 134
LEAF SPOT (FUNGAL) 12, 13, 143
MEALYBUGS 32, 37, 147
POWDERY MILDEWS 49, 165
TARSONEMID MITES 27, 181
VINE WEEVIL 53, 186
WATERLOGGING 34, 188
WESTERN FLOWER THRIPS 42, 189

## D

**Dracaena**
LEAF SPOT (FUNGAL) 12, 13, 143

## E

**Echeveria**
MEALYBUGS 32, 37, 147
RUSTS 173
VINE WEEVIL 53, 186

**Epiphyllum**
APHIDS 99
MEALYBUGS 32, 37, 147

# F

### Ferns
APHIDS 99
FOOT AND ROOT ROTS **55**, 128
HEMISPHERICAL SCALE **16**, 135
LEAF AND BUD EELWORM **25**, 142
LEAF SCORCH 143
MEALYBUGS **32**, **37**, 147
RUSTS 173
SMUTS **35**, 176
SOFT SCALE 177

### Frangipani (*Plumeria*)
GLASSHOUSE RED SPIDER MITE **22**, **23**, 132
MEALYBUGS **32**, **37**, 147

### Fuchsia
APHIDS 99
FOOT AND ROOT ROTS **55**, 128
FUCHSIA RUST **18**, 129
GLASSHOUSE LEAFHOPPER **20**, 132
GLASSHOUSE RED SPIDER MITE **22**, **23**, 132
GLASSHOUSE WHITEFLY 132
LEAF SPOT (FUNGAL) **12**, **13**, 143
ROOT MEALYBUGS **55**, 170
VINE WEEVIL **53**, 186

# G

### Gardenia
APHIDS 99
BUD, DRY 112
GLASSHOUSE RED SPIDER MITE **22**, **23**, 132
GREY MOULD **34**, 134
IRON DEFICIENCY AND LIME-INDUCED CHLOROSIS **24**, 146
MAGNESIUM DEFICIENCY **22**, 146
MEALYBUGS **32**, **37**, 147
SOFT SCALE 177

### Gazania
APHIDS 99
CHRYSANTHEMUM LEAF MINER **25**, 117
GLASSHOUSE RED SPIDER MITE **22**, **23**, 132
GLASSHOUSE WHITEFLY 132

### Gerbera
APHIDS 99
CHRYSANTHEMUM LEAF MINER **25**, 117
FOOT AND ROOT ROTS **55**, 128
GLASSHOUSE WHITEFLY 132
LEAF SPOT (FUNGAL) **12**, **13**, 143
VERTICILLIUM WILTS **29**, **33**, 185

### Gloxinia (*Sinningia*)
FOOT AND ROOT ROTS **55**, 128
GLASSHOUSE LEAFHOPPER **20**, 132
POWDERY MILDEWS **49**, 165
VINE WEEVIL **53**, 186
WESTERN FLOWER THRIPS **42**, 189

# H

### Hibiscus
APHIDS 99
GLASSHOUSE RED SPIDER MITE **22**, **23**, 132
VIRUSES **23**, 187

### Hot water plant (*Achimenes*)
APHIDS 99
GLASSHOUSE LEAFHOPPER **20**, 132
WESTERN FLOWER THRIPS **42**, 189

# K

### Kalanchoe
APHIDS 99
FOOT AND ROOT ROT **55**, 128
POWDERY MILDEWS **49**, 165
VINE WEEVIL **53**, 186

### Kangaroo vine (*Rhoicissus*)
MEALYBUGS **32**, **37**, 147
POWDERY MILDEWS **49**, 165
VINE WEEVIL **53**, 186

# L

### Lantana
GLASSHOUSE WHITEFLY 132

### Lapageria
GLASSHOUSE RED SPIDER MITE **22**, **23**, 132
MEALYBUGS **32**, **37**, 147

# M

### Marguerite (*Chrysanthemum frutescens*)
APHIDS 99
CHRYSANTHEMUM LEAF MINER **25**, 117
FUSARIUM WILT 130
LEAF SPOT (FUNGAL) **12**, **13**, 143
POWDERY MILDEW **49**, 165

### Mimosa (*Mimosa*)
CARNATION TORTRIX MOTH **20**, 114

### Glasshouse Red Spider Mite
GLASSHOUSE RED SPIDER MITE **22**, **23**, 132
MEALYBUGS **32**, **37**, 147

### Money tree (*Crassula*)
CARNATION TORTRIX MOTH **20**, 114
GREY MOULD **34**, 134
MEALYBUGS **32**, **37**, 147
VINE WEEVIL **53**, 186

# O

### Oleander (*Nerium*)
APHIDS 99
CARNATION TORTRIX MOTH **20**, 114
GLASSHOUSE RED SPIDER MITE **22**, **23**, 132
GREY MOULD **34**, 134
HEMISPHERICAL SCALE **16**, 135
MEALYBUGS **32**, **37**, 147
OLEANDER SCALE 152

### Orchids
APHIDS 99
FOOT AND ROOT ROTS **55**, 128
GLASSHOUSE RED SPIDER MITE **22**, **23**, 132
LEAF SPOT (FUNGAL) **12**, **13**, 143
SCALE INSECTS 173
SLUGS **29**, **43**, 176
VIRUSES **23**, 187

# P

### Palms
BANDED PALM THRIPS **22**, 104
FOOT AND ROOT ROTS **55**, 128
GLASSHOUSE RED SPIDER MITE **22**, **23**, 132
HEMISPHERICAL SCALE **16**, 135
LEAF SPOT (FUNGAL) **12**, **13**, 143
MEALYBUGS **32**, **37**, 147
OLEANDER SCALE 152
SOFT SCALE 177
WATERLOGGING **34**, 188

### Papyrus (*Cyperus*)
MEALYBUGS **32**, **37**, 147

### Passion flower (*Passiflora*)
CUCUMBER MOSAIC VIRUS 120
GLASSHOUSE LEAFHOPPER **20**, 132
GLASSHOUSE RED SPIDER MITE **22**, **23**, 132

### Magnesium Deficiency
MAGNESIUM DEFICIENCY **22**, 146
MEALYBUGS **32**, **37**, 147
VIRUSES **23**, 187

### Pelargonium
APHIDS 99
BLACKLEG OF CUTTINGS 108
CARNATION TORTRIX MOTH **20**, 114
CROWN GALL **38**, 119
GLASSHOUSE LEAFHOPPER **20**, 132
GLASSHOUSE WHITEFLY 132
LEAF SPOT (FUNGAL) **12**, **13**, 143
LEAFY GALL **39**, 143
MEALYBUGS **32**, **37**, 147
OEDEMA **19**, 152
ROOT MEALYBUGS **55**, 170
RUSTS 173
TARSONEMID MITES **27**, 181
VERTICILLIUM WILTS **29**, **33**, 185
VINE WEEVIL **53**, 186
WESTERN FLOWER THRIPS **42**, 189

### Pepper, ornamental (*Capsicum*)
APHIDS 99
FOOT AND ROOT ROTS **55**, 128
GLASSHOUSE RED SPIDER MITE **22**, **23**, 132
GLASSHOUSE WHITEFLY 132
VIRUSES **23**, 187

### Plumbago
CARNATION TORTRIX MOTH **20**, 114
GLASSHOUSE WHITEFLY 132
MEALYBUGS **32**, **37**, 147

### Poinsettia (*Euphorbia pulcherrima*)
GLASSHOUSE WHITEFLY 132
LEAFY GALL **39**, 143
SOFT SCALE 177

# R

### Rubber plant (*Ficus elastica*)
BANDED PALM THRIPS **22**, 104
GLASSHOUSE RED SPIDER MITE **22**, **23**, 132
IRREGULAR WATERING **36**, 140
LEAF SCORCH 143
MEALYBUGS **32**, **37**, 147
SOFT SCALE 177
WATERLOGGING **34**, 188

# S

### Schefflera
BANDED PALM THRIPS **22**, 104

### Glasshouse Red Spider Mite
GLASSHOUSE RED SPIDER MITE **22**, **23**, 132
IRREGULAR WATERING **36**, 140
MEALYBUGS **32**, **37**, 147
SOFT SCALE 177
WATERLOGGING **34**, 188

### Spider plant (*Chlorophytum*)
HEMISPHERICAL SCALE **16**, 135
SOFT SCALE 177

### Stephanotis
GLASSHOUSE RED SPIDER MITE **22**, **23**, 132
HEMISPHERICAL SCALE **16**, 135
MEALYBUGS **32**, **37**, 147
SOFT SCALE 177
VIRUSES **23**, 187

### Streptocarpus
APHIDS 99
GLASSHOUSE LEAFHOPPER **20**, 132
GREY MOULD **34**, 134
MEALYBUGS **32**, **37**, 147
POWDERY MILDEWS **49**, 165
ROOT MEALYBUGS **55**, 170
TARSONEMID MITES **27**, 181
VINE WEEVIL **53**, 186
WATERLOGGING **34**, 188
WESTERN FLOWER THRIPS **42**, 189

# W

### Wandering Jew (*Tradescantia*)
GLASSHOUSE LEAFHOPPER **20**, 132
VINE WEEVIL **53**, 186

### Wax plant (*Hoya*)
GLASSHOUSE RED SPIDER MITE **22**, **23**, 132
MEALYBUGS **32**, **37**, 147
VIRUSES **23**, 187

### Weeping fig (*Ficus benjamina*)
BANDED PALM THRIPS **22**, 104
GLASSHOUSE RED SPIDER MITE **22**, **23**, 132
IRREGULAR WATERING **36**, 140
LEAF SCORCH 143
MEALYBUGS **32**, **37**, 147
SOFT SCALE 177
WATERLOGGING **34**, 188

### Winter cherry (*Solanum capsicastrum*)
APHIDS 99
GLASSHOUSE RED SPIDER MITE **22**, **23**, 132
GLASSHOUSE WHITEFLY 132
MAGNESIUM DEFICIENCY **22**, 146

# FRUITS AND NUTS

MANY GARDENERS are happy to use chemicals on ornamentals, but when it comes to cropping plants, most prefer to avoid them wherever possible. Good cultivation, preventive measures taken at the correct time, and the use of resistant cultivars will all help. Some problems affect many fruits (*see right*). Others are more host-specific; plants so affected are listed below. More resistant species or cultivars may be suggested in the entry for the problem in question in the *A–Z of Pests, Diseases and Disorders*.

## COMMON PROBLEMS

The most common problems affecting plants and trees bearing edible fruits and nuts are:

APHIDS 99
BIRD DAMAGE ON FRUITS 107
CORAL SPOT 119
GREY MOULD 134
HONEY FUNGUS 136
MAGNESIUM DEFICIENCY 146
POWDERY MILDEWS 165

RABBITS 166
SQUIRRELS 178
WASPS 188

## A

### Apple
APPLE BITTER PIT **48**, 100
APPLE BLOSSOM WEEVIL **46**, 100
APPLE CANKER **36**, 100
APPLE CAPSID **47**, 100
APPLE LEAF MINER **19**, 101
APPLE SAWFLY **47**, **48**, 101
APPLE SCAB **16**, 101
APPLE SOOTY BLOTCH 102
APPLE WATER CORE 102
BLOSSOM WILT **41**, 108
BROWN ROT **47**, **49**, 111
CODLING MOTH **47**, 118
CORAL SPOT **41**, 119
CROWN GALL **38**, 119
FIREBLIGHT **35**, 127
FRUIT TREE RED SPIDER MITE **20**, **40**, 129
HIGH TEMPERATURE INJURY 135
HONEY FUNGUS **34**, 136
IRON DEFICIENCY AND LIME-INDUCED CHLOROSIS **24**, 139
LEOPARD MOTH 144
MAGNESIUM DEFICIENCY **22**, 146
MUSSEL SCALE **36**, 149
NITROGEN DEFICIENCY 151
PAPERY BARK **40**, 154
PHYTOPHTHORA 159
POTASSIUM DEFICIENCY **24**, 162
POWDERY MILDEWS **49**, 165
PROLIFERATION **44**, 166
ROSY APPLE APHID **48**, 173
SILVER LEAF 175
SPLITTING 178
VIRUSES **23**, 187
WINTER MOTH **31**, 191
WOOLLY APHID **35**, 192

### Apricot *see* Peach

## B

### Blackberry
BLACKBERRY CANE SPOT 108

CROWN GALL **38**, 119
FROST DAMAGE (BUDS AND FLOWERS) **44**, 129
FROST DAMAGE (FOLIAGE AND STEMS) **28**, 129
GREY MOULD **50**, 134
LEAFHOPPERS 142
LEAF SPOT (FUNGAL) **12**, **13**, 143
POWDERY MILDEWS **49**, 165
RASPBERRY BEETLE **50**, 167
RED BERRY MITE **49**, 168
RUSTS 173
VIRUSES **23**, 187

## C

### Cherry
APPLE LEAF MINER **19**, 101
BACTERIAL CANKER **35**, 104
BLOSSOM WILT **41**, 108
BROWN ROT **47**, 111
CHERRY BLACKFLY **24**, 116
CROWN GALL **38**, 119
FROST DAMAGE (BUDS AND FLOWERS) **44**, 129
FROST DAMAGE (FOLIAGE AND STEMS) **28**, 129
HONEY FUNGUS **34**, 136
IRON DEFICIENCY AND LIME-INDUCED CHLOROSIS **24**, 139
LEAF SPOT (FUNGAL) **12**, **13**, 143
MAGNESIUM DEFICIENCY **22**, 146
NITROGEN DEFICIENCY 151
PEAR AND CHERRY SLUGWORM **19**, 157
PHOSPHORUS DEFICIENCY 159
POTASSIUM DEFICIENCY 162
POWDERY MILDEWS **49**, 165
SHOTHOLE **18**, **30**, 175
SHOT-HOLE BORERS **33**, 175
SILVER LEAF 175
SPLITTING 178
VIRUSES **23**, 187
WINTER MOTH **31**, 191
WITCHES' BROOMS **41**, 191

### Citrus *see* GREENHOUSE & HOUSE PLANTS 202–3

### Currant, black
AMERICAN GOOSEBERRY MILDEW **25**, **49**, 98
BIG BUD MITES **41**, 107
BLACKCURRANT GALL MIDGE **26**, 108
CAPSID BUGS **29**, **46**, 114
CORAL SPOT **41**, 119
CURRANT BLISTER APHID 121
CURRANT CLEARWING MOTH **33**, 121
GLASSHOUSE RED SPIDER MITE **22**, **23**, 132
GOOSEBERRY DIEBACK 133
GREY MOULD **50**, 134
HONEY FUNGUS **34**, 136
LEAF SPOT (BACTERIAL) **13**, **29**, 143
LEAF SPOT (FUNGAL) **12**, **13**, 143
VIRUSES **23**, 187
WOOLLY VINE SCALE **39**, 192

### Currant, red/white
BROWN SCALE **36**, 111
CAPSID BUGS **29**, **46**, 114
CORAL SPOT **41**, 119
CURRANT AND GOOSEBERRY LEAF SPOT 120
CURRANT BLISTER APHID **27**, 121
CURRANT CLEARWING MOTH **33**, 121
DIEBACK 123
GOOSEBERRY SAWFLY **31**, 133
GREY MOULD **50**, 134
HONEY FUNGUS **34**, 136
LEAF SPOT (FUNGAL) **12**, **13**, 143
VIRUSES **23**, 187
WOOLLY VINE SCALE **39**, 192

## F

### Fig
CORAL SPOT **41**, 119
DIEBACK 123
FROST DAMAGE **28**, **44**, 129
GLASSHOUSE RED SPIDER MITE **22**, **23**, 132
GREY MOULD **50**, 134
LEAF SPOT (FUNGAL) **12**, **13**, 143

MEALYBUGS **32**, **37**, 147
SOFT SCALE 177
VIRUSES **23**, 187

## G

### Gooseberry
AMERICAN GOOSEBERRY MILDEW **25**, **49**, 98
APHIDS 99
BROWN SCALE **36**, 111
BULLFINCHES **46**, 112
CROWN GALL **38**, 119
CURRANT AND GOOSEBERRY LEAF SPOT 120
DIEBACK 123
FROST DAMAGE (BUDS AND FLOWERS) **44**, 129
FROST DAMAGE (FOLIAGE AND STEMS) **28**, 129
GOOSEBERRY SAWFLY **31**, 133
GREY MOULD **50**, 134
HONEY FUNGUS **34**, 136
IRON DEFICIENCY AND LIME-INDUCED CHLOROSIS **24**, 139
RUSTS 173
VIRUSES **23**, 187

### Grape *see* Vine

## H

### Hazelnut
APHIDS 99
BIG BUD MITES **41**, 107
LEAF SPOT (FUNGAL) **12**, **13**, 143
NUT WEEVIL **52**, 151
POWDERY MILDEWS **49**, 165
SQUIRRELS **52**, 178
VIRUSES **23**, 187
WINTER MOTH **31**, 191

## M

### Mulberry
MULBERRY BACTERIAL BLIGHT 149
MULBERRY CANKER **38**, 149

## P

### Peach, nectarine, apricot
BACTERIAL CANKER **35**, 104
BLOSSOM WILT **41**, 108
BROWN SCALE **36**, 111
CROWN GALL **38**, 119
FROST DAMAGE (BUDS AND FLOWERS) **44**, 129
FROST DAMAGE (FOLIAGE AND STEMS) **28**, 129
GLASSHOUSE RED SPIDER MITE **22**, **23**, 132
HONEY FUNGUS **34**, 136
IRON DEFICIENCY AND LIME-INDUCED CHLOROSIS **24**, 139
NITROGEN DEFICIENCY 151
PEACH APHIDS 156
PEACH LEAF CURL **26**, 156
PHOSPHORUS DEFICIENCY 159
PHYTOPHTHORA 159
PLUM RUST **22**, 161
POTASSIUM DEFICIENCY 162
POWDERY MILDEWS **49**, 165
SHOTHOLE **18**, **30**, 175
SILVER LEAF 175

### Pear
APPLE AND PEAR CANKER **36**, 100
APPLE POWDERY MILDEW **25**, 101
APPLE SCAB **16**, 101
BLOSSOM WILT **41**, 108
BORON DEFICIENCY 109
BROWN ROT **47**, **49**, 111
BULLFINCHES **46**, 112
CODLING MOTH **47**, 118
CORAL SPOT **41**, 119
FIREBLIGHT **35**, 127
FROST DAMAGE (BUDS AND FLOWERS) **44**, 129
HONEY FUNGUS **34**, 136
MAGNESIUM DEFICIENCY **22**, 146
NITROGEN DEFICIENCY 151
PEAR AND CHERRY SLUGWORM **19**, 157
PEAR BEDSTRAW APHID 157
PEAR LEAF BLISTER MITE **16**, 157
PEAR MIDGE **48**, 157

PEAR STONY PIT VIRUS **48**, 157
PEAR SUCKER 157
PHOSPHORUS DEFICIENCY 159
POTASSIUM DEFICIENCY 162
SOOTY MOULD 177
VIRUSES **23**, 187
WINTER MOTH **31**, 191

**Plum, damson, gage**
BACTERIAL CANKER **35**, 104
BLOSSOM WILT **41**, 108
BROWN SCALE **36**, 111
BULLFINCHES **46**, 112
CROWN GALL **38**, 119
FROST DAMAGE **28**, **44**, 129
FRUIT TREE RED SPIDER MITE **20**, **40**, 129
HONEY FUNGUS **34**, 136
IRON DEFICIENCY AND LIME-INDUCED CHLOROSIS **24**, 139
MEALY PLUM APHID 146
NITROGEN DEFICIENCY 151
PHOSPHORUS DEFICIENCY 159
PHYTOPHTHORA 159
PLUM LEAF-CURLING APHID **27**, 161
PLUM LEAF GALL MITE **16**, 161
PLUM MOTH **48**, 161

PLUM RUST **22**, 161
PLUM SAWFLY 161
POTASSIUM DEFICIENCY 162
POWDERY MILDEWS **49**, 165
SHOTHOLE **18**, **30**, 175
SHOT-HOLE BORERS **33**, 175
SILVER LEAF 175
WINTER MOTH **31**, 191

**Q**

**Quince**
BROWN ROT **47**, **49**, 111
CROWN GALL **38**, 119
FIREBLIGHT **35**, 127
HONEY FUNGUS **34**, 136
POWDERY MILDEWS **49**, 165
QUINCE LEAF BLIGHT **48**, 166
SILVER LEAF 175
WINTER MOTH **31**, 191

**R**

**Raspberry, loganberry, other hybrid berries**
APHIDS 99

BROWN SCALE **36**, 111
CROWN GALL **38**, 119
FROST DAMAGE **28**, **44**, 129
GLASSHOUSE RED SPIDER MITE **22**, **23**, 132
GREY MOULD **50**, 134
HONEY FUNGUS **34**, 136
PHYTOPHTHORA 159
POOR POLLINATION 162
RASPBERRY BEETLE **50**, 167
RASPBERRY CANE BLIGHT 167
RASPBERRY CANE SPOT **34**, 167
RASPBERRY LEAF AND BUD MITE **18**, 167
RASPBERRY SPUR BLIGHT 168
RUSTS 173
VIRUSES **23**, 187
WEEDKILLER DAMAGE – BROADLEAVED **27**, 188
WEEDKILLER DAMAGE – GLYPHOSATE 189

**S**

**Strawberry**
APHIDS 99

GLASSHOUSE RED SPIDER MITE **22**, **23**, 132
GREY MOULD **34**, 134
LEAF SPOT (FUNGAL) **12**, **13**, 143
MILLIPEDES 148
POOR POLLINATION 162
SLUGS **29**, 176
STRAWBERRY BLACK EYE 178
STRAWBERRY GREEN PETAL 179
STRAWBERRY GREY MOULD **50**, 179
STRAWBERRY LEAF SPOT **12**, **17**, 179
STRAWBERRY MITE 179
STRAWBERRY RED CORE 179
STRAWBERRY SEED BEETLES 179
STRAWBERRY VIRUSES 180
VINE WEEVIL **29**, **53**, 186

**V**

**Vine (grape)**
BROWN SCALE **36**, 111
CROWN GALL **38**, 119

DOWNY MILDEW 124
GLASSHOUSE RED SPIDER MITE **22**, **23**, 132
GREY MOULD **50**, 134
HIGH TEMPERATURE INJURY 135
MEALYBUGS **32**, **37**, 147
OEDEMA **19**, 152
POWDERY MILDEWS **49**, 165
SHANKING OF GRAPES 174
VINE ERINOSE MITE **15**, 186
WOOLLY VINE SCALE **39**, 192

**W**

**Walnut**
CORAL SPOT **41**, 119
FROST DAMAGE (FOLIAGE AND STEMS) **28**, 129
SLIME FLUX AND WETWOOD 175
SQUIRRELS **52**, 178
VIRUSES **23**, 187
WALNUT BLISTER MITE 187
WALNUT BLOTCH 187
WALNUT SOFT SHELL 187

# VEGETABLES AND HERBS

SOME PROBLEMS affect a wide range of vegetables (*right*); others are more host-specific (*below*). Even more so than with fruit, the choice of locally disease-resistant cultivars can help ensure healthy crops and minimize the need for chemicals. Resistant cultivars may be suggested in the entry for the problem in question in the *A–Z of Pests, Diseases and Disorders*.

## COMMON PROBLEMS

The most common non-host-specific ailments affecting vegetable crops are:

APHIDS 99
GREY MOULD 134
LEAF SPOT (FUNGAL) 143

POWDERY MILDEWS 165
SLUGS 176
VIRUSES 187

**A**

**Artichoke, globe**
BLACKFLY 108
GREY MOULD **50**, 134
LEAF SPOT (FUNGAL) **12**, **13**, 143
LETTUCE DOWNY MILDEW **22**, 144

**Artichoke, Jerusalem**
ROOT APHIDS **55**, 170
SCLEROTINIA **36**, 174
SLUGS **36**, **57**, 176

**Aubergine**
APHIDS 99
FOOT AND ROOT ROTS **55**, 128
GLASSHOUSE RED SPIDER MITE **22**, **23**, 132
GLASSHOUSE WHITEFLY 132
VIRUSES **23**, 187

**B**

**Basil**
FOOT AND ROOT ROTS **55**, 128
GREY MOULD **50**, 134

**Bay** *see* **Sweet bay** *see* SHRUBS AND CLIMBERS 196–7

**Bean, broad**
BEAN SEED BEETLES **51**, 105
BLACKFLY 108
CHOCOLATE SPOT 117
DOWNY MILDEW 124
FOOT AND ROOT ROTS **55**, 128
GREY MOULD **50**, 134
IRON DEFICIENCY AND LIME-INDUCED CHLOROSIS **24**, 139
MICE **51**, 147
PEA AND BEAN WEEVIL **31**, 155

RUSTS 173
SCLEROTINIA **36**, 174
VIRUSES **23**, 187
WILTS 191

**Bean, dwarf/French**
BEAN ANTHRACNOSE 105
BEAN SEED FLY **51**, 105
BLACKFLY 108
FOOT AND ROOT ROTS **55**, 128
GLASSHOUSE RED SPIDER MITE **22**, **23**, 132
GREY MOULD **50**, 134
HALO BLIGHT 135
ROOT APHIDS **55**, 170
RUSTS 173
VIRUSES **23**, 187
WILTS 191

**Bean, runner**
BEAN ANTHRACNOSE 105
BEAN SEED FLY **51**, 105
BLACKFLY 108
FOOT AND ROOT ROTS **55**, 128

FROST DAMAGE (FOLIAGE AND STEMS) **28**, 129
GLASSHOUSE RED SPIDER MITE **22**, **23**, 132
GREY MOULD **50**, 134
HALO BLIGHT 135
IRON DEFICIENCY AND LIME-INDUCED CHLOROSIS **24**, 139
NECTAR ROBBING **46**, 151,
ROOT APHIDS **43**, 170
RUSTS 173
SCLEROTINIA **36**, 174
SLUGS **36**, **57**, 176
VIRUSES **23**, 187
WILTS 191

**Beetroot**
BEET LEAF MINER **17**, 106
BORON DEFICIENCY 109
DOWNY MILDEW 124
FOOT AND ROOT ROTS **55**, 128
FUSARIUM ROT 130
IRON DEFICIENCY AND LIME-INDUCED CHLOROSIS **24**, 139

LEAF SPOT (FUNGAL) **12**, **13**, 143
RUSTS 173
VIOLET ROOT ROT **57**, 186
VIRUSES **23**, 187

**Brassicas**
BOLTING **44**, 109
BORON DEFICIENCY 109
BRASSICA DARK LEAF SPOT **13**, 110
BRASSICA LIGHT LEAF SPOT 113
CABBAGE MOTH **30**, 112
CABBAGE ROOT FLY 112
CABBAGE WHITE BUTTERFLY 113
CABBAGE WHITEFLY **32**, 113
CLUBROOT **55**, 118
DOWNY MILDEW 124
FLEA BEETLES **29**, 127
GREY MOULD **50**, 134
IRON DEFICIENCY AND LIME-INDUCED CHLOROSIS **24**, 139
MAGNESIUM DEFICIENCY **22**, 146

MEALY CABBAGE APHID **32**, 146
NITROGEN DEFICIENCY 151
OEDEMA **19**, 152
PIGEONS **30**, 160
POWDERY MILDEWS **49**, 165
SOFT ROT 177
TURNIP GALL WEEVIL 184
VIRUSES **23**, 187
WEEDKILLER DAMAGE –
  BROADLEAVED **39**, 188
WHIPTAIL OF BRASSICAS 189
WHITE BLISTER **14**, 189

## C

**Capsicum (peppers)**
APHIDS 99
BLOSSOM END ROT 108
FOOT AND ROOT ROTS **55**, 128
GLASSHOUSE RED SPIDER
  MITE **22**, **23**, 132
GLASSHOUSE WHITEFLY 132
OEDEMA **19**, 152
SCLEROTINIA **36**, 174
VIRUSES **23**, 187

**Carrot**
CARROT CAVITY SPOT 114
CARROT FLY **51**, 114
CARROT MOTLEY DWARF
  VIRUS **25**, 114
DOWNY MILDEW 124
POWDERY MILDEWS **49**, 165
ROOT APHIDS **55**, 170
SCLEROTINIA **36**, 174
SPLITTING 178
VIOLET ROOT ROT **57**, 186
VIRUSES **23**, 187

**Celery, celeriac**
BOLTING **44**, 109
BORON DEFICIENCY 109
CARROT FLY **51**, 114
CELERY LEAF MINER **19**, 115
CELERY LEAF SPOT **12**, 115
DOWNY MILDEW 124
FOOT AND ROOT ROT **55**, 128
GREY MOULD **50**, 134
SCLEROTINIA **36**, 174
SLUGS **36**, **57**, 176
VIOLET ROOT ROT **57**, 186
VIRUSES **23**, 187

**Courgette** *see* **Marrow**

**Cucumber, melon**
APHIDS 99
CUCUMBER GUMMOSIS 120
CUCUMBER MOSAIC VIRUS
  **21**, **25**, 120
DOWNY MILDEW 124
FOOT AND ROOT ROT **55**, 128
FUSARIUM WILT 130
GLASSHOUSE RED SPIDER
  MITE **22**, **23**, 132
GLASSHOUSE WHITEFLY 132
GREY MOULD **50**, 134

LEAF SPOT (BACTERIAL) **13**, **29**, 143
LEAF SPOT (FUNGAL) **12**, **13**, 143
POWDERY MILDEWS **49**, 165
SCLEROTINIA **36**, 174
VERTICILLIUM WILTS **33**, 185
VIRUSES **23**, 187
WEEDKILLER DAMAGE –
  BROADLEAVED **39**, 188

## L

**Leek**
FOOT AND ROOT ROTS **55**, 128 ·
LEAF SPOT (FUNGAL) **12**, **13**, 143
LEEK MOTH **23**, 144
LEEK RUST **17**, 144
ONION FLY 153
ONION THRIPS 153
ONION WHITE ROT **52**, 153
SMUTS **35**, 176

**Lettuce**
APHIDS 99
BACTERIAL SOFT ROT 104
BOLTING **44**, 109
CUTWORMS **53**, 121
FOOT AND ROOT ROT **55**, 128
GREY MOULD **50**, 134
LEAF SPOT (FUNGAL) **12**, **13**, 143
LETTUCE DOWNY MILDEW
  **22**, 144
LETTUCE ROOT APHID 145
SLUGS **36**, **57**, 176
VIRUSES **23**, 187

## M

**Marrow, courgette, pumpkin**
APHIDS 99
CUCUMBER MOSAIC VIRUS
  **21**, **25**, 120
FOOT AND ROOT ROT **55**, 128
GREY MOULD **50**, 134
POWDERY MILDEWS **49**, 165
SLUGS **36**, **57**, 176
VIRUSES **23**, 187

**Melon** *see* **Cucumber**

**Mint**
CATERPILLARS 115
LEAFHOPPERS 142
MINT BEETLE **31**, 148
MINT RUST 148

## O

**Onion, shallot, garlic**
BACTERIAL SOFT ROT 104
BOLTING 109

CABBAGE MOTH **30**, 112
LEEK MOTH **23**, 144
ONION EELWORM **52**, 153
ONION FLY 153
ONION NECK ROT **52**, 153
ONION THRIPS 153
ONION WHITE ROT **52**, 153
RUSTS 173
SMUTS **35**, 176
VIRUSES **23**, 187

## P

**Parsley**
CARROT FLY **51**, 114
CARROT MOTLEY DWARF
  VIRUS **25**, 114
CELERY LEAF MINER **19**, 115
DOWNY MILDEW 124
LEAF SPOT (FUNGAL) **12**, **13**, 143
MAGNESIUM DEFICIENCY **22**, 146
POWDERY MILDEWS **49**, 165
VIRUSES **23**, 187

**Parsnip**
CARROT FLY **51**, 114
CELERY LEAF MINER **19**, 115
DOWNY MILDEW 124
LEAF SPOT (FUNGAL) **12**, **13**, 143
PARSNIP CANKER **57**, 154
POWDERY MILDEWS **49**, 165
ROOT APHIDS **55**, 170
SCLEROTINIA **36**, 174
SPLITTING 178
VIOLET ROOT ROT **57**, 186
VIRUSES **23**, 187

**Pea**
APHIDS 99
DOWNY MILDEW 124
FOOT AND ROOT ROT **55**, 128
FUSARIUM WILT 130
GREY MOULD **50**, 134
MICE **51**, 147
PEA AND BEAN WEEVIL **31**, 155
PEA LEAF AND POD SPOT **14**, **51**, 155
PEA MOTH **51**, 155
PEA THRIPS 155
POWDERY MILDEWS **49**, 165
SLUGS **36**, **57**, 176
VIRUSES **23**, 187

**Potato**
APHIDS 99
FROST DAMAGE (FOLIAGE
  AND STEMS) **28**, 129
LEAF SPOT (FUNGAL) **12**, **13**, 143
MAGNESIUM DEFICIENCY **22**, 146
MILLIPEDES 148
POTATO BLACK LEG 163
POTATO BLIGHT **28**, **56**, 163
POTATO COMMON SCAB **56**, 163

POTATO CYST EELWORMS **55**, 163
POTATO DRY ROT **57**, 164
POTATO EARLY BLIGHT 164
POTATO GANGRENE **57**, 164
POTATO HOLLOW HEART 164
POTATO INTERNAL RUST SPOT 164
POTATO POWDERY SCAB **56**, 164
POTATO SILVER SCURF 165
POTATO SPRAING **56**, 165
POTATO VIRUSES 165
POTATO WART DISEASE 165
SCLEROTINIA **36**, 174
SLUGS **36**, **57**, 176
SPLITTING 178
VERTICILLIUM WILTS **33**, 185
VIOLET ROOT ROT **57**, 186
WIREWORMS **54**, **56**, 191

**Pumpkin** *see* **Marrow**

## R

**Radish**
CABBAGE ROOT FLY 112
DAMPING OFF **34**, 122
DOWNY MILDEW 124
FLEA BEETLES **29**, 127
LEAF SPOT (FUNGAL) **12**, **13**, 143
SLUGS **36**, **57**, 176
VIRUSES **23**, 187
WHITE BLISTER **14**, 189

**Rhubarb**
BLACKFLY 108
CROWN ROT 120
CUCUMBER GUMMOSIS 120
DOWNY MILDEW 124
HONEY FUNGUS **34**, 136
LEAF SPOT (FUNGAL) **12**, **13**, 143
RHUBARB GUMMING 169
SLUGS **36**, **57**, 176
SNAILS **29**, 177
VIRUSES **23**, 187

**Rocket, salad**
FLEA BEETLES **29**, 127

**Rosemary**
FROST DAMAGE **28**, 129
HONEY FUNGUS **34**, 136
LEAFHOPPERS 142

## S

**Spinach, perpetual**
BOLTING **44**, 109
CATERPILLARS 115
DOWNY MILDEW 124
IRON DEFICIENCY AND LIME-
  INDUCED CHLOROSIS **24**, 139
LEAF SPOT (FUNGAL) **12**, **13**, 143
MAGNESIUM DEFICIENCY **22**, 146

POWDERY MILDEWS **49**, 165
RUSTS 173
SLUGS **36**, **57**, 176
VIRUSES **23**, 187

**Swede**
BORON DEFICIENCY 109
BRASSICA DARK LEAF SPOT
  **13**, 110
CABBAGE ROOT FLY 112
CLUBROOT **55**, 118
DOWNY MILDEW 124
FLEA BEETLES **29**, 127
LEAF SPOT (FUNGAL) **12**, **13**, 143
POWDERY MILDEWS **49**, 165
VIOLET ROOT ROT **57**, 186
VIRUSES **23**, 187
WHITE BLISTER **14**, 189

**Sweetcorn**
BORON DEFICIENCY 109
MICE **51**, 147
SWEETCORN SMUT **51**, 180
VIRUSES **23**, 187

## T

**Tomato**
APHIDS 99
BLOSSOM END ROT **50**, 108
FUSARIUM WILT 130
GLASSHOUSE LEAFHOPPER **20**, 132
GLASSHOUSE RED SPIDER
  MITE **22**, **23**, 132
GLASSHOUSE WHITEFLY 132
GREY MOULD **50**, 134
OEDEMA **19**, 152
SLUGS **36**, **57**, 176
SNAILS **29**, 177
SPLITTING 178
TOMATO BLIGHT **51**, 181
TOMATO BLOTCHY RIPENING
  **51**, 181
TOMATO GREENBACK **50**, 182
TOMATO LEAF MOULD 182
TOMATO MOTH 182
TOMATO PITH NECROSIS 182
VERTICILLIUM WILTS **33**, 185
VIRUSES **23**, 187
WEEDKILLER DAMAGE –
  BROADLEAVED **27**, 188

**Turnip**
BORON DEFICIENCY 109
BRASSICA DARK LEAF SPOT
  **13**, 110
CABBAGE ROOT FLY 112
CLUBROOT **55**, 118
DOWNY MILDEW 124
FLEA BEETLES **29**, 127
LEAF SPOT (FUNGAL) **12**, **13**, 143
POWDERY MILDEWS **49**, 165
TURNIP GALL WEEVIL 184
VIOLET ROOT ROT **57**, 186
VIRUSES **23**, 187
WHITE BLISTER **14**, 189

# REFERENCE
# SECTION

IN ADDITION TO a comprehensive index, conversion charts and glossary of terms, this section contains clear, tabular information on the chemical controls formulated for use in the garden. The active ingredients, methods of application and uses of all commonly available garden insecticides and fungicides are given, with notes on safety and any special conditions for use.

# GARDEN INSECTICIDES

BEFORE BUYING or using any garden insecticide, read the manufacturer's instructions carefully and note any limitations, such as the types of plants on which the chemical can be used, the way in which it is applied, whether it can be used on food plants and, if so, the safe harvest interval. Follow the general safety precautions on pp.84–5 when applying any chemical. Avoid contaminating ponds or other waterways and safeguard bees by spraying plants either when not in bloom, or at dusk.

| ACTIVE INGREDIENT(S) | ACTION(S) | FORMULATIONS AVAILABLE | USABLE ON EDIBLE CROPS* | HARVEST INTERVAL |
|---|---|---|---|---|
| Aluminium sulphate | Contact | Soluble powder | Yes | None when applied to soil, not directly onto food plants |
| Bendiocarb | Contact | Dust and bait (for ants) | No | |
| Bifenthrin | Contact | Spray concentrate and ready-to-use spray | Yes | I day |
| Bioallethrin with permethrin | Contact | Ready-to-use spray | Yes | I day |
| Borax | Poison bait | Sweet liquid | No | |
| Butoxycarboxim | Systemic | Impregnated cardboard strips for insertion in the compost of pot plants | No | |
| Chlorpyrifos | Contact | Ready-to-use spray in or near buildings | No | |
| Coumatetralyl | Poison bait | Cereal-based | No | |
| Cypermethrin | Contact | Ready-to-use spray | No | |
| Deltamethrin | Contact | Ready-to-use spray | No | |
| Derris see Rotenone | | | | |
| Diazinon | Contact | Aerosol spray | No | |
| Difenacoum | Poison bait | Cereal-based | No | |
| Dimethoate | Systemic | Spray concentrate | Yes | 7 days |
| Fatty Acids see Insecticidal soap | | | | |
| Fenitrothion | Contact | Spray concentrate | Yes | 14 days (7 for raspberry) |
| Gamma-BHC see Lindane | | | | |
| HCH see Lindane | | | | |
| Heptenophos with permethrin | Systemic and contact | Spray concentrate | Yes | I day |
| Insecticidal soap | Contact | Spray concentrate and ready-to-use spray | Yes | I day |
| Lindane | Contact | Dust | Yes | 14 days |
| Malathion | Contact | Spray concentrate | Yes | 4 days |
| Metaldehyde | Poison bait and liquid spray | Pellets and spray concentrate | Yes | None for pellets placed near (not on) plants; 10 days for liquid spray |
| Methiocarb | Poison bait | Pellets | Yes | None for pellets placed near (not on) plants |
| Permethrin (see also Bioallethrin with permethrin; Heptenophos with permethrin) | Contact | Spray concentrate, aerosol dusts, fumigant | Yes | I day |
| Pirimicarb | Partially systemic | Spray concentrate and aerosol | Yes | 3 days (lettuce under glass 14 days) |
| Pirimiphos-methyl | Contact | Spray concentrate, dust, aerosol, fumigant | Yes | 7 days (fumigant 2 days) |
| Pyrethrum | Contact | Spray concentrate, ready-to-use sprays, dust, aerosol | Yes | I day |
| Rotenone (derris) | Contact | Spray concentrate, dust | Yes | I day |
| Sulphur | Fumigant | Smoke generators | No | |
| Tar oil | Contact | Spray concentrate | Yes ** | |
| Trichlorphon | Poison bait | Sweet liquid | No | |

* Many insecticides are suitable for use on certain crops only; these will be specified on the label.

** Tar oil is sprayed on deciduous fruit trees and bushes in winter, not during the cropping period, so there is no harvest interval.

## CONTROLS

| Aphids | Whitefly | Mealybugs | Caterpillars | Thrips | Capsid Bugs | Red Spider Mite | Leafhoppers | ALSO CONTROLS | COMMENTS/LIMITATIONS |
|---|---|---|---|---|---|---|---|---|---|
| | | | | | | | | Slugs and snails | |
| | | | | | | | | Ants, wasp nests, woodlice, earwigs | Dangerous to fish, aquatic life, pets, birds and bees. |
| • | • | • | • | | | • | | Beetles | Dangerous to pets and birds; extremely dangerous to bees. |
| • | • | • | • | | | • | | | Dangerous to bees; extremely dangerous to fish and aquatic life. |
| | | | | | | | | Ants in or near buildings | |
| • | | | | | | • | | Other sap-feeding pests | |
| | | | | | | | | Ants and other household crawling pests | Dangerous to bees; extremely dangerous to fish and aquatic life. |
| | | | | | | | | Rats and mice | Prevent access to bait by children and domestic animals; burn or bury dead rodents and uneaten bait. |
| | | | | | | | | Ants, earwigs, wasps in or near houses | Extremely dangerous to fish and aquatic life. |
| | | | | | | | | Ants and other crawling insects in or near houses | Dangerous to bees; extremely dangerous to fish and aquatic life. |
| | | | | | | | | Ants and other crawling insects in buildings | Dangerous to fish, aquatic life and bees. |
| | | | | | | | | Rats and mice | Prevent access to bait by children and domestic animals; burn or bury dead rodents and uneaten bait. |
| • | • | • | | • | • | • | • | Suckers, leaf miners, scale insects | Apply away from birds, fish and bees. |
| • | | | • | | • | | | Apple sawfly, pea moth, codling moth, raspberry beetle | Dangerous to fish, aquatic life and bees. |
| • | • | | • | • | | | | Pea moth | Dangerous to bees; extremely dangerous to fish and aquatic life. |
| • | • | • | | • | | • | • | Scale insects | Apply away from birds, fish and bees. |
| | | | | | • | | | Ants, woodlice, earwigs, flea beetle, cabbage root fly, carrot fly, onion fly, chafers, millipedes, wireworm | Apply away from birds, fish and bees. |
| • | • | • | • | • | | • | • | Leaf miners, scale insects | Dangerous to bees; extremely dangerous to fish and aquatic life. |
| | | | | | | | | Slugs and snails | Dangerous to fish and aquatic life and to pets. |
| | | | | | | | | Slugs and snails | Apply away from birds, fish and bees. |
| • | • | | • | • | | | • | Ants | Dangerous to bees; extremely dangerous to fish and aquatic life. |
| • | | | | | | | | | Dangerous to fish and aquatic life. |
| • | • | | | • | • | • | • | Sawflies, fungus gnats, leaf miners. Dust formulation: carrot fly, cabbage root fly, onion fly, flea beetle, ants, wireworm, earwigs, woodlice | Dangerous to fish and aquatic life (smoke extremely so) and to bees. |
| • | • | | • | • | • | | • | Ants | Apply away from birds, fish and bees. |
| • | | | • | • | | • | | Raspberry beetle, flea beetle | Dangerous to fish and aquatic life. |
| | | | | | | | | Moles | |
| | | | | | | | | Overwintering stages of fruit aphids, apple sucker, scale insects | Dangerous to fish and aquatic life. |
| | | | | | | | | Ants in or near buildings | |

# GARDEN FUNGICIDES

BEFORE BUYING or using any garden insecticide, read the manufacturer's instructions carefully and note any limitations, such as the types of plants on which the chemical can be used, the way in which it is applied, whether it can be used on food plants and, if so, the safe harvest interval. Follow the general safety precautions on pp.84–5 when applying any chemical. Avoid contaminating ponds or other waterways and safeguard bees by spraying plants either when not in bloom, or at dusk.

| ACTIVE INGREDIENT(S) | ACTION(S) | FORMULATIONS AVAILABLE | USABLE ON EDIBLE CROPS* | HARVEST INTERVAL |
|---|---|---|---|---|
| Ammonium hydroxide with copper sulphate | Contact, preventive | Ready-to-use spray | Yes | I day |
| Bordeaux mixture (copper sulphate and hydrated lime) | Contact, preventive | Wettable powder for spraying/dusting | Yes | None |
| Bupirimate with triforine | Systemic | Spray concentrate | Yes | 7 days for apples, 14 days for gooseberries and currants |
| Carbendazim | Systemic | Wettable powder, liquid for spraying or sometimes used as a dip | Yes | None, except lettuce: 14 days |
| Cheshunt compound (copper sulphate and ammonium carbonate) | Preventive | Soluble powder, drench | Yes (as seedlings) | |
| Copper oxychloride | Preventive | Wettable powder, spray or drench | Yes | None |
| Copper sulphate see Ammonium hydroxide and copper sulphate | | | | |
| Mancozeb | Contact, preventive | Wettable powder, spray | Yes | Seven days, 21 days for lettuce |
| Myclobutanil | Systemic | Wettable powder, spray | Yes | None |
| Penconazole | Systemic | Spray concentrate, ready-to-use spray | No | |
| Sulphur (colloidal) | Preventive | Ready-to-use spray | Yes (one formulation only) | I day |
| Sulphur (dust) | Preventive | Dust | Yes | None |
| Sulphur with triforine | Preventive, systemic | Wettable powder spray | No | |
| Tar acids and oils (phenols) | Preventive | Liquid, spray or drench | On fruit trees as winter wash only. | |
| Thiophanate-methyl | Systemic | Liquid, as a root dip | Only on brassicas, against clubroot | |
| Triforine see Bupirimate with triforine, Sulphur with triforine | | | | |

* Many fungicides are suitable for use on certain crops only; these will be specified on the label

** Spray concentrate only

**CONTROLS**

| Powdery Mildew | Rust | Downy Mildew | Grey Mould | Rose Blackspot | Potato/Tomato Blight | Leaf Spot | Apple/Pear Scab | Bacterial Canker | ALSO CONTROLS | COMMENTS/LIMITATIONS |
|---|---|---|---|---|---|---|---|---|---|---|
| • | | | • | • | | | | | | Powdery mildew control on roses, chrysanthemums and gooseberries. Harmful to fish. |
| | • | | | • | • | | | • | Apple canker, peach leaf curl | Rust control on blackcurrants, gooseberries. Leaf spot control on celery. Canker control on edible fruit trees. Harmful to livestock (keep them out of treated areas for at least three weeks). Toxic to fish. Some pear cultivars may be damaged by it, e.g. 'Doyenné du Comice'. |
| • | • | | • | | • | • | | | | Powdery mildew control on ornamentals plus apples, blackcurrants, gooseberries. Rust control on ornamentals. Apple scab control on apples. Harmful to fish. |
| • | | | • | • | • | • | • | | Tulip fire, peony wilt, raspberry cane spot and spur blight, tomato leaf mould, certain bulb and corm diseases, certain turf diseases, clubroot (as a dip), apple canker | Powdery mildew control on apples, pears, vines, soft fruit and ornamentals. Grey mould control on soft fruit, vines, tomatoes, beans, other vegetables, ornamentals and pot plants. Leaf spot control on currants, gooseberries, celery. Resistant strains of fungi have developed and more may develop, rendering this fungicide ineffective. Harmful to fish. Do not use on strawberry beds. |
| | | | | | | | | | Prevents damping off disease of seedlings | Apply away from fish and do not mix up solution in vessels of iron, tin or zinc. |
| • | | | | | | | | • | Peach leaf curl, damping off disease of seedlings and common fruit diseases | |
| | | | | | | | | | | |
| • | • | • | • | • | • | | • | | Tomato leaf mould | Will not control white rust on chrysanthemums. |
| • | • | | • | | | | • | | | Powdery mildew control on roses and apples. Rust control on roses. |
| • | •** | | •** | | | | | | | Powdery mildew and rust control on ornamentals, but do not use on fuchsias. Harmful to fish and other aquatic life. |
| | | | | | | | | | | Apply away from fish. Sulphur may taint jams and wines made from treated fruits. Do not apply to ferns, plants under stress, or gooseberry or grape cultivars known to be "sulphur shy" i.e. damaged by sulphur. |
| • | | | | | | | | | Storage rots of bulbs, corms and dahlia tubers | Apply away from fish. Sulphur may taint jams and wines made from treated fruits. Do not apply to ferns, plants under stress, or gooseberry or grape cultivars known to be "sulphur shy", i.e. damaged by sulphur. |
| • | | | • | | | | | | | For use specifically on roses. Dangerous to bees and extremely dangerous to fish. |
| | | | | | | | | | For partial sterilization of soil or implements. On fruit trees (as winter wash) as a partial preventive against certain diseases e.g. blossom wilt | Should be kept off the skin and away from eyes, food and food containers. Dangerous to fish. Wear protective gloves when handling the concentrates. |
| | | | | | | | | | | Resistant strains of the clubroot organism may develop, so this chemical (and also the closely related carbendazim) would be ineffective. |

# BRAND NAMES

THE TABLE BELOW lists the most commonly available garden chemicals, giving their general use, formulation and active ingredients; on the facing page, these commercial preparations are listed by their active ingredients, as recommended for use in the *A–Z of Pests, Diseases and Disorders*.

| KEY TO ABBREVIATIONS | |
|---|---|
| F – Fungicide | Fm – Fumigant |
| I – Insecticide or other pest control | ICS – Impregnated cardboard strips |
| A – Aerosol | RTU – Ready-to-use spray |
| B – Baits | SC – Spray concentrate (may also |
| D – Dust and/or Wettable powder | be used as a dip) |

| BRAND NAME | TYPE | ACTIVE INGREDIENT(S) |
|---|---|---|
| 6 X Slug and Snail Killer (D) | I | Aluminium sulphate |
| Bio Anti-Ant Duster (D) | I | Pyrethrum |
| Bio Cheshunt Compound (D) | F | Cheshunt compound |
| Bio Dithane 945 (D) | F | Mancozeb |
| Bio Fenitrothion (SC) | I | Fenitrothion |
| Bio Flydown (SC) | I | Permethrin |
| Bio Friendly Pest Pistol (RTU) | I | Insecticidal soap |
| Bio Liquid Clubroot Control (dip only) | F | Thiophanate-methyl |
| Bio Liquid Derris (SC) | I | Rotenone (derris) |
| Bio Malathion Greenfly Killer (SC) | I | Malathion |
| Bio Multirose (SC) | F, I | Sulphur, triforine, permethrin and a foliar feed |
| Bio Racumin (B) | I | Coumatetralyl |
| Bio Slug Gard (B) | I | Methiocarb |
| Bio Slug Mini-Pellets (B) | I | Metaldehyde |
| Bio Sprayday (SC) | I | Permethrin |
| Bio Spraydex Ant and Crawling Insect Killer (RTU) | I | Bioallethrin, permethrin |
| Bio Spraydex Greenfly Killer (RTU) | I | Bioallethrin, permethrin |
| Bio Spraydex Houseplant Spray (RTU) | I | Bioallethrin, permethrin |
| Bio Supercarb (D) | F | Carbendazim |
| Bio Systhane (D) | F | Myclobutanil |
| Doff All in One Insecticide Spray (RTU) | I | Pyrethrum |
| Doff Ant Control Powder (D) | I | Bendiocarb |
| Doff Ant Killer Spray (RTU) | I | Deltamethrin |
| Doff Antlak (A) | I | Diazinon |
| Doff Derris Dust (D) | I | Rotenone (derris) |
| Doff Economy Ant Killer Powder (D) | I | Lindane |
| Doff Gamma-BHC Dust (D) | I | Lindane |
| Doff Houseplant Insecticide Spray (RTU) | I | Pyrethrum |
| Doff Slugoids Slug Killer (B) | I | Metaldehyde |
| Doff Systemic Insecticide (SC) | I | Dimethoate |
| Doff Wasp Nest Killer (D) | I | Bendiocarb |
| Doff Woodlice Killer (D) | I | Bendiocarb |
| Fertosan Slug and Snail Powder (D) | I | Aluminium sulphate |
| Fumite General Purpose Smoke (Fm) | I | Pirimiphos-methyl |
| Fumite Whitefly Smoke (Fm) | I | Permethrin |
| Levington Insect Spray for Houseplants (A) | I | Permethrin |
| Levington Nature's Answer Fungicide and Insect Killer (RTU) | F, I | Colloidal sulphur |
| Levington Nature's Answer to Insect Pests (RTU) | I | Pyrethrum |
| Miracle Garden Care Ant and Crawling Insect Gun (RTU) | I | Deltamethrin |
| Miracle Garden Care Bug Gun for Fruit and Vegetables (RTU) | I | Pyrethrum |
| Miracle Garden Care Bug Gun for Roses and Flowers (RTU) | I | Pyrethrum |
| Miracle Garden Care Keri Insect Spray (RTU) | I | Pyrethrum |
| Miracle Garden Care Longer Lasting Bug Gun (RTU) | I | Bioallethrin, permethrin |
| Miracle Garden Care Nimrod T (SC) | F | Bupirimate, triforine |
| Miracle Garden Care Picket (SC) | I | Permethrin |
| Miracle Garden Care Rapid (SC, A) | I | Pirimicarb |
| Miracle Garden Care Ratak (B) | I | Difenacoum |
| Miracle Garden Care Slug Xtra (B) | I | Metaldehyde |
| Miracle Garden Care Sybol (D) | I | Pirimiphos-methyl |
| Miracle Garden Care Sybol (SC) | I | Pirimiphos-methyl |
| Murphy Bug Master (RTU) | I | Pyrethrum |
| Murphy Derris Dust (D) | I | Rotenone (derris) |
| Murphy Gamma-BHC Dust (D) | I | Lindane |
| Murphy Kil-Ant (D) | I | Bendiocarb |
| Murphy Kil-Ant (RTU) | I | Cypermethrin |
| Murphy Liquid Malathion (SC) | I | Malathion |
| Murphy Liquid Slugit (SC) | I | Metaldehyde |
| Murphy Mole Smokes (Fm) | I | Sulphur |
| Murphy Mortegg (SC) | I | Tar oil |
| Murphy Slugits (B) | I | Metaldehyde |
| Murphy Systemic Action Insecticide (SC) | I | Heptenophos, permethrin |
| Murphy Traditional Copper Fungicide (D) | F | Copper oxychloride |
| Murphy Tumbleblite II (SC) | F | Penconazole |
| Murphy Tumblebug (SC) | I | Heptenophos, permethrin |
| Nippon Ant Killer Liquid (B) | I | Borax |
| Nippon Ant Killer Powder (D) | I | Permethrin |
| Phostrogen Plant Pin (ICS) | I | Butoxycarboxim |
| Phostrogen Safer's Fruit and Vegetable Insecticide (RTU) | I | Insecticidal soap |
| Phostrogen Safer's Garden Fungicide (RTU) | F | Colloidal sulphur |
| Phostrogen Safer's Garden Insecticide Concentrate (SC) | I | Insecticidal soap |
| Phostrogen Safer's House Plant Insecticide (RTU) | I | Insecticidal soap |
| Phostrogen Safer's Rose and Flower Insecticide (RTU) | I | Insecticidal soap |
| Py Garden Insecticide (A) | I | Pyrethrum |
| Py Powder Garden and Household Insect Killer (D) | I | Pyrethrum |
| Py Spray Garden Insect Killer Concentrate (SC) | I | Pyrethrum |
| Py Spray Insect Killer (RTU) | I | Pyrethrum |
| Rentokil Ant Trap (B) | I | Trichlorphon |
| Secto Ant and Crawling Insect Lacquer (A) | I | Diazinon |
| Secto Ant and Crawling Insect Powder (D) | I | Bendiocarb |
| Secto Ant Bait (B) | I | Bendiocarb |
| Secto Extra Strength Insect Killer (D) | I | Permethrin |
| Secto Nature Care Garden Insect Powder (D) | I | Pyrethrum |
| Secto Nature Care Garden Insect Spray (RTU) | I | Pyrethrum |
| Secto Nature Care Houseplant Insect Spray (RTU) | I | Pyrethrum |
| Secto Wasp Killer Powder (D) | I | Bendiocarb |
| Solaris Ant Stop (RTU) | I | Chlorpyrifos |
| Solaris Polysect (RTU) | I | Bifenthrin |
| Solaris Polysect Concentrate (SC) | I | Bifenthrin |
| Spraydex General Purpose Fungicide (RTU) | F | Ammonium hydroxide, copper sulphate |
| Vapona Ant Trap (B) | I | Trichlorphon |
| Vapona Woodlice Killer Powder (D) | I | Permethrin |
| Vitax Bordeaux Mixture (D) | F | Bordeaux mixture |
| Vitax Derris Dust (D) | I | Rotenone (derris) |
| Vitax Green Sulphur (D) | F | Sulphur |
| Vitax Yellow Sulphur (D) | F | Sulphur |

| ACTIVE INGREDIENT(S) | TYPE | BRANDS |
|---|---|---|
| Aluminium sulphate | I | 6 X Slug and Snail Killer (D) |
| | | Fertosan Slug and Snail Powder (D) |
| Ammonium hydroxide with copper sulphate | F | Spraydex General Purpose Fungicide (RTU) |
| Bendiocarb | I | Doff Ant Control Powder (D) |
| | | Doff Woodlice Killer (D) |
| | | Doff Wasp Nest Killer (D) |
| | | Murphy Kil-Ant (D) |
| | | Secto Ant and Crawling Insect Powder (D) |
| | | Secto Ant Bait (B) |
| | | Secto Wasp Killer Powder (D) |
| Bifenthrin | I | Solaris Polysect Concentrate (SC) |
| | | Solaris Polysect (RTU) |
| Bioallethrin with permethrin | I | Bio Spraydex Ant and Crawling Insect Killer (RTU) |
| | | Bio Spraydex Greenfly Killer (RTU) |
| | | Bio Spraydex Houseplant Spray (RTU) |
| | | Miracle Garden Care Longer Lasting Bug Gun (RTU) |
| Borax | I | Nippon Ant Killer Liquid (B) |
| Bordeaux mixture | F | Vitax Bordeaux Mixture (D) |
| Bupirimate with triforine | F | Miracle Garden Care Nimrod T (SC) |
| Butoxycarboxim | I | Phostrogen Plant Pin (ICS) |
| Carbendazim | F | Bio Supercarb (D) |
| Cheshunt compound | F | Bio Cheshunt Compound (D) |
| Chlorpyrifos | I | Solaris Ant Stop (RTU) |
| Copper oxychloride | F | Murphy Traditional Copper Fungicide (D) |
| Copper sulphate with ammonium hydroxide | F | Spraydex General Purpose Fungicide (RTU) |
| Coumatetralyl | I | Bio Racumin (B) |
| Cypermethrin | I | Murphy Kil-Ant (RTU) |
| Deltamethrin | I | Miracle Garden Care Ant and Crawling Insect Gun (RTU) |
| | | Doff Ant Killer Spray (RTU) |
| Derris **see** Rotenone | | |
| Diazinon | I | Doff Antlak (A) |
| | | Secto Ant and Crawling Insect Lacquer (A) |
| Difenacoum | I | Miracle Garden Care Ratak (B) |
| Dimethoate | I | Doff Systemic Insecticide (SC) |
| Fenitrothion | I | Bio Fenitrothion (SC) |
| Gamma-BHC **see** Lindane | | |
| HCH **see** Lindane | | |
| Heptenophos with permethrin | I | Murphy Systemic Action Insecticide (SC) |
| | | Murphy Tumblebug (SC) |
| Insecticidal soap | I | Bio Friendly Pest Pistol (RTU) |
| | | Phostrogen Safer's Fruit and Vegetable Insecticide (RTU) |
| | | Phostrogen Safer's Garden Insecticide Concentrate (SC) |
| | | Phostrogen Safer's House Plant Insecticide (RTU) |
| | | Phostrogen Safer's Rose and Flower Insecticide (RTU) |
| Lindane | I | Doff Economy Ant Killer Powder (D) |
| | | Doff Gamma-BHC Dust (D) |
| | | Murphy Gamma-BHC Dust (D) |
| Malathion | I | Bio Malathion Greenfly Killer (SC) |
| | | Murphy Liquid Malathion (SC) |
| Mancozeb | F | Bio Dithane 945 (D) |

| ACTIVE INGREDIENT(S) | TYPE | BRANDS |
|---|---|---|
| Metaldehyde | I | Bio Slug Mini-Pellets (B) |
| | | Doff Slugoids Slug Killer (B) |
| | | Miracle Garden Care Slug Xtra (B) |
| | | Murphy Liquid Slugit (SC) |
| | | Murphy Slugits (B) |
| Methiocarb | I | Bio Slug Gard (B) |
| Myclobutanil | F | Bio Systhane (D) |
| Penconazole | F | Murphy Tumbleblite II (SC) |
| Permethrin | I | Bio Flydown (SC) |
| (**see also** Bioallethrin with permethrin; Heptenophos with permethrin; Sulphur with triforine and permethrin) | | Bio Sprayday (SC) |
| | | Fumite Whitefly Smoke (Fm) |
| | | Levington Insect Spray for Houseplants (A) |
| | | Miracle Garden Care Picket (SC) |
| | | Nippon Ant Killer Powder (D) |
| | | Secto Extra Strength Insect Killer (D) |
| | | Vapona Woodlice Killer Powder (D) |
| Pirimicarb | I | Miracle Garden Care Rapid (SC, A) |
| Pirimiphos-methyl | I | Miracle Garden Care Sybol (SC) |
| | | Miracle Garden Care Sybol (D) |
| | | Fumite General Purpose Smoke (Fm) |
| Pyrethrum | I | Bio Anti-Ant Duster (D) |
| | | Doff All in One Insecticide Spray (RTU) |
| | | Doff Houseplant Insecticide Spray (RTU) |
| | | Py Spray Garden Insect Killer Concentrate (SC) |
| | | Py Spray Insect Killer (RTU) |
| | | Miracle Garden Care Bug Gun for Fruit and Vegetables (RTU) |
| | | Miracle Garden Care Bug Gun for Roses and Flowers (RTU) |
| | | Murphy Bug Master (RTU) |
| | | Levington Nature's Answer to Insect Pests (RTU) |
| | | Secto Nature Care Garden Insect Spray (RTU) |
| | | Secto Nature Care Houseplant Insect Spray (RTU) |
| | | Miracle Garden Care Keri Insect Spray (RTU) |
| | | Py Garden Insecticide (RTU) |
| | | Py Powder Garden and Household Insect Killer (D) |
| | | Secto Nature Care Garden Insect Powder (D) |
| Rotenone | I | Bio Liquid Derris (SC) |
| | | Doff Derris Dust (D) |
| | | Murphy Derris Dust (D) |
| | | Vitax Derris Dust (D) |
| Sulphur (**see also** Sulphur with triforine and permethrin) | F, I | Levington Nature's Answer Fungicide and Insect Killer (RTU) |
| | | Murphy Mole Smokes (Fm) |
| | | Vitax Green Sulphur (D) |
| | | Vitax Yellow Sulphur (D) |
| | | Phostrogen Safer's Garden Fungicide (RTU) |
| Sulphur with triforine and permethrin | F, I | Bio Multirose (SC) |
| Tar oil | I | Murphy Mortegg (SC) |
| Thiophanate-methyl | F | Bio Liquid Clubroot Control (dip only) |
| Trichlorphon | I | Rentokil Ant Trap (B) |
| | | Vapona Ant Trap (B) |
| Triforine with bupirimate | F | Miracle Garden Care Nimrod T (SC) |
| Triforine with sulphur and permethrin | F | Bio Multirose (SC) |

REFERENCE SECTION

# GLOSSARY

**ABDOMEN** The hind part of an insect's body.

**ACARICIDE** A pesticide effective against mites.

**ACID** (of soil) With a pH value of less than 7; see also *Alkaline* and *Neutral*.

**ADVENTITIOUS** (of roots) Arising above ground from a stem or leaf.

**AERATE** (of soil) 1. Loosen by mechanical means in order to allow air to enter. 2. Alter soil texture and structure by creating more air spaces.

**ALKALINE** (of soil) With a pH value of more than 7; see also *Acid* and *Neutral*.

**ALTERNATE HOST** One of two host plants needed for some pathogens or pests to complete their life cycles. The alternate host is often unrelated to the other host: most pathogens and pests are able to complete their life cycles using a single host plant.

**ANAEROBIC** (of an organism) Able to live and grow without oxygen.

**ANNUAL** A plant or fungus that completes its life cycle, from germination through to death, in one year.

**ANTENNA** (*pl.* antennae) A jointed sensory organ on the head of insects and mites.

**ANTHER** The part of a stamen that produces pollen; it is usually borne on a filament.

**ANTHRACNOSE** A term used to describe several unrelated fungal diseases which are seen as dark spots or lesions on foliage, stems, pods or fruits.

**ARTHROPODS** Invertebrate animals which have jointed legs, e.g. insects, mites, woodlice, centipedes and millipedes.

**ASEXUAL REPRODUCTION** A form of reproduction not involving fertilization.

**AXIL** The upper angle between a leaf and stem, between a main stem and a lateral branch, or between a stem and a bract; where an axillary bud develops.

**BACTERIUM** (*pl.* bacteria) A microscopic single-celled organism. Some are beneficial to plants, others may be pathogenic or cause secondary rotting.

**BACTERICIDE** A chemical capable of controlling bacteria.

**BARK** The surface layer of the trunk and branches of woody plants, protecting the tissue within; usually composed of dead corky cells.

**BARK-RINGING** The controlled removal of a partial ring of bark from the trunk or branches of certain fruit trees, to reduce vigorous growth and encourage fruit cropping. Complete removal or girdling of a trunk, as in damage by rabbits, squirrels, voles and vine weevil grubs,

can lead to the death of woody plants. See also *Girdling*.

**BASAL PLATE** The flattened base of a bulb, from which the roots grow.

**BIENNIAL** A plant that produces leafy growth in the first year, and then flowers, sets seed and dies in the next year.

**BIOLOGICAL CONTROL** The control of pests, diseases and weeds by the use of natural enemies such as predators, parasites and pathogens.

**BLEEDING** The oozing of sap through a cut or wound.

**BLIGHT** A term loosely used to describe a fungal, or occasionally an insect, attack.

**BLIND SHOOT** A shoot that does not form a terminal flower bud, or one where the growing point has been destroyed.

**BOLT** To produce flowers and seed prematurely.

**BORDEAUX MIXTURE** A fungicide mixture containing copper sulphate and lime.

**BRACT** A modified leaf at the base of a flower or flower cluster, sometimes resembling a flower petal or leaves. Bracts may also be reduced to scale-like structures.

**BROADLEAVED** Of trees or shrubs that bear broad, flat leaves, rather than needle-like foliage as in conifers.

**BROAD-SPECTRUM** Used to describe a chemical which has an effect on a wide range of often unrelated organisms.

**BUD** A condensed shoot containing an embryonic leaf, leaf cluster or flower.

**BULB** A storage organ consisting mainly of fleshy scales and swollen, modified leaf-bases on a much reduced stem. Bulbs usually, but not always, grow from underground.

**CALLUS** The corky wound tissue formed by a plant in response to wounding. Callus frequently grows in from the edges of a wound created when a tree limb is removed, or at the base of a cutting.

**CATERPILLAR** The larval stage of moths and butterflies.

**CERTIFIED STOCK** Plants sold guaranteed to be of good quality and free from pests and diseases, especially virus infection, at the time of sale.

**CHEMICAL** In a garden situation, a fungicide, pesticide, weedkiller, repellent or fertilizer used to control pathogens, pests or weeds, or to increase the soil fertility.

**CHESHUNT COMPOUND** A fungicide mixture of ammonium carbonate and copper sulphate.

**CHLOROPHYLL** The green pigment in plants that absorbs light, providing the energy for photosynthesis.

**CHLOROSIS** Yellowing of plant tissue due to a deficiency or loss of chlorophyll (the green pigment). It may be brought about by a range of physiological disorders, nutrient deficiencies, pests and diseases.

**CHLOROTIC** Showing the symptoms of chlorosis.

**CHRYSALIS** The resting stage between a fully-fed caterpillar and the adult moth or butterfly. See also *Pupa*.

**COCOON** A protective outer covering, usually of silk, produced by caterpillars and some other larvae before they pupate.

**COLOUR-BREAKING** A term used to describe a patterned change in flower colour, where the original colour is "broken" into elaborate feathered patterns. It is usually caused by virus infections.

**COMPANION PLANTING** Positioning plants together that are reputed to have a beneficial effect on neighbouring plants by discouraging pests and diseases or improving growth.

**COMPLETE METAMORPHOSIS** See *Metamorphosis*.

**COMPOST** 1. A potting medium comprising a mixture of loam, sand, peat, leaf mould, coir, bark or other ingredients. 2. An organic material, rich in humus, formed by decomposed plant remains and other organic matter, including material from the kitchen and garden, used as a soil improver or mulch. 3. A growing medium available in different forms.

**COMPOUND** Made up of several or many parts, e.g. a leaf that is divided into two or more leaflets.

**CONTACT** Used to describe a chemical that remains on the outside of a plant and is absorbed by a pest or pathogen when it is sprayed or comes into contact with a treated surface.

**CORM** A bulb-like underground storage organ, consisting mainly of a swollen stem base and often surrounded by a papery tunic.

**COTYLEDON** A seed leaf; the first leaf or leaves to emerge from some seeds after germination, often markedly different from mature leaves. Flowering plants (angiosperms) are classified into monocotyledons (one) and dicotyledons (two) depending on how many cotyledons are contained in the mature seed. In gymnosperms (conifers) they are often produced in whorls.

**CRUCIFEROUS** Of the family Cruciferae, to which brassicas belong.

**CULTIVAR** A type of plant which has developed from the original species in cultivation (rather than in the wild), either by accident or by deliberate breeding; a contraction of "cultivated variety". See also *Variety*.

**DECIDUOUS** Plants that shed their leaves at the end of the growing season (in autumn and winter) and renew them at the beginning of the next season.

**DEFOLIATION** Loss of leaves.

**DERRIS** An insecticide derived from the roots of *Derris* and *Lonchocarpus* spp. Its active chemical ingredient is rotenone.

**DIAPAUSE** A resting period in the life cycle of invertebrate animals, the onset and duration of which is controlled by environmental factors such as day length and temperature. It enables them to survive adverse conditions, such as winter, and has similarities to hibernation in vertebrate animals.

**DIEBACK** The death of shoots, often spreading down from the tip of the stem, caused by damage or disease.

**DORMANCY** 1. The state of temporary cessation of growth in plants, and slowing down of other activities, usually during winter. 2. Seed dormancy: non-germination of seed when placed in conditions suitable for germination, due to physical, chemical, or other factors inherent in the seed.

**DRUPES, DRUPELETS** See *Stone fruits*.

**DRENCH** Pest or disease control applied as a liquid to the roots of a plant to control soil-borne pests or pathogens.

**DUST** Pest or disease control applied as a fine dust, usually to the plant, but occasionally to the soil round it.

**ECDYSIS** The process in invertebrate animals of shedding their outer skin as they grow and increase in size.

**ECTOPARASITE** A parasite that feeds on the outside of a plant or animal's body.

**ELYTRON** (*pl.* elytra) A beetle's wing case; when not in flight these are folded back and form a covering for the top of the abdomen.

**ENDOPARASITE** A parasite that feeds inside a plant or animal's body.

**EYE** On a stem or, more commonly, on a tuber such as a potato, a dormant or latent growth bud that is visible. Of a fruit, e.g. apple, the end opposite the stalk.

**f.** [Lat.] *forma*; a variant within a species.

**FAMILY** A category in plant classification, grouping together related genera. All these genera have characteristics which are all constant, or which are clearly different from those of other families.

**FERTILIZATION** The fusion of a pollen grain nucleus (male) with an ovule (female) to form a fertile seed.

**FERTILIZER** A plant food (usually in concentrated form), naturally or synthetically produced, applied to the soil or to plant foliage. A balanced fertilizer contains a balance of nitrogen, phosphate and potash. A complete fertilizer contains both these and other nutrients.

**FLEECE, HORTICULTURAL** Synthetic material composed of very fine fibres woven or compressed together into a fabric which can be draped over plants to protect them from weather extremes and certain pests.

**FOLIAR FEED** Liquid fertilizer formulated for application direct to healthy plant foliage. Nutrients applied in this manner are very rapidly taken up by the plant.

**FRASS** Insect excrement pellets, especially of caterpillars.

**FREE-LIVING** An organism such as a nematode which is associated with but not attached to its host.

**FROST HARDY** A plant able to withstand temperatures down to −5˚C (23˚F).

**FROST TENDER** A plant vulnerable to frost damage.

**FRUITING BODIES** (of fungi) Spore-bearing structures of larger fungi, commonly known as toadstools, mushrooms or bracket fungi. May be annual or perennial.

**FULLY HARDY** See *Hardy*.

**FUNGICIDE** A pesticide capable of controlling a fungus.

**FUNGUS** (*pl.* fungi) A very variable group of organisms, which obtain their food materials from living or dead organic matter (see also *Parasite* and *Saprophyte*). They vary greatly in size, from microscopic to clearly visible with the naked eye.

**GALL** An abnormal growth produced by a plant in response to chemicals secreted by an animal, fungus or bacterium that lives within the galled tissue.

**GENUS** (*pl.* genera) A category in plant classification, grouping related species. Denoted by the first part of the scientific name. See also *Species*.

**GERMINATION** The physical and chemical changes that take place as a seed starts to grow and develop into a plant.

**GIRDLING** The removal of bark all round a trunk, stem or branch, by animals or physical injury. See also *Bark-ringing*.

**HABIT** 1. Characteristic, natural form of growth of a plant – upright, prostrate, weeping, etc. 2. The characteristic growth or general appearance of a plant.

**HALF HARDY** 1. Plants not able to tolerate frost, but generally able to withstand lower temperatures than frost-tender plants. 2. Not tolerating frost, but withstanding temperatures down to 0˚C (32˚F).

**HARDY** Plants tolerating year-round climatic conditions in temperate regions, including frost, without protection.

**HAULM** Used to describe the top-growth of plants such as potatoes and legumes.

**HEAVY** (of soil) Having a high proportion of clay, and often poorly drained.

**HERBACEOUS** 1. A non-woody plant in which the above-ground part dies down to the rootstock at the end of the growing season. 2. Dying down at the end of the growing season.

**HERBICIDE** A chemical used to control or kill weeds.

**HERBIVORE** A plant-feeding animal.

**HERMAPHRODITE** Animals with both female and male sexual organs, e.g. slugs, snails and earthworms.

**HONEYDEW** A sugary liquid excreted by sap-feeding insects including many, but not all, aphids, whiteflies, suckers, mealybugs and scale insects.

**HOST** The plant on which an insect, pest or pathogen develops or feeds; also the insect or other animal in which parasitic animals develop.

**HOST RANGE** The range of host plants that a given pest or pathogen will use. The host range is often restricted to closely-related plants.

**HYPHA** (*pl.* hyphae) The minute thread-like growths formed by fungi.

**IMMUNE** A plant with characteristics that prevent a given pest or pathogen from attacking or colonizing it. A plant immune to one disease or pest may, however, be susceptible to others.

**INCOMPLETE METAMORPHOSIS** See *Metamorphosis*.

**INORGANIC** 1. Used generally, not of plant or animal origin, i.e. a mineral or synthetically-produced material. 2. Of a chemical compound, one that does not contain carbon. Many man-made pesticides contain carbon-based chemicals and so are organic compounds as defined by a chemist, but not in the sense in which gardeners use the word organic. Inorganic fertilizers are refined from naturally-occurring minerals or produced artificially. See also *Organic*.

**INSECT** Invertebrate animal whose body is divided into a head, thorax and abdomen. Adult insects have three pairs of jointed legs on the thorax and usually two pairs of wings, e.g. moths, butterflies, beetles, ants, bees, wasps, sawflies, some aphids, whitefly and earwigs.

**INSECTICIDAL SOAP** A pesticide containing fatty acids, used against some pests, diseases and weeds.

**INSECTICIDE** A pesticide capable of controlling insects.

**INSTAR** The stages between ecdysis in the development of larvae and nymphs of insects and mites.

**INTEGRATED CONTROL** A combination of chemical and biological controls with good cultivation techniques, aimed at keeping pests and diseases below the level at which damage occurs.

**INVERTEBRATE** Animals without backbones, e.g. insects, mites, woodlice, molluscs, nematodes, earthworms and millipedes.

**LARVA** (*pl.* larvae) The immature feeding-stage in the life cycle of an insect undergoing complete metamorphosis, e.g. moth caterpillars, beetle grubs, fly maggots. See also *Nymph*.

**LATENT INFECTION** Infection by a pathogen which does not result in any visible symptoms on the plant.

**LEACHING** The loss of nutrients when they are washed down through the soil to areas out of reach of plant roots, often caused by excessive watering or rain.

**LEGUME** A plant of the family Leguminosae, to which peas and beans belong, or of one of its three sub-families, Caesalpiniaceae, Mimosaceae and Papilionaceae.

**LENTICEL** A breathing pore found in the stems of woody plants. Often enlarged and clearly visible to the naked eye.

**LESION** Point or area of damage caused by pests, pathogens or physical injury.

**LIME** Compounds of calcium; the amount of lime in soil determines whether it is alkaline, neutral or acid.

**MANDIBLES** The biting mouthparts of insects, especially leaf-eating types.

**METAMORPHOSIS** The changes undergone during an insect or mite's development. Incomplete metamorphosis, as shown by aphids, capsid bugs, earwigs and mites, involves little more than an increase in size and the gradual development of sexual organs and sometimes wings. Complete metamorphosis involves a dramatic change of form with the larva bearing no resemblance to the adult, e.g. in moths, flies, beetles and sawflies.

**MOISTURE SHELLS** Layers of humid air which accumulate or build up round a plant or its leaves as they transpire and increase humidity locally.

**MOLLUSCICIDE** A pesticide capable of controlling slugs and snails.

**MULCH** 1. A material applied in a layer to the soil surface to suppress weeds, conserve moisture, and maintain a cool, even root temperature. In addition to organic materials such as manure, bark and garden compost, which also enrich the soil, polythene, foil and gravel may also be used. 2. A layer of organic matter applied to the soil over or around a plant to conserve moisture, protect the roots from frost, reduce the growth of weeds, and enrich the soil.

**MULTI-STEMMED** Tree or shrub with several main stems arising either directly from the ground or from a short main stem.

**MYCELIUM** A compacted mass of fungal hyphae, often developed into a distinct "sheet" of fungal growth.

**MYCOPLASMA** Microscopic organism closely related and with similar characteristics to a virus.

**MYCORRHIZA** (*pl.* mycorrhizae) Soil fungi that live in mutually beneficial association with plant roots, helping them to absorb nutrients from the soil.

**NECROSIS** The deterioration and death of plant tissues.

**NECROTIC** Dead, frequently brown or black, areas of plant tissue.

**NECTAR** A sweet, sugary liquid secreted by nectary glandular tissue, usually found in the flower, but sometimes on the leaves or stem.

**NEMATICIDE** A pesticide capable of controlling nematodes or eelworms.

**NEMATODES** Microscopic, unsegmented, worm-like animals, also known as eelworms. Some are used as pest predators, e.g. of slugs.

**NEUTRAL** (of soil) With a pH value of 7, the point at which soil is neither acid nor alkaline. See also *Acid* and *Alkaline*.

**NITRATE** A salt of nitric acid, having a high nitrogen content that is available to plants. Either produced by the activity of bacteria in the soil or manufactured.

**NITRITE** A salt of nitric acid in which the nitrogen is not readily available to plants.

**NODE** The point on a stem where one or more buds, leaves, shoots or flowers are attached.

**NODULE** A small swelling.

**NOSE** (of a bulb or corm) The pointed upper end.

**NOTIFIABLE** A pest or disease is described as notifiable when it is obligatory to notify the Ministry of Agriculture, Fisheries and Food (MAFF) as soon as it is discovered.

**NUTRIENTS** Minerals (mineral ions) used to develop proteins and other compounds required for plant growth.

**NYMPH** The immature stage of an insect or mite that undergoes incomplete metamorphosis. See also *Larva*.

**ORGANIC** 1. Of garden chemicals, referring to compounds containing carbon derived from plant or animal organisms. 2. Loosely, mulches, composts, or similar materials derived from plant or animal-derived materials. 3. A system of crop production and gardening without using synthetic or non-organic materials.

**OVICIDE** A pesticide capable of controlling the egg stage of pests.

**OVIPOSITOR** The egg-laying organ of an insect.

**PARASITE** An organism that gets all or part of its food from another plant or animal over a period of time, often without killing its host. Parasitic insects that eventually kill the insect in which they have developed are more correctly termed parasitoids.

**PARTHENOGENESIS** The ability of some female pests to reproduce without fertilization. Many aphids have this ability, as have vine weevils, some sawflies, scale insects and mealybugs. Males are rare or non-existent in these species.

**PATHOGEN** An organism which is parasitic and causes disease in another organism. Pathogens include certain bacteria, fungi, viruses and mycoplasmas.

**PATHOVAR** A distinct strain or type of a named pathogen.

**PEDICEL** The stalk of an individual flower.

**PERENNIAL** Living for at least three seasons. In this book the term when used as a noun, and unless qualified, denotes a herbaceous perennial. A woody-based perennial dies down only partially, leaving a woody stem at the base.

**PERSISTENCE** (of chemicals) The length of time a chemical persists in an active form either on the plant or in the soil or compost after it has been applied.

**PESTICIDE** A chemical substance; a term usually applied to chemicals that control pests, but in its broader definition also includes fungicides, herbicides, animal repellents and wood preservatives.

**PETIOLE** The stalk of a leaf.

**pH** The scale by which the acidity or alkalinity of soil is measured. See also *Acid, Alkaline, Neutral.*

**PHEROMONE** Volatile chemicals produced by insects as a means of communicating with others of the same species. Males often locate virgin females by tracking down the source of the pheromone or scent released by the female.

**PHLOEM** Conducting elements within plant tissue, largely associated with the transport of food materials in solution, produced in the leaves and distributed to the rest of the plant.

**PHOTOSYNTHESIS** The synthesis of carbohydrates in green plants from carbon dioxide and water, using light energy from sunshine or artificial light absorbed by the green pigment chlorophyll.

**PHYTOTOXIC** Harmful to plants. Usually used to describe a chemical (usually an insecticide or fungicide) which can damage a plant onto which it has been sprayed.

**POLLEN** The male cells of a plant, formed in the anthers.

**POLLINATION** The transfer of pollen from the anthers to the stigma of the same or different flowers, often carried out by insects, especially bees, or by air movement, resulting in fertilization of the embryonic seeds in the ovary.

**POLYPHAGOUS** Pests that feed on a wide range of plants which may be botanically unrelated.

**POME FRUIT** A firm, fleshy fruit, with seeds enclosed in a central core, as in apples and pears.

**PROBOSCIS** Insect mouthparts, used for sucking up liquids, especially nectar.

**PROLEGS** The clasping legs on the abdomen of a caterpillar or sawfly larva.

**PUPA** (*pl.* pupae) The resting, non-feeding stage in the life cycle of an insect undergoing complete metamorphosis. Also known as a chrysalis.

**PUPATION** The period of time during which larval tissues are dissolved and reconstructed into the form of the adult insect.

**Pycnidium** (*pl.* pycnidia) Minute raised, rounded fungal fruiting body, produced on the surface of infected plant tissue.

**Race** (of fungi) A specific strain of a named fungus which may vary in its physical or pathogenic characteristics from other races of that fungus.

**Repellents** Substances that deter pests without causing them harm.

**Resistant** A plant is described as resistant to a pest or disease when it is attacked by it, but shows no ill effects. A fungus may also develop resistance to a fungicide, or to a closely-related group of fungicides. Fungicide resistance occurs most commonly when a particular fungicide or closely-related ones have been used repeatedly for many years. Similar resistance problems also occur with some pests and weeds that are no longer controlled by insecticides and herbicides that were formerly effective.

**Rhizome** A specialized, usually horizontally creeping, swollen or slender underground stem that acts as a storage organ and produces aerial shoots at its apex and along its length.

**Rhizomorph** A cord-like fungal structure, which is very resilient and withstands extremes of temperature and moisture. Honey fungus rhizomorphs are dark and tough, and grow from host to host.

**Root** The part of a plant, normally underground, that anchors it and through which water and nutrients are absorbed.

**Rotation** Planning where plants are grown, to ensure that different types of plant are grown on a given piece of land from season to season. By rotating plants some pest and disease outbreaks can be avoided. Rotation is a particularly effective method of avoiding problems with soil-borne pests and pathogens.

**Russeting** A discoloration, generally brown, of a plant surface (usually the skin of fruits) which is often roughened.

**Saprophyte** An organism (most commonly a fungus) which lives on dead or decaying organic material but does not attack living material.

**Sclerotium** (*pl.* sclerotia) The resilient fungal resting bodies formed by some fungi to allow them to overwinter successfully and withstand extremes of weather.

**Shoot** The aerial part of a plant which bears leaves. A side-shoot arises from a bud along the length of a main shoot.

**Species** A category in plant and animal classification, comprising individuals in the same genus that are alike and naturally breed with each other. Denoted by the second part of the scientific name. See also *Genus*.

**Spore** Microscopic reproductive structure of fungi and bacteria. Certain non-flowering plants such as ferns may also produce spores.

**Spur** A short fruiting branch on fruit bushes and trees, particularly apples and pears.

**Ssp.** Subspecies; a higher category in plant classification than *forma* or variety.

**Stamen** The male reproductive organ in a plant, comprising the pollen-producing anther and usually its supporting filament or stalk.

**Stem** The main axis of a plant, usually above ground and supporting leaves, flowers and fruits.

**Stolon** A horizontally spreading or arching stem, usually above ground, which roots at its tip to produce a new plant, e.g. strawberry.

**Stone fruits** Fruits, also known as "drupes", with one or more seeds ("stones") surrounded by fleshy, usually edible, tissue. They are common in the genus *Prunus* (e.g. apricots, plums, cherries).

**Stool** A group of shoots emerging from the base of a single plant. In propagation the term refers to the roots of a plant such as a chrysanthemum after it has finished flowering.

**Strain** (of bacteria, etc.) A distinct form of a named pathogen.

**Stylets** The mouthparts of sap-feeding insects and mites, used for piercing plant tissues.

**Symbiosis** A relationship between two organisms, where neither harms the other and both derive benefit from the relationship, e.g. nitrogen-fixing bacteria in root nodules.

**Systemic** 1. Insecticides and fungicides that are absorbed into a plant's sap and translocated within the plant. Systemic fungicides are, however, rarely able to move great distances but may move from one leaf surface through the leaf to the lower surface. 2. Of diseases, some pathogens, in particular the viruses, that are systemic within the host plant, moving with the plant sap and being found in all or a large part of the host.

**Tender** Vulnerable to low temperatures. Tender plants may be categorized as: cool-growing, down to a minimum temperature of 10°C (50°F); intermediate, minimum 13°C (55°F); or warm-growing, minimum 18°C (64°F). See also *Frost tender*.

**Thorax** The mid-part of an insect's body, between the head and abdomen.

**Tolerant** Having the ability to tolerate an attack by a pest or a pathogen. A pest or pathogen may be described as tolerant if able to withstand the effects of a pesticide.

**Trace Elements** Chemical elements needed by plants, but only in very small amounts, e.g. boron, copper, manganese, iron, molybdenum and zinc.

**Transpiration** Loss of water by evaporation from the leaves and stems of plants.

**Tuber** A swollen, usually underground, organ derived from a stem or root, used for food storage.

**Variegated** Irregularly marked with various colours; particularly of leaves patterned with markings in white, yellow or other colours.

**Variety** A plant which has developed with a slight difference from a wild plant species. Plant varieties are often brought into cultivation once developed.

**Vector** See *Virus vector*.

**Virus** A sub-microscopic particle often responsible for causing plant virus diseases.

**Virus Vector** An insect, mite or nematode that transmits plant viruses, usually on its mouthparts or in its saliva. Some fungal and bacterial diseases can also be spread by plant pests.

**Viviparous** Giving birth to live young. Most aphid reproduction is by this method.

**Wetting Agent** A material used to lower the surface tension of liquids on a plant surface, so allowing a spray to form a film over the surface rather than forming into droplets. Many garden pesticide formulations contain wetting agents (also known as wetters or spreaders).

**Wind Snap** Wind causing a tree to be broken off at ground level; this can be made more likely by decay round the base of the trunk.

**Wind Throw** Wind causing a tree to be uprooted. Decay in the tree roots can make the tree unstable and therefore susceptible to wind damage.

**Worm** Various animals with elongate, cylindrical bodies. Eelworms or nematodes have simple, unsegmented bodies; earthworms are more advanced animals with bodies divided into many segments.

**Xylem** Vascular tissue under bark taking water and nutrients up a stem.

**Zoospore** A motile stage, capable of self-propulsion, in the life cycle of certain fungi.

# INDEX

References to illustrations and tables are in **bold**. Sub-headings are arranged with symptoms first.

## A

*Abies alba see* fir, silver
abutilon 202
acacia, false (*Robinia*) **39**, 194
*Acer see* maple, sycamore
*Acer pseudoplatanus see* sycamore
acer gall mite **17**, 98
acer tar spot **13**, 180–1
*Achimenes see* hot water plant
acidanthera 200
acid-loving plants **24**, 139–40
acorn gall wasp **52**, 98
adelgids 75, 98
*Aesculus hippocastanum see* horse chestnut
African marigold (*Tagetes*) 201
African violet (*Saintpaulia*) 202
agapanthus 198
*Alcea see* hollyhock
alder (*Alnus*) 98, 194
  leaf symptoms **32**
alder sucker **32**, 98
algae 98
almond (*Prunus dulcis*) 156, 175, 194
  symptoms **26**, **33**
*Alnus see* alder
alstroemeria 198
alternaria leaf spot 98
alyssum 198
amaryllis (*Hippeastrum*) 112, 150, 202
  symptoms **29**, **54**
*Ambleyseius* spp. **91**, 148
amelanchier 196
American gooseberry mildew **25**, **49**, 98–9
anchusa 198
anemone 99, 161, 198
  leaf symptoms **22**
anemone rust *see* plum rust
anemone smut 99
*Anemone* hybrids *see* Japanese anemone
angels' trumpets (*Datura*) 202
animals
  beneficial **60–3**, 79, 90, **90**, 141
  pesticide injury 83, **211**
  plant relationships 66–7, **67**, 68–9
anthocorid bugs **60**
anthracnose 99
*Antirrhinum see* snapdragon
antirrhinum rust **15**, 99
ants **32**, **59**, 83, 87, 99
  chemical control **208–9**
*Aphidius* spp. **91**
*Aphidoletes aphidimyza* **91**, 130
aphids 73, 74, 75, 99–100
  biological control **60–1**, 90-1, **91**
  chemical control 81, **208–9**

life cycle **74**, 77, **77**
apple 100–2, 118, 129, 133, 144, 156, 173, 204
  flower symptoms **43**, **44**, **46**
  fruit symptoms **47**, **48**
  leaf symptoms **16**, **19**, **20**, **25**, **31**
  stem symptoms **35**, **36**, **40**, **41**
  fruit protection 88
  replanting 96, 172
  resistant cultivars 100, 101, 156
apple and pear canker **36**, 100, 156
  chemical control **210–11**
apple bitter pit **48**, 100
apple blossom weevil **46**, 75, 100
apple capsid **47**, 100
apple frost damage 100–1
apple fruit split **48**, 101
apple leaf miner **19**, 101
apple powdery mildew **25**, 101
apple sawfly **47**, **48**, 75, 101
  chemical control **208–9**
apple scab **16**, 81, 101
  chemical control **210–11**
apple scald **47**, 101
apple sooty blotch 102
apple sucker **43**, 102
apple water core 102
apricot *see* peach
*Aquilegia see* columbine
aquilegia sawfly 102
*Arabis see* rock cress
arabis mosaic virus 102
*Araucaria araucana see* monkey puzzle
arbutus leaf spot **14**, 102
*Arbutus unedo see* strawberry tree
artichoke, globe 205
artichoke, Jerusalem 205
arum lily (*Zantedeschia*) 202
*Aruncus dioicus see* goatsbeard
aruncus sawfly **30**, 77, 102
ash (*Fraxinus*) 102–3, 194
ash catkin gall mite 102
ash sucker 103
asparagus beetle **40**, 103
asparagus fern 202
aspidistra 202
*Aster see* Michaelmas daisy
aster wilt 103
astilbe 198
aubergine 120, 123, 205
  stem symptoms **33**
  resistant cultivars 120
aubrieta 198
auricula (*Primula*) 198
autumn crocus (*Colchicum*) 200
azalea *see* rhododendron
azalea gall 103
azalea whitefly 103

## B

*Bacillus thuringiensis* 91, **91**
bacteria 67, 72–3, 103–4
  nitrogen-fixing **68**, 69
  as pest control 91, **91**

bacterial canker **35**, 104
  chemical control **210–11**
bacterial leaf spot **13**, 29, 143
bacterial soft rot **73**, 104
bait **82**, 83
  chemicals **208–9**, **212**, **213**
banana (*Musa*) 202
banded palm thrips **22**, 104
barberry (*Berberis*) 196
bark beetles **33**, 75, 104
barriers **78**, 86, 88–9, **88**, **89**
basal rots *see* foot rots
basil 205
bay, sweet bay (*Laurus*) 104–5, 194, 197, 205
  leaf symptoms **26**
bay sucker **26**, 104–5
bean, broad 69, 105, 107, 147, 151, 205
  flower symptoms **46**
  leaf symptoms **12**, **31**, **32**
  seed symptoms **51**
bean, dwarf *see* bean, French
bean, French 105–6, 147, 205
  seed symptoms **51**
  nitrogen fixation 69
  resistant cultivars 105
bean, runner 105–6, 147, 151, 205
  symptoms **46**, **51**
  nitrogen fixation 69
  resistant cultivars 105
bean anthracnose 105
bean chocolate spot **12**, 105
bean halo blight 105
bean rust 105
bean seed beetle **51**, 105
bean seed fly **51**, 105–6
beech (*Fagus*) 106, 194
  bark symptoms **35**, **39**
  leaf symptoms **24**, **32**
beech bark disease **35**, 106
beech bark scale **35**, 106
beech gall midge 106
beech leaf-mining weevil **24**, 106
beech woolly aphid **32**, 106
beefsteak fungus 127
bees
  beneficial **62**
  burrowing **58**, 112
  leaf-cutting **30**, 75, 142
  nectar robbing **46**, 151
  pesticide injury **208–11**
beet leaf miner **17**, 106
beetles 75, 106–7
  chemical control **208–9**
beetroot 106, 109, 205
  symptoms **17**, **44**
begonia 202, 107, 202
begonia powdery mildew **23**, 107
*Bellis perennis see* daisy
beneficial creatures **60–3**, 79, 90, **90**, 141
*Berberis see* barberry
*Bergenia see* elephant's ears
*Betula see* birch
biennials 201
big bud mites **41**, 107
biological control 79, 90–1, **91**

birch (*Betula*) 107, 194
birds 46, **47**, 75, 107
  beneficial **63**, **90**
  chemical injury 83, **83**
  seed dispersal 67
bitter rot *see* gleosporium rot
black bean aphid **32**, 107–8
black leg of cuttings 108
blackberry 108, 168, 204
  symptoms **34**, **49**
blackberry cane spot **34**, 108
blackcurrant *see* currant, black
blackcurrant gall midge **26**, 108
blackfly 108
bleeding canker 108
blind shoots **40**, 108
blindness of bulbs **41**, 108
blossom end rot **50**, 108
blossom wilt **41**, 108–9
blotch *see* leaf spot
blue mould on bulbs **54**, 109
bluebell (*Hyacinthoides*) 109, 200
bluebell rust 109
blushing bracket *see* *Daedaleopsis confragosa*
bolting **44**, 109
boron deficiency 109
*Botrytis see* grey mould
box (*Buxus*) **26**, 109, 196
box sucker **26**, 109-10
bracket fungus **37**, **38**, 110
brassica bacterial soft rot 110
brassica dark leaf spot **13**, 110
brassica downy mildew 110
brassica light leaf spot 110
brassica ring spot 110
brassica wire stem 110
brassicas 110, 112-13, 127–8, 146, 160, 189, 205-6
  leaf symptoms **13**, **14**, **29**, **30**, **32**
  root symptoms **55**
  stem symptoms **44**
  chemical control **210–11**
  plant protection 89, **89**
  resistant cultivars **92**, 118
broom (*Cytisus*) 110–11, 196
  stem symptoms **39**
broom gall mite **39**, 110–11
brown core of primula 111
brown rot (*Sclerotinia*) **47**, **49**, 111
brown scale **36**, 111
brown-tail moth 111
*Brugmansia see Datura*
bryobia mite 111
bud, dry 112
bud drop 111–12
buddleia (*Buddleja*) 196
buff-tip moth 112
bulb scale mite 112
bulbs 92, 108, 109, 124, 200
  symptoms **41**, **54–5**
  fungicides **210–11**
  hot water treatment 89
bullfinch **46**, 74, 75, 112
busy Lizzie (*Impatiens*) 125, 201
  leaf symptoms **30**
*Buxus see* box

## C

cabbage *see* brassicas
cabbage moth **30**, 112
cabbage root fly 75, 112–13
  chemical control **208–9**
  plant protection 89, **89**
cabbage white butterfly 74, **76**, 113
cabbage whitefly **32**, 113
cacti **71**, 113, 147, 202
  plant symptoms **22**
cactus corky scab **22**, 113
calceolaria 202
calcium 69
  deficiency 113
*Calendula see* marigold
Californian poppy 198
*Callistephus see* China aster
  resistant cultivars 103
*Calluna see* heather
camellia 113–14, 121, 196
  leaf symptoms **15**, **22**, **26**
camellia gall **26**, 113
camellia yellow mottle **22**, 114
campanula 198
*Campsis see* trumpet vine
candytuft (*Iberis*) 198
canker **73**, 114
  chemical control **210–11**
canna 198
*Capsicum* (ornamental pepper) 203
*Capsicum* (edible pepper) **50**, 108, 206
capsid bugs
  symptoms **29**, **46**, 75
  chemical control **208–9**
carabid beetle **60**
carnation (*Dianthus*) 123, 198, 202
carnation tortrix moth **20**, 86, 114
carpenteria 196
*Carpinus see* hornbeam
carrot 98, 114–15, 206
  symptoms **25**, **51**
carrot cavity spot 114
carrot fly **51**, 75, 89, 114
  chemical control **208–9**
carrot motley dwarf virus **25**, 114-15
caryopteris 196
*Castanea sativa see* sweet chestnut
castor-oil plant (*Ricinus communis*) 201
catalpa 194
caterpillars 74, 75, 91, **91**, 115
  chemical control **208–9**
cats 89, 115
cavity decay 115
ceanothus 196
cedar (*Cedrus*) 115, 194
  leaf symptoms **32**
cedar aphid **32**, 115
celeriac **12**, 206
celery 115–16, 206
  symptoms **12**, **19**, **36**
celery leaf miner **19**, 115
celery leaf spot **12**, 115–16
centaurea 198
centipedes **61**, 116
*Cercidiphyllum japonicum* 194
*Cercis siliquastrum see* Judas tree

chaenomeles 197
chafer grubs **53**, **58**, 75, 116
chemical control **208–9**
*Cheiranthus see* wallflower
chemicals 78–9, 80-3
damage 160, 188–9, **211**
ingredient tables **212**, **213**
trials 81
*see also* pesticides
cherry 101, 116, 157, 175, 204
bark symptoms **33**, **41**
leaf symptoms **19**, **24**
resistant cultivars 104
cherry, flowering *see* prunus, ornamental
cherry blackfly **24**, 116
cherry laurel *see* laurel, cherry
chicken of the woods *see Laetiporus sulphureus*
China aster (*Callistephus*) 201
resistant cultivars 103
China aster fusarium wilt 103
china mark moth **30**, 116
chironomid midges 116–17
chives **17**, 144
*Chlorophytum see* spider plant
chlorosis 117
*Choisya see* Mexican orange
Christmas cactus (*Schlumbergera*) 202
chrysanthemum (*Dendranthema*) 117, 147, 198, 202
flower symptoms **43**, **45**
leaf symptoms **14**, **25**
hot water treatment 89, 117
chrysanthemum eelworm 89, 117
*Chrysanthemum frutescens see* marguerite
chrysanthemum leaf miner **25**, 117
chrysanthemum petal blight **45**, 117
chrysanthemum ray blight 117
chrysanthemum white rust **14**, 117
cineraria (*Senecio*) 98, 202
*Cistus see* sun rose
citrus 202
clematis 117–18, 125, 196
flower symptoms **46**
clematis green petal 117–18
clematis wilt 118
click beetle 118
climbing plants 78, **93**, 196-7
clivia 202
cloches 88, **89**
clubroot **55**, 118
chemical control **210–11**
*Codiaeum see* croton
codling moth **47**, 75, 86, 118
chemical control **208–9**
*Colchicum see* autumn crocus
coleus (*Solenostemon*) 202
collar rot *see* phytophthora
*Colletotrichum* 118
Colorado beetle 118
columbine (*Aquilegia*) 198
companion planting 78, **86**

composting **78**, 79, 96, **96**
conifer aphid 119, 121
conifer needle rust 151
conifer red spider mite 119
conifer woolly aphid *see* adelgids
conifers 98, 119, 121, 128, 151, 174
symptoms **37**
contact pesticides 82, **82**, **83**, **208**, **210**
chemical tables **212**, **213**
*Convallaria see* lily-of-the-valley
convolvulus 198
copper deficiency 119
coral flower (*Heuchera*) 198
coral spot **41**, 119
corms 124, 200
corm symptoms **55**
chemical control **210–11**
*Cornus see* dogwood
*Corylus see* hazel
cosmos 198
cotinus 196
cotoneaster 119, 192, 196
stem symptoms **35**, **37**
cotoneaster webber moths **37**, 119
courgette *see* marrow
covers, protective 88–9, **88**, **89**
crab apple (*Malus*) 101, 194
leaf symptoms **16**
crane flies (leatherjackets) 119, 143-4
cranesbill (*Geranium*) 131, 198
leaf symptoms **25**
*Crassula see* money tree
*Crataegus see* hawthorn
crinum **29**, 150, 200
*Crocosmia see* montbretia
crocus 147, 177–8, 200
symptoms **46**, **55**
crop rotation 78, 89, 96
croton (*Codiaeum*) 202
crown gall **38**, 119–20
crown rot 120
*Cryptolaemus montrouzieri* **91**, 141
cuckoo spit **39**, **43**, 75, 120
cucumber 120, 128, 147, 206
fruit symptoms **55**
leaf symptoms **21**, **25**, **43**
resistant cultivars 120
cucumber foot & root rot **55**, 128
cucumber gummosis 120
cucumber mosaic virus **21**, **25**, **72**, **73**, 120
*Cupressus see* cypress
currant, black 98–9, 108, 121, 204
bud symptoms **41**
leaf symptoms **25**, **26**
stem symptoms **33**
resistant cultivars 108
*see also* currants
currant blister aphid **27**, 121
currant clearwing moth **33**, 75, 121
currant, flowering (*Ribes*) 196
currant, red **31**, 204
*see also* currants
currant reversion 121
currant, white **31**, 204

currant & gooseberry leaf spot 120–1
*see also* currants
currants 119, 120–1, 133–4, 192
symptoms **27**, **39**
plant protection 88, **88**
*see also* currant, black; currant, red/white
cushion scale **15**, 121
cutworms **53**, 75, 121
cyclamen 202
cyclamen, hardy 200
*Cyperus see* papyrus
cypress (*Cupressus*) **35**, 121, 194
susceptible spp. 174
cypress aphid 119, 121
*Cytisus see* broom

# D

*Daedaleopsis confragosa* **37**, 122
daffodil (*Narcissus*) 43, 112, 149–51, 200
symptoms **23**, **29**, **43**, **54**
hot water treatment 89
daffodil bulb scale mite 112
dahlia 107–8, 122, 125, 147, 200
symptoms **32**, **43**, **46**
chemical injury **210–11**
dahlia smut 122
daisy (*Bellis perennis*) 198
damping off **34**, 122
chemical control **210–11**
damson *see* plum
*Datura see* angels' trumpets
daylily (*Hemerocallis*) 135, 198
flower symptoms **45**
deer **41**, **70**, 75, 88–9, 122
delphinium 122–3, 198
leaf symptoms **12**, **18**
delphinium black blotch **12**, 122
delphinium leaf miner **18**, 122
delphinium moth 123
delphinium viruses 123
*Dendranthema see* chrysanthemum
devil's coach horse **62**
*Dianthus see* carnation
dianthus anther smut 123
*Dianthus barbatus see* sweet William
dianthus fusarium wilt 123
didymella stem rot **33**, 123
dieback 123
*Digitalis see* foxglove
diseases 72–3, **72**, **73**, 74
resistant spp. 137
dog lichens 58, **59**
dogs 123
dogwood (*Cornus*) 123–4, 194, 196
dogwood anthracnose 123–4
Douglas fir (*Pseudotsuga menziesii*) 124, 194
Douglas fir adelgid 124
downy mildew **73**, 124
chemical control **210–11**
dracaena 202
drought **28**, **44**, 70–1, **70**, 124
bud disorders 111–12
dry rot of bulbs & corms 124
Dutch elm disease 124–5

earthworms **58**, **61**, **66**, 67, 125
earwigs 46, 74, 75, 125
chemical control **208–9**
traps 87, 88, **88**
echeveria 203
*Echinops see* globe thistle
ecology in the garden 66–7
eelworm 73, 75, 96, 125
biological control 91, **91**
hot water treatment 89
elaeagnus 119, 196
elder (*Sambucus*) 196
elephant hawk moth **30**, 125
elephant's ears (*Bergenia*) 198
elm bark beetle **33**, 104
elm gall mite **16**, 125–6
elm (*Ulmus*) 104, 124–5, 125–6, 194
symptoms **16**, **33**
*Encarsia formosa* **91**
enchytraeid worms 126
epimedium 198
epiphyllum 113, 203
plant symptoms **22**
*Eranthis see* winter aconite
*Erica see* heather
erythronium 200
escallonia 196
eucalyptus 126, 194
eucalyptus sucker 126
euonymus 126, 196
leaf symptoms **21**
euonymus scale **21**, 126
euphorbia 198
*Euphorbia pulcherrima* 203
evening primrose (*Oenothera*) 201

# F

*Fagus see* beech
fairy rings 183–4
false acacia (*Robinia*) 39, 194
fasciation **43**, 126
fatsia 196
felt gall mites **19**, 126
felted beech coccus *see* beech bark scale
ferns 203, 211
fertilizers 68, 93
*Ficus benjamina see* weeping fig
*Ficus elastica see* rubber plant
fig (*Ficus*) 204
weeping (*Ficus benjamina*) 203
figwort weevil 126–7
fir, Douglas 124, 194
fir, silver (*Abies alba*) 98, 175, 194
fireblight **35**, 127
firethorn (*Pyracantha*) 166, 192, 196
symptoms **35**, **49**
resistant cultivars 166
fish 83, **83**
harmful chemicals **210–11**
*Fistulina hepatica* 127
flatworms 127
flea beetles **29**, 74, 75, 89, 127–8
chemical control **208–9**
sticky traps 87
flies 128
flower damage **42–6**, **73**, 158, 166

frost **44**, 129
pests 74-5
*Fomes pomaceous* 39, 128
fomes root & butt rot 37, 128
food chains 67–8, **83**
foot and root rots **55**, 128
forget-me-not (*Myosotis*) 201
forsythia 128, 196
stem symptoms **40**
forsythia gall **40**, 128
*Fortunella see* citrus
foxes 128–9
foxglove (*Digitalis*) 198
frangipani (*Plumeria*) 203
*Fraxinus see* ash
freesia 200
French marigold *see* African marigold
fritillary (*Fritillaria*) 200
froghopper *see* cuckoo spit
frogs 79, 90
frost damage **28**, **44**, **70**, 71, 129
apple 100–1
fruit cages 88, **88**
fruit damage **47–51**, 111, 178
pests 75
fruit plants 204–5
fruit tree red spider mite **20**, **40**, 129
fuchsia 114, 129, 203
symptoms **18**, **30**, **46**
fuchsia, hardy **30**, 114, 125, 196
fuchsia rust **18**, 129
fumigation 80, **208–9**
fungi 67, 72–3, **72**, **73**
mycorrhizal associations 69
pathogen resistance **211**
fungicides **210–11**
chemical tables **212**, **213**
phytoxicity 73, 81, 160
*see also* pesticides
fungus gnat **53**, 87, 129–30
biological control **91**
chemical control **208–9**
fusarium bulb rot *see* gladiolus corm rot; narcissus basal rot
fusarium wilt 130

# G

gage *see* plum
*Galanthus see* snowdrop
gall midges 75, 130
gall mites 75, 130
gall wasps 75, 77, 130
gall-forming pests 74, 75, 77, 130
ganoderma 130–1
garden, gardening 66–7
hygiene 79, 92, **94**, 96, **96**
techniques 78-9, 86-9, 92-3
tool disinfection 73, **73**
*see also* pruning
gardenia 203
garlic 206
gazania 203
gentian 198
*Geranium see* cranesbill
geranium downy mildew **25**, 131
geranium sawfly 131
gerbera 203
geum 198

REFERENCE SECTION

geum sawfly 131
ghost spot *see* grey mould
gladiolus 131–2, 200
  symptoms **45**, **55**
gladiolus core rot 131
gladiolus corm rot **55**, 131
gladiolus dry rot 131-2
gladiolus hard rot 132
gladiolus scab & neck rot
  132
gladiolus thrips **45**, 132
glasshouse leafhopper **20**,
  132
glasshouse plants *see*
  greenhouse plants
glasshouse red spider mite
  **22**, **23**, **43**, 77, 81, 132
  biological control **91**
glasshouse whitefly 77, 81,
  87, **91**, 132–3
gleditsia gall midge **27**, 133
*Gleditsia see* honey locust
gleosporium rot 133
globe thistle (*Echinops*) 198
gloxinia (*Sinningia*) 203
glyphosate damage 189
goatsbeard (*Aruncus*) 102,
  198
  leaf symptoms **30**
golden rod (*Solidago*) 198
gooseberry 98–9, 120–1,
  133–4, 204
  symptoms **25**, **31**, **49**
  fungicide damage **210-11**
  resistant cultivars 99
gooseberry dieback 133
gooseberry mildew *see*
  American gooseberry
  mildew
gooseberry rust 133
gooseberry sawfly **31**, 133–4
grape *see* vine (grape)
grape hyacinth (*Muscari*)
  109, 200
grapefruit *see* citrus
grease bands 87
green spruce aphid 134
green tiger beetle **63**
greenfly *see* aphids
greenhouse plants 77,
  132–3, 135, 147, 181,
  202–3
  flower symptoms **42**, **43**
  leaf symptoms **16**, **20**,
    **22**, **23**, **27**, **32**
  high temperature injury
    135–6
  pest control **79**, 81, 85,
    87
grey mould (*Botrytis*) 134
  symptoms **34**, **42**, **45**, **50**
  chemical control 81,
    **210-11**
*Grifola frondosa* 134
ground beetles 134
gummosis 120
gypsy moth 134–5

## H

hail damage 71
halo blight 105
hardenbergia 203
hawthorn (*Crataegus*) 100,
  119, 135, 194
  symptoms **19**, **37**
hawthorn gall midge 135
hazel (*Corylus*) 107, 151,
  178, 204
  symptoms **41**, **52**
heather (*Erica*) 196
hebe 135, 196

hebe downy mildew 135
*Hedera see* ivy
hedgehog **63**, **67**, 90
*Helianthemum see* rock
  rose
hellebore leaf blotch **20**, 135
hellebore (*Helleborus*) 135,
  198
*Helleborus see* hellebore
  leaf symptoms **20**
  resistant spp. 135
*Hemerocallis see* daylily
hemerocallis gall midge **45**,
  75, 135
hemispherical scale **16**, 135
herbaceous plants 114, 143,
  198-9
  symptoms **20**, **39**
*Heterobasidion annosum
  see* fomes root & butt
  rot
*Heterorhabditis megidis* 91,
  **91**
*Heuchera see* coral flower
hibiscus 203
high temperature injury 71,
  135-6
Himalayan poppy
  (*Meconopsis*) 199
*Hippeastrum see* amaryllis
  symptoms **29**, **54**
holly (*Ilex*) 121, 136, 194
  leaf symptoms **15**, **17**
holly leaf miner **17**, 136
hollyhock (*Alcea*) 136, 199
  leaf symptoms **17**
  resistant spp. 136
hollyhock rust **72**, 136
holm oak leaf miner 136
honesty (*Lunaria*) 199
honey fungus **34**, **73**, 136–7
honey locust 133, 194–5
  leaf symptoms **27**
honeydew **32**, 137
honeysuckle (*Lonicera*)
  137, 197
honeysuckle aphid 137
hornbeam (*Carpinus*) 137,
  195
hornbeam witches' broom
  137
horse chestnut (*Aesculus*)
  137–8, 195
  symptoms **20**, **40**
horse chestnut leaf blotch
  **20**, 137
horse chestnut marginal
  leaf scorch 137
horse chestnut scale **40**,
  137–8
horticultural fleece **88**, 89
hosta 199
hot water plant
  (*Achimenes*) 203
house leek (*Sempervivum*)
  199
house plants 104, 129–30,
  147, 202–3
  symptoms **22**, **37**, **53**
hoverflies **61**, 90, 138
*Hoya see* wax plant
hyacinth (*Hyacinthus*)
  200
*Hyacinthoides see* bluebell
*Hyacinthus see* hyacinth
hydrangea 138, 197
  leaf symptoms **15**
hydrangea leaf spot 138
hydrangea powdery
  mildew 138
hydrangea scale **15**, 138
hypericum 197
*Hypoaspis miles* **91**

## I

*Iberis see* candytuft
ice plant (*Sedum*) 199
*Ilex see* holly
*Impatiens see* busy Lizzie
injury, physical **40**, **70**
*Inonotus hispidus* 138
insecticides **208–9**
  chemical tables **212**, **213**
  organic 83, 90
  phytoxicity 73, 81, 160
  *see also* pesticides
insects 74–75, **74**, **75**, **77**
  beneficial **60–3**, **60**, 141
  life cycles 76–7, **76**
integrated control **78–9**
  complementary practices
    86–9, **86**, **87**, **88**, **89**
*Ipomoea see* morning glory
iris 138–9, 200
  symptoms **15**, **29**, **55**
iris ink disease 138–9
iris leaf spot 139
iris rhizome rot **55**, 139
iris rust **15**, 139
iris sawfly **29**, 139
iris scorch 139
iron 69
  deficiency **24**, 139–40
ivy (*Hedera*) 140, 197
ivy leaf spot 140
ivy on tree trunks 140

## J

jade plant (*Crassula*) 203
Japanese anemone
  (*Anemone*) 199
Japanese quince
  (*Chaenomeles*) 197
jasmine (*Jasminum*) 197
Jerusalem sage (*Phlomis*)
  197
Judas tree (*Cercis
  siliquastrum*) 195
*Juglans see* walnut
juniper (*Juniperus*) 140, 195
  symptoms **32**, **37**
juniper aphid 119, 140
juniper scale **37**, 140
juniper webber moth **32**,
  140
*Juniperus see* juniper

## K

kabatina shoot blight 140–1
kalanchoe 203
kangaroo vine 203
keithia thujina needle blight
  **28**, 141
*Kniphofia see* red-hot poker
knopper galls *see* acorn gall
  wasp
kumquat *see* citrus

## L

laburnum 141, 195
  leaf symptoms **18**
laburnum leaf miners **18**,
  141
lacebug **21**, 168–9
lacewing **63**, 141
lackey moth **30**, 141
ladybird **60**, 77, **91**, 141
*Laetiporus sulphureus* **38**,
  141–2
lantana 203
lapageria 203
larch (*Larix*) 142, 195
  needle symptoms **32**

larch adelgid **32**, 98, 142
*Larix see* larch
*Lathyrus see* sweet pea
laurel (*Laurus*) 197
laurel, cherry (*Prunus
  laurocerasus*) 197
*Laurus see* bay, sweet bay
lavatera 142, 197
lavatera leaf & stem rot 142
lavender (*Lavandula*) 142,
  197
lavender shab 142
lawns 78, 116, 125, 183–4,
  199
  symptoms **58**, **59**
  disease resistance 183
  fungicides **210–11**
leaf & bud eelworm **25**, 142
leaf curl **26**, 156
leaf damage 12–32, 75,
  142-3, 160
  frost **28**, **44**, **70**, 129
  weedkillers **34**, 188–9
leaf miners 142
  chemical control **208–9**
leaf scorch 143
leaf spot (bacterial) **13**, **29**,
  143
leaf spot (fungal) **73**, 143
  symptoms **12**, **13**
  chemical control **210–11**
leaf weevils 143
leaf-cutting bees **30**, 75,
  142
leafhoppers 74, 75, 142
  chemical control **208–9**
leafy gall **39**, 143
leatherjackets **53**, **58**, 75,
  143-4
  *see also* crane flies
leek 144, 206
  symptoms **17**, **23**, **52**
  resistant cultivars 144
leek moth **23**, 144
leek rust **17**, 144
leek white tip 144
lemon *see* citrus
leopard moth 75, 144
lettuce 109, 144–5, 206
  symptoms **22**, **28**, **44**, **55**
  resistant cultivars 144–5
lettuce downy mildew **22**,
  144
lettuce grey mould **28**,
  144–5
lettuce root aphid 145
lettuce viruses 145
lightning damage **70**
lilac (*Syringa*) 145, 165–6,
  197
  symptoms **19**, **22**, **33**
lilac blight **33**, 145
lilac leaf miner **19**, 145
*Lilium see* lily
lily (*Lilium*) 145–6, 200
  resistant spp. 145
lily beetle 145
lily disease 145
lily viruses 146
lily-of-the-valley
  (*Convallaria*) 199
lime (*Tilia*) 146, 195
  leaf symptoms **14**
lime nail gall mite **14**, 146
lime-induced chlorosis **24**,
  68, 139–40
liquidamber 195
*Liriodendron* 195
lobelia 199, 201
loganberry *see* raspberry
*Lonicera see* honeysuckle
*Lunaria see* honesty
lupin (*Lupinus*) 146, 199

stem symptoms **41**
lupin aphid **41**, 146
*Lupinus see* lupin
lychnis 199

## M

magnesium 69
  deficiency **22**, 146
magnolia 119, 195
mahonia 146, 197
  leaf symptoms **12**, **24**
mahonia rust **12**, **24**, 146
*Malus spp.* 101, 194
  leaf symptoms **16**
manganese 69
  deficiency **24**, 139–40
maple (*Acer*) 98, 119,
  180–1, 195
  leaf symptoms **13**, **17**
marguerite 203
marigold (*Calendula*) 201
marrow 206
  resistant cultivars 120
*Matthiola see* stock
mealy cabbage aphid **32**,
  146
mealy plum aphid 146–7
mealybugs **32**, **37**, 74, 75,
  147
  biological control **91**
  chemical control **208-9**
*Meconopsis see* Himalayan
  poppy
melon 147, 206
  plant symptoms **43**
melon cotton aphid **43**, 147
*Meripilus giganteus* **39**, 147
metamorphosis 76–7, **76**, **77**
*Metaphycus helvolus* **91**
Mexican orange (*Choisya*)
  197
mice **51**, **79**, 89, 147
  chemical control **208–9**
Michaelmas daisy (*Aster*)
  147, 199
  flower symptoms **45**
  susceptible cultivars 147
Michaelmas daisy mite **45**,
  74, 75, 147
Michaelmas daisy wilt 147
midges *see* chironomid
  midges; gall midges
millipedes **54**, 148
  chemical control **208–9**
mimosa 203
mint (*Mentha*) **31**, 148, 206
mint beetle **31**, 148
mint rust 148
mistletoe **67**
mites 148
mites, bryobia 111
mock orange
  (*Philadelphus*) 107, 197
moles **59**, 89, 148
  fumigants **208–9**
molybdenum deficiency 110
money tree (*Crassula*) 203
monkey puzzle (*Araucaria
  araucana*) 195
montbretia (*Crocosmia*)
  124, 200
morning glory (*Ipomoea*)
  197
moths 148–9
mountain ash (*Sorbus
  aucuparia*) 149, 195
mountain ash blister mite
  149
mulberry **38**, 149, 204
mulberry bacterial blight
  149
mulberry canker **38**, 149

mullein moth 149
*Musa see* banana
*Muscari see* grape hyacinth
mussel scale **36**, 149
mycoplasmas 72, 149
mycorrhiza 69
*Myosotis see* forget-me-not

# N

*Narcissus see* daffodil
narcissus basal rot 149–50
narcissus bulb fly **54**, 89, 150
narcissus eelworm **54**, 89, 150
narcissus leaf scorch 29, 150
narcissus smoulder 150
narcissus viruses **23**, 150–1
nasturtium (*Tropaeolum*) 107, 201
    leaf symptoms **32**
nectar robbing 46, 151
nectarine *see* peach
nectria canker 151
needle blights of pines 151
needle rusts of conifers 151
nematodes *see* eelworm
nemesia 201
*Nerium see* oleander
New Zealand flax (*Phormium*) 159, 197
*Nicotiana see* tobacco plant
nitrogen 93
    deficiency 151
    fixation 68, **68**, 69
nut weevil **52**, 75, 151
*Nymphaea see* water lily

# O

oak (*Quercus*) 152, 195
    symptoms **17**, **24**, **39**, **52**
oak gall wasp **17**, 152
oak phylloxera 152
oak powdery mildew **24**, 152
oedema **19**, 152
*Oenothera see* evening primrose
oleander (*Nerium*) 152, 203
oleander scale 152
onion 144, 152-4, 206
    leaf symptoms **17**, **23**, **52**
    resistant cultivars 153–4
onion bolting 152–3
onion downy mildew 153
onion eelworm **52**, 153
onion fly 75, 153
    chemical control **208–9**
onion neck rot **52**, 153
onion thrips 153
onion white rot **52**, 153–4
*Opuntia* **22**, 113
orange *see* citrus
orchid **12**, 154, 203
orchid viruses **12**, 154
organic gardening 79
oribatid mites 154
ornithogalum 200
overwintering of pests 77

# P

*Paeonia see* peony
palms 203
pansy (*Viola*) 154, 201
    leaf symptoms **13**
pansy downy mildew 154
pansy leaf spot **13**, 154
pansy sickness 154
*Papaver see* poppy

papery bark **40**, 154
papyrus (*Cyperus*) 203
parsley 114–15, 206
    symptoms **19**, **25**, **51**
parsnip 114–15, 154–5, 206
    symptoms **19**, **51**, **57**
    resistant cultivars 154–5
parsnip canker **57**, 154–5
parsnip viruses 155
parthenogenesis 76–7
*Passiflora see* passion flower
passion flower (*Passiflora*) 203
pea 105, 147, 155–6, 206
    leaf symptoms **14**, **28**, **30**, **31**
    pod symptoms **51**
    nitrogen fixation 69
    resistant cultivars 155, 156
pea & bean weevil **31**, 155
pea downy mildew 155
pea leaf & pod spot **14**, **51**, 155
pea marsh spot 155
pea moth **51**, 75, 155
    chemical control **208–9**
pea powdery mildew 155
pea thrips 155
pea viruses 156
pea wilt **28**, 156
peach 156, 204
    leaf symptoms **26**
peach aphids 156
peach leaf curl **26**, 156
    chemical control **210–11**
pear 118, 127, 156–8, 204–5
    flower symptoms **41**, **44**
    fruit symptoms **47**, **48**
    leaf symptoms **16**, **19**, **31**
    stem symptoms **35**, **36**
    fruit protection 88
    fungicide damage **211**
    ornamental (*Pyrus salicifolia*) 195
    resistant cultivars 157
pear & apple canker **36**, 100, 156
pear & cherry slugworm **19**, 157
pear bedstraw aphid 157
pear boron deficiency 157
pear leaf blister mite **16**, 157
pear midge **48**, 157
pear scab **48**, 157
    chemical control **210–11**
pear stony pit virus **48**, 157
pear sucker 157–8
pedicel necrosis 158
pelargonium 152, 158, 203
    leaf symptoms **15**, **19**, **21**
pelargonium rust **15**, 158
pelargonium viruses **21**, 158
penstemon 199
peony (*Paeonia*) 158, 199
    flower symptoms **43**
peony wilt **43**, 158
    chemical control **210–11**
pepper, edible (*Capsicum*) **50**, 108, 206
pepper, ornamental (*Capsicum*) 203
perennials, herbaceous 198-9
periwinkle (*Vinca*) **15**, 158, 199
periwinkle rust **15**, 158
pestalotiopsis 158
pesticides 78–9, 80–3
    application **80**, **82**, 83–5, **84**, **85**, **208–211**
    pathogen resistance to 80–1, **211**
    selective 83, 90
    *see also* chemicals

pests 74–5, **74**, **75**
    chemical control 78–9, 80–5, **208–211**
    complementary control 86–9
    life cycles 76–7, **76**
petunia 159, 201
petunia foot rot 159
petunia viruses 159
*Phasmarhabditis bermaphrodita* 91, **91**
*Phellinus ignarius* 159
*Phellinus pomaceous see Fomes pomaceous*
pheromone traps 86–7, **87**, 118
*Philadelphus see* mock orange
*Phlomis see* Jerusalem sage
phlox **33**, 89, 159, 199
phlox eelworm **33**, 89, 159
phlox powdery mildew 159
phomopsis canker 159
*Phormium* (New Zealand flax) 159, 197
phormium mealybug 159
phosphorus 69, 93
    deficiency 159
photosynthesis 66–7
phygelius 199
phytophthora 73, **73**, 159–60
*Phytoseiulus persimilis* **91**, 148
phytotoxicity 73, 81, 160
*Picea see* spruce
pigeons **30**, 89, 160
pine (*Pinus*) 98, 119, 151, 160, 195
    symptoms **39**
pine, needle blight 151
pine adelgid **39**, 98, 160
pine aphid 98, 119
pineapple gall adelgid **38**, 160
pink *see* carnation
*Pinus see* pine
plane (*Platanus*) 160, 195
plane anthracnose 160
plant associations 67, **67**
plant metabolism 66–7
plant nutrition 67, 68–9, 93
plant quality 73, 92, **92**
plant raising 92–3, **93**, 122
plant requirements 67, 70
*Platanus see* plane
plum 146-7, 160–2, 175, 205
    bark symptoms **33**, **40**, **41**
    fruit symptoms **48**
    leaf symptoms **16**, **19**, **20**, **22**, **27**
    resistant cultivars 104, 175
    susceptible cultivars 161
plum blossom wilt 160–1
plum leaf gall mite **16**, 161
plum leaf-curling aphid **27**, 161
plum moth **48**, 75, 86, 161
plum pox (sharka) 161
plum rust **22**, 161
plum sawfly 75, 161
*Plumeria see* frangipani
pocket plum 162
poinsettia (*Euphorbia pulcherrima*) 203
pollen beetle **43**, 74, 75, 162
pollination 60, 67
    nectar robbing bees **46**, 151
    poor 71, 162
polyanthus (*Primula*) 143, 199

symptoms **13**, **46**
*Polygonatum see* Solomon's seal
*Polygonum baldschuanicum see* Russian vine
pome fruits, fireblight **35**, 127
pond plants 116
    *see also* water lily
ponds **83**, 90
    harmful chemicals **210–11**
poplar (*Populus*) 100, 162, 195
poplar canker 162
poplar yellow blister 162
poppy (*Papaver*) 201
*Populus see* poplar
pot plants **54**, **55**, 202–3
potash deficiency **24**, 162–3
potassium 69
    deficiency **24**, 162-3
potato 98, 163–5, 176, 206
    leaf symptoms **28**
    tuber symptoms **51**, **54**, **55**, **56**, **57**
    resistant cultivars 163–5, 176
potato black leg 163
potato blight **28**, **51**, **56**, 163
    fungicides **210–11**
potato common scab **56**, 163
potato cyst eelworm **55**, 163–4
potato dry rot **57**, 164
potato early blight 164
potato gangrene **57**, 164
potato hollow heart 164
potato internal rust spot 164
potato powdery scab **56**, 164–5
potato silver scurf **56**, 165
potato spraing **56**, 165
potato viruses 165
potato wart disease 165
powdery mildew **25**, **49**, **73**, 165
    fungicides **210–11**
primrose (*Primula*) 143, 199
*Primula see* primrose, polyanthus
    symptoms **13**, **46**
*Primula auricula* 198
privet (*Ligustrum*) 165–6, 197
    leaf symptoms **22**
privet thrips **22**, 165–6
proliferation **44**, 166
pruning 94–5, **94**, **95**
    poor **35**, **40**, 71
*Prunus dulcis see* almond
*Prunus laurocerasus see* laurel, cherry
prunus, ornamental 175, 195
    leaf symptoms **18**, **30**
*Pseudotsuga menziesii see* Douglas fir
pulmonaria 199
pumpkin *see* marrow
*Pyracantha see* firethorn
pyracantha leaf miner 166
pyracantha scab **49**, 166
pyrethrum (*Tanacetum*) 199
*Pyrus salicifolia see* ornamental pear
pythium root rot 166

# Q

*Quercus see* oak
quince 166, 205
    fruit symptoms **48**
quince leaf blight **48**, 166
quince powdery mildew 166

# R

rabbits 70, 75, 88–9, 166–7
radish 206
raspberry 162, 167–8, 205
    cane symptoms **34**, **40**
    fruit symptoms **50**
    leaf symptoms **18**
    resistant cultivars 167–8
raspberry beetle **50**, 75, 167
    chemical control **208–9**
raspberry cane blight 167
raspberry cane spot & spur blight **34**, 167
    chemical control **210–11**
raspberry leaf & bud mite **18**, 167
raspberry rust **40**, 168
raspberry spur blight 168
raspberry viruses **18**, 168
rats 83, 168
    chemical control **208–9**
red berry mite **49**, 168
red spider mite 75, 119, 129, 132, 148
    symptoms **20**, **22**, **23**, **40**, **43**
    chemical control 81, **208–9**
red thread *see* turf red thread
redcurrant *see* currant, red
red-hot poker (*Kniphofia*) 199
repellents 89
replant problems *see* rose sickness
replanting 96, 172
resistant cultivars 92
    *see also individual pathogens and pests*
rhododendron 111–12, 168–9, 197
    flower symptoms **41**, **42**
    leaf symptoms **14**, **15**, **21**
    susceptible spp. 169
rhododendron bud blast **41**, 168
rhododendron lacebug **21**, 168-9
rhododendron leaf spot **14**, 169
rhododendron leafhopper 169
rhododendron petal blight **42**, 169
rhododendron powdery mildew **15**, 169
rhododendron rust 169
*Rhoicissus see* kangaroo vine
rhubarb 169–70, 206
rhubarb gumming 169–70
*Rhus see* sumach
*Ribes see* flowering currant
*Ricinus communis see* castor-oil plant
*Robinia pseudoacacia* **39**, 194
Robin's pin cushion **39**, 170
rock cress (*Arabis*) 199
rock rose (*Helianthemum*) 197
*Romneya* 198

root aphids **55**, 75, 170
root damage **53–7**, 128, 170
root feeding pests 74-5
root knot eelworm 170
root mealybug **55**, 75, 170
root rots **55**, 96, 128
rose (*Rosa*) 39, 127, 170–3, 197
    flower symptoms **43**, **44**
    leaf symptoms **13**, **16**, **18**, **20**, **21**, **27**
    stem symptoms **35**, **39**, **40**
    replanting 96, 172
    resistant cultivars 171, 172, 185
rose aphid **43**, 170
rose balling **44**, 170–1
rose black spot **13**, 81, 171
    chemical control **210–11**
rose canker & die back **40**, 171
rose grey mould *see* grey mould; rose canker
rose leafhopper **21**, 171
rose leaf-rolling sawfly **27**, 171
rose powdery mildew **20**, 81, 171–2
rose root aphid **35**, 172
rose rust **16**, 172
rose sickness/replant problems 96, 172
rose slugworm **18**, 172
rose viruses **21**, 172–3
rosemary (*Rosmarinus*) 197, 206
*Rosmarinus see* rosemary
rosy apple aphid **48**, 173
rotation of crops 78, 89, 96
rowan (*Sorbus aucuparia*) 149, 195
rubber plant (*Ficus elastica*) 203
Russian vine (*Solanum*) 197
rusts **73**, 173
    chemical control **210–11**

**S**

sage (*Salvia*) 199
*Saintpaulia see* African violet
salad rocket 206
*Salix see* willow
salt damage 173
*Salvia see* sage, salvia
salvia (*Salvia*) 201
*Sambucus see* elder
sap-feeding pests 74, 75
    chemical control **208–9**
sawflies 74, 75, 173
saxifrage 199
scale insects 74, 75, 173–4
    chemical control **208–9**
schefflera **22**, 104, 203
*Schlumbergera* 202
sciarid flies *see* fungus gnat
sclerotinia **36**, 174
    *see also* brown rot
scorch 174
*Sedum see* ice plant
seedlings 92–3, **93**, 122
    symptoms **34**, **51**, **53**, **54**
seeds 67
    damage **51**, **52**
seiridium canker **35**, 174
*Sempervivum see* houseleek
*Senecio see* cineraria
shallot 144, 153, 206
    symptoms **17**, **52**
shanking of grapes 174
sharka *see* plum pox
shieldbugs 174–5

shothole **18**, **30**, 175
shot-hole borers **33**, 175
shrew **62**, 90
shrubs 196–7
    resistant spp. 137
silver fir *see* fir, silver
silver fir adelgid 175
silver leaf 175
*Sinningia see* gloxinia
slime flux & wetwood 175–6
slime mould **58**, 176
slugs **57**, 75, 77, 176
    symptoms **29**, **36**, **43**
    biological control 91, **91**
    chemical control **208–9**
    trap **82**, 83, 86, 88
small bulb fly *see* narcissus bulb fly
small ermine moths **38**, 176
smut **35**, **44**, 176–7
snail 74, 75, 77, **90**, 177
    symptoms **29**, **43**
    chemical control 208-9
    trap 86, 88
snapdragon (*Antirrhinum*) 99, 199
    leaf symptoms **15**
    resistant cultivars 99
snow 71
snowdrop (*Galanthus*) 150, 177, 200
snowdrop grey mould 177
soft scale **91**, 177
soils 68–9, **68**, **69**, 93
    problems **53–7**
    sterilisation **210–11**
*Solanum capsicastrum see* winter cherry
*Solenostemon see* coleus
*Solidago see* golden rod
Solomon's seal (*Polygonatum*) 177, 200
    leaf symptoms **31**
Solomon's seal sawfly **31**, 177
sooty mould 177
*Sorbus aucuparia see* mountain ash, rowan
sparrow **46**, 74, 75, 177–8
spider plant (*Chlorophytum*) 203
spinach beet 106
    leaf symptoms **17**
spinach, perpetual 206
splitting 178
spruce (*Picea*) 98, 119, 134, 160, 195
    stem symptoms **38**
squirrels **52**, 75, 178
stachys 178, 199
stachys case-bearing caterpillar 178
statice 201
*Steinernema carpocapsae* 91, **91**
stem & bulb eelworm 178
stem damage 33–41, 129, 178
stephanotis 203
stock (*Matthiola*) 201
storage rots 178
stranvaesia 197
strawberry 178–80, 205
    fruit symptoms **50**
    leaf symptoms **12**, **17**
    plant protection 88, **88**
    susceptible cultivars 180
strawberry black eye 178–9
strawberry green petal 179
strawberry grey mould **50**, 179

strawberry leaf spot **12**, **17**, 179
strawberry mite 179
strawberry red core 179
strawberry seed beetle 179
strawberry tree (*Arbutus*) 102, 185–6, 195
    leaf symptoms **15**
strawberry viruses 180
strawberry yellows 180
streptocarpus 203
succulents **22**, 113, 147
sucker 75, 180
    chemical control **208–9**
sumach (*Rhus*) 197
sun rose (*Cistus*) 197
swede 89, 146, 206
    leaf symptoms **32**
    resistant cultivars 118
sweet chestnut (*Castanea sativa*) 195
sweet pea (*Lathyrus*) 201
sweet pepper *see* Capsicum
sweet William (*Dianthus*) 201
sweetcorn 147, 180, 206
    seed symptoms **51**
sweetcorn smut **51**, 180
swift moth **53**, 75, 180
sycamore (*Acer*) 98, 180–1, 195
    symptoms **13**, **17**, **36**
sycamore sooty bark disease **36**, 180
systemic pesticides 82, **82**, **83**, **208**, **210**, **212**, **213**

**T**

tachinid fly **61**
*Tagetes see* African marigold
tar spot of acer **13**, 180–1
tarsonemid mites **27**, 181
*Taxus see* yew
tayberry *see* raspberry
temperature extremes 71
thrips 74, 75, 87, 181
    chemical control **208–9**
thrushes **63**, **90**
thuja **28**, 141, 195
thunderworms 181
*Tilia see* lime
toads **63**, **79**, 90
toadstools 69, 181
tobacco plant (*Nicotiana*) 201
tomato 123, 181–3, 206
    symptoms **28**, **33**, **50**, **51**, **55**
    resistant cultivars 182, 183
tomato blight **28**, **51**, 181
    chemical control **210–11**
tomato blotchy ripening **51**, 181
tomato cyst eelworm *see* potato cyst eelworm
tomato fruit splitting **50**, 182
tomato ghost spot 182
tomato greenback **50**, 182
tomato leaf mould 182
    chemical control **210–11**
tomato moth 182
tomato pith necrosis 182
tomato viruses 183
tortrix moth *see* carnation tortrix moth
*Tradescantia see* wandering Jew
traps 86–8, **86**, 88
    sticky **78**, **79**, 87, **87**
trees 194–5
    bark symptoms **33**, **34**, **37**, **38**, **39**, **40**, **41**

leaf symptoms **19**, **30**, **31**
    resistant spp. 136–7
*Tropaeolum see* nasturtium
trumpet vine (*Campsis*) 197
tulip (*Tulipa*) 183, 200
    symptoms **22**, **55**
tulip fire **22**, **55**, 183
    chemical control **210–11**
tulip grey bulb rot 183
tulip tree 195
turf *see* lawns
turf dollar spot **58**, 183
turf fairy ring **59**, 183–4
turf ophiobolus patch 184
turf red thread **59**, 184
turf snow mould **59**, 184
turf thatch fungus mycelium **58**, 184
turf toadstool 184
turnip 206
turnip gall weevil 184–5
two-spotted mite *see* glasshouse red spider mite

**U**

*Ulmus see* elm

**V**

vapourer moth **31**, 185
vegetables 109, 113, 205
    symptoms **50–2**, **53–4**, **55–7**
    crop rotation 78, 89, 96
    *see also individual plants*
verticillium wilt **29**, **33**, 185
viburnum 107, 185–6, 197
    leaf symptoms **15**
viburnum beetle 185
viburnum scale 185
viburnum whitefly **15**, 185–6
*Vinca see* periwinkle
vine (grape) 165, 174, 186, 205
    symptoms **15**, **29**, **47**, **49**, **53**
    fungicide damage **210–11**
vine erinose mite **15**, 186
vine weevil **29**, **53**, 75, 77, 186
    control 87, **91**
violet (*Viola*) 186, 199
    symptoms **27**, **57**
violet gall midge **27**, 186
violet root rot **57**, 186
viruses 72, **72**, 73, **73**, 187
    flower symptoms **42**, **73**
    leaf symptoms **23**, **25**, **27**
    vectors 73, 186–7
voles **36**, 75, 187

**W**

wallflower (*Cheiranthus*) 201
walnut 178, 187–8, 205
    nut symptoms **52**
walnut blister mite 187
walnut blotch 187
walnut soft shell 187–8
wandering Jew (*Tradescantia*) 203
wasps **47**, **49**, **61**, 75, 188
    beer trap **86**
    chemical control **208–9**
    parasitic **60**, **62**
water core 102
water lily (*Nymphaea*) 116, 188, 199
    leaf symptoms **30**
water lily aphid 188

water lily beetle **30**, 188
water lily leaf spot 188
watering 93
    irregular **26**, **36**, **57**, 71, 140
waterlogging **34**, 71, **71**, 94, 188
wax plant (*Hoya*) 203
weather 70–1, **70**, **71**
    disease development 73
    pest populations 77
weed control 94
weedkillers, damage **23**, **34**, **39**, 188–9
weeping fig (*Ficus*) 203
weevils 189
western flower thrips **42**, **91**, 189
whiptail of brassicas 189
white blister **14**, 189
whitecurrant *see* currant, white
whitefly 87, 189–90
    chemical control 81, **208–9**
wildlife, pest control 83
willow (*Salix*) 190–1, 195
    symptoms **17**, **18**, **36**, **41**
    resistant spp. 195
willow anthracnose 190
willow bark aphid **36**, 190
willow bean gall sawfly **17**, 190
willow black canker 190
willow leaf beetle **18**, 190
willow scab **41**, 190
willow watermark disease 190–1
wilt **73**, 191
wind damage **45**, 71, 191
winter aconite (*Eranthis*) 191, 200
winter aconite smut 191
winter cherry (*Solanum*) 203
winter moth **31**, 87, 191
wireworms **54**, **56**, 75, 77, 191
    chemical control **208–9**
    traps 88
wisteria 197
witches' brooms **41**, 191
wolf spider **62**
woodlice **53**, 67, 192
    chemical control **208–9**
woolly aphid **35**, 192
woolly vine scale **39**, 192
worm casts *see* earthworms
worms
    cutworms **53**, 75, 121
    earthworms **58**, **61**, **66**, 67, 125
    eelworms 73, 75, 96, 125
    enchytraeid worms 126
    flatworms 127
    thunderworms 181
    *see also* wireworms
wound paints 95

**Y**

yellow-tail moth 192
yew (*Taxus*) 107, 195
yucca **18**, 192, 197
yucca leaf spot **18**, 192

**Z**

*Zantedeschia see* arum lily
*Zelkova* 124-5, 195
zinnia 192, 201
zucchini yellow mosaic virus 192

# CONVERSION TABLES

METRIC MEASUREMENTS have been used throughout this book, so that the tiny sizes of some of the insects and mites mentioned can be given accurately. In addition, metric measurements are now given exclusively on most commercial packaging, and so must be used to calculate dosage and dilution rates for chemicals: however, the imperial conversions given here, although approximate, are suitable in most ordinary garden situations.

Five conversion tables are given here, which should cover all possible measurements needed when gardening.

### WEIGHT

| Metric | Imperial |
|---|---|
| 5g | ⅛oz |
| 16g | ½oz |
| 25g | 1oz |
| 50g | 2oz |
| 85g | 3oz |
| 110g | 4oz |
| 140g | 5oz |
| 180g | 6oz |
| 200g | 7oz |
| 225g | 8oz |
| 250g | 9oz |
| 450g | 16oz |
| 1kg | 2.2lb |
| 5kg | 11lb |
| 10kg | 22lb |

### POT SIZES

| Metric | Imperial |
|---|---|
| 5cm | 2in |
| 6cm | 3in |
| 9cm | 3½in |
| 10cm | 4in |
| 12cm | 4½in |
| 13cm | 5in |
| 14cm | 5½in |
| 15cm | 6in |
| 19cm | 7in |
| 21cm | 8in |
| 24cm | 9in |
| 25cm | 10in |
| 30cm | 12in |
| 38cm | 15in |
| 45cm | 18in |

### LENGTH

| Metric | Imperial |
|---|---|
| 1mm | 1/16in |
| 2mm | 1/16in |
| 3mm | ⅛in |
| 4mm | ⅛in |
| 5mm/0,5cm | ¼in |
| 6mm | ¼in |
| 7mm | ¼in |
| 8mm | ⅜in |
| 9mm | ⅜in |
| 1cm | ½in |
| 2cm | ¾in |
| 2.5cm | 1in |
| 3cm | 1¼in |
| 3.5cm | 1⅛in |
| 4cm | 1½in |
| 4.5cm | 1¾in |
| 5cm | 2in |
| 6cm | 2½in |
| 7cm | 3in |
| 8cm | 3in |
| 9cm | 3½in |
| 10cm | 4in |
| 11cm | 4½in |
| 12cm | 5in |
| 13cm | 5in |
| 14cm | 5½in |
| 15cm | 6in |
| 16cm | 6in |
| 18cm | 7in |

### LENGTH (CONTINUED)

| Metric | Imperial |
|---|---|
| 20cm | 8in |
| 22cm | 9in |
| 25cm | 10in |
| 30cm | 12in |
| 35cm | 14in |
| 40cm | 16in |
| 45cm | 18in |
| 50cm | 20in |
| 55cm | 22in |
| 60cm | 24in |
| 65cm | 26in |
| 70cm | 28in |
| 75cm | 30in |
| 80cm | 32in |
| 85cm | 34in |
| 90cm | 1ft |
| 1m | 3ft |
| 1.1m | 3½ft |
| 1.2m | 4ft |
| 1.35m | 4½ft |
| 1.5m | 5ft |
| 1.7m | 5½ft |
| 2m | 6ft |
| 2.2m | 7ft |
| 2.5m | 8ft |
| 3m | 10ft |
| 3.5m | 11ft |
| 4m | 12ft |
| 5m | 15ft |

### VOLUME

| Metric | Imperial |
|---|---|
| 5ml | ⅛fl oz |
| 50ml | 2fl oz |
| 100ml | 4fl oz |
| 150ml | 5fl oz |
| 300ml | 10fl oz |
| 600ml | 1 pint |
| 1 litre | 1.76 pints |
| 2 litres | 3.52 pints |
| 3 litres | 5.28 pints |
| 4 litres | 7.04 pints |
| 5 litres | 1.1 gallons |
| 10 litres | 2.2 gallons |
| 50 litres | 11 gallons |
| 100 litres (1 hectolitre) | 22 gallons |

### AREA

| Metric | Imperial |
|---|---|
| 1 sq cm | 0.16 sq in |
| 1 sq m | 10 sq ft |
| 2 sq m | 20 sq ft |
| 3 sq m | 30 sq ft |
| 4 sq m | 40 sq ft |
| 5 sq m | 50 sq ft |
| 6 sq m | 60 sq ft |
| 7 sq m | 70 sq ft |
| 8 sq m | 80 sq ft |
| 1 sq km | 0.3 sq mile |
| 1 hectare (1,000 sq km) | 2.47 acres |

# ACKNOWLEDGMENTS

## AUTHORS' ACKNOWLEDGMENTS

**Pippa Greenwood** would like to thank all the colleagues past and present, and friends and family, who helped her to retrieve obscure but useful information and continued to put up with her whilst she was writing this book. In particular, Dr John Fletcher, her plant pathology "guru", who taught her so much and who really does share her fascination with diseased plants! She would also like to thank Justine for her word-processing help, and the numerous people at Dorling Kindersley who have been involved in this project and who have all been so tirelessly enthusiastic and such fun to work with.

**Andrew Halstead** is indebted to Dorothy Gibson for her speedy and accurate conversion of hand-written pages to a word-processed script. He would also like to thank the companies and colleagues who have given assistance and encouragement in writing this book.

## PUBLISHER'S ACKNOWLEDGMENTS
### PICTURE RESEARCH
Melanie Simmonds, Louise Thomas and Victoria Walker

### DESIGN ASSISTANCE
Fay Singer and Heather Dunleavy

### EDITORIAL ASSISTANCE
James Nugent

### ADDITIONAL ILLUSTRATIONS
Karen Cochrane

### CREEPY CRAWLY FONT
Mark Bracey

### INDEX
Ella Skene

Dorling Kindersley would also like to thank Susanne Mitchell, Karen Wilson and Barbara Haynes at the Royal Horticultural Society and Ian Hodgson at *The Garden* magazine.

## PHOTOGRAPHY CREDITS
The publisher would like to thank the following for their kind permission to reproduce photographs.

t=top, b=bottom, c=centre, l=left, r=right.

**A–Z Botanical** Front Jacket bc, 39br. DW Bevan: 47c, 60tr. Martin Stankewitz: 67tr.

**Biofotos/Heather Angel** Front Jacket tr. 10 title page, 37bl, 38tl, 39cl, 43cb, 44cr, 59br, 61tl, 63tc, 63tr, 63cl, 72bl, 75tl, 92tl.

**Bruce Coleman** Jane Burton: 62bl. Kim Taylor: 63tl, 90tr.

**Forestry Commission** 36bl, 38br, 39tl, 40cr, 41tc, 70cl.

*The Garden* magazine Tim Sandall: Front Jacket bcr, bcl, cbl, cr, tcr. Back Jacket bl, tl. 5bl, 12tc, 12cr, 13tl, 13cr, 13tr, 14bc, 14tr, 15tc, 15cr, 15bl, 16tl, 16tr, 16bl, 16bc, 16br, 17tr, 17tc, 17cl, 17cr, 17br, 18cr, 18br, 18tl, 18tc, 19bc, 19bl, 19tl, 20cr, 20tl, 21tl, 21tc, 21bl, 22tr, 22cl, 22cr, 22bl, 22bc, 22tl, 23bl, 23bc, 23tr, 23cr, 23br, 24tc, 24tr, 24cl, 24br, 25tr, 25cl, 25c, 26cr, 26bl, 27tl, 27tr, 27bc, 27tc, 27cl, 27c, 27cr, 27bl, 28tr, 28cl, 29bc, 29cr, 29cr, 29br, 30tc, 30tr, 30bl, 30cr, 31tr, 31cr, 31tl, 31br, 32tc, 32tr, 32cl, 32tl, 33tr, 34tl, 34bl, 34cr, 35c, 35tl, 35bc, 36cl, 37tl, 37tr, 38cr, 38bl, 39tr, 39c, 41bc, 43tr, 43br, 44bl, 45bl, 46bl, 47c, 48tc, 48bl, 48c, 49tl, 50tr, 51tc, 51tr, 52cl, 53tr, 53tc, 53bl, 54tl, 54cr, 54br, 54tr, 55cl, 55cr, 56bc, 57br, 58bl, 58br, 75cr. Derek St Romaine: 15tl, 27br, 42bl, 55tl.

**Garden/Wildlife Matters** 48cr, 48bc.

**The Garden Picture Library** John Glover: 20tr, 26tr, 74bl.

**Jerry Harpur** 83bl.

**Holt Studios** John Adams: 28bl, 44tr. Gordon Roberts: 45tr. Inga Spence: 51bl. Nigel Cattlin: Front Jacket cr, br, bl, cl, tl, tcl, tc. Back Jacket cr, br, clb, cl, tcl, tc. Inside Jacket flap bl. p6, 12tr, 12bl, 12bc, 13tc, 14tl, 14cr, 17bc, 21br, 23tl, 24tl, 25tc, 25br, 25bl, 27br, 28tl, 28cr, 28bc, 29tl, 29tc, 29tr, 33bl, 33bc, 35bl, 37br, 38tr, 40tc, 41cr, 42bc, 42tr, 43c, 43tl, 43bl, 44tl, 47tr, 47br, 48tr, 49bc, 50tl, 50bl, 51cr, 52tl, 52br, 55tc, 55tr, 56br, 56tl, 56cr, 57cl, 58bc, 59tc, 59cl, 60bl, 60bc, 61cl, 62tl, 72bl, 80cl, 91tr, 96br.

**Oxford Scientific Films** 68cl. Marshall Black: 75cbr. Scott Camazine: 72br. Harry Fox: Back Jacket cal, 61tc. Harold Taylor: 60br, 77bc, 77br. Terry Heathcote: 70bl. John McCammon: 81tr. Colin Milkins: 62bl, 77tl. James Robinson: 75tr. Tony Tilford: 83tr. Thorsten Klapp/Okapia: 81bl, 81bc.

**RHS Wisley** Front Jacket cra, cbr. Spine tc. 14br, 15br, 17c, 18bl, 19tr, 19cr, 20bc, 21cl, 22tc, 26bc, 27bl, 30tl, 30bc, 30c, 30br, 31bl, 31bc, 32bl, 32br, 32cr, 32bc, 33br, 33tc, 33c, 33cr, 36bc, 36bl, 36tr, 38bc, 39cr, 45tl, 45cl, 46tc, 46cl, 46br, 46tl, 47bl, 48tl, 49c, 51c, 51cl, 51bc, 52tc, 52bl, 53bc, 53cr, 54bc, 55br, 55bc, 58cr, 58c, 58tc, 61bl.

**Photos Horticultural** 15tc, 28br, 41cl, 43cr, 44bc, 44br, 45tc, 55bl, 62cr, p64, 71br, 75crb, 75br, 90bl. BT: 34br, 40c, 40cl, 47cr. RB: 49tr.

**Harry Smith Collection** 50cr.

**Unwins Seeds Ltd** Calabrese 'Trixie', 92c.

**Jerry Young** 63bl, 67br.

Additional photography by **Peter Anderson**.

Every effort has been made to trace the copyright holders. Dorling Kindersley apologises for any unintentional omissions and would be pleased, in such cases, to add an acknowledgment in future editions.

# USEFUL ADDRESSES

**Alpine Garden Society**, The AGS Centre, Avonbank, Pershore, Worcs WR10 3JP

**Arboricultural Association**, Ampfield House, Ampfield, Romsey, Hants SO51 9PA

**British Agrochemical Association**, 4 Lincoln Court, Lincoln Road, Peterborough PE1 2RP

**Henry Doubleday Research Association**, Ryton Organic Gardens, Ryton-on-Dunsmore, Coventry CV8 3LG

**National Council for the Conservation of Plants and Gardens**, The Pines, RHS Garden, Wisley, Woking, Surrey GU23 6QB

**National Gardens Scheme Charitable Trust**, Hatchlands Park, East Clandon, Guildford, Surrey GU4 7RT

**National Vegetable Society**, 56 Waun-y-Groes Avenue, Rhiwbina, Cardiff, South Glamorgan

**Northern Horticultural Society**, Harlow Carr Botanical Gardens, Crag Lane, Harrogate HG3 1QB

**Royal Caledonian Horticultural Society**, 28 Silverknowes, Southway, Edinburgh EH4 5PX

**Royal Horticultural Society**, 80 Vincent Square, London SW1P 2PE

**Royal Horticultural Society of Ireland**, Swanbrook House, Bloomfield Avenue, Donnybrook, Dublin 4, Eire

**Scotland's Garden Scheme**, 31 Castle Terrace, Edinburgh EH1 2EL

## SUPPLIERS OF BIOLOGICAL CONTROLS AND/OR PHEROMONE TRAPS AND HORTICULTURAL FLEECES

**Agralan Ltd**, The Old Brickyard, Ashton Keynes, Swindon, Wilts SN6 6QR

**Chase Organics Ltd**, Coombelands Lane, Addlestone, Weybridge, Surrey KT15 1HY

**Defenders Ltd**, Occupation Road, Wye, Ashford, Kent TN25 5EN

**English Woodlands Biocontrol**, Hoyle Depot, Graffham, Petworth, West Sussex GU28 0LR

**Green Gardener**, 41 Strumpshaw Road, Brundall, Norfolk NR13 5PG

**Natural Pest Control Ltd**, Yapton Road, Barnham, Bognor Regis, West Sussex PO22 0AY

**Plant Sales Centre**, RHS Garden, Wisley, Woking, Surrey GU23 6QB

**Scarletts**, Nayland Road, West Bergholt, Colchester, Essex CO6 3DH

40–783